GALILEO ENGINEER

BOSTON STUDIES IN THE PHILOSOPHY OF SCIENCE

VOLUME 269

For further volumes:
http://www.springer.com/series/5710

GALILEO ENGINEER

by

MATTEO VALLERIANI
Max Planck Institute for the History of Science
Berlin, Germany

Springer

Matteo Valleriani
Max Planck Institute for the History of Science
Boltzmannstr. 22
14195 Berlin
Germany
valleriani@mpiwg-berlin.mpg.de

Dissertation zur Erlangung des Doktorgrades an der Philosophischen Fakultät I der Humboldt Universität zu Berlin.

ISBN 978-90-481-8644-0 e-ISBN 978-90-481-8645-7
DOI 10.1007/978-90-481-8645-7
Springer Dordrecht Heidelberg London New York

Library of Congress Control Number: 2010922899

Printed on acid-free paper

Springer is part of Springer Science+Business Media (www.springer.com)

Carlo de Bonardis (17th century). *Scientist undertaking an experiment*. Oil painting (Bona Castellotti, Gamba et al. 1999/2000, 141)

Contents

Foreword: The Historical Epistemology of Mechanics

Jürgen Renn

The historical epistemology of mechanics studies the long-term development of mechanical knowledge. Mechanical knowledge concerns material bodies in time and space, their motions, and the forces that cause or resist such motions. Mechanical knowledge enables us to predict how bodies change their position with time as long as we know their current state and the forces acting upon them. Mechanical knowledge of this kind played a special role in the process of transformation from natural philosophy to modern science. Natural philosophy from its very inception in the works of Aristotle constructed conceptual systems to represent pictures of the world as a whole. But, in contrast to such global intentions, the origins of mechanical knowledge have to be sought in the much more down-to-earth practical activities of achieving the specific tasks of everyday life.

Over a long historical period, the development of mechanical knowledge and its transmission from one generation to the next remained an inherent dimension of such activities, unrelated to any cognitive endeavors aimed at constructing a mechanical worldview. It was only after the first attempts in classical antiquity to include mechanical knowledge in the conceptual systems of natural philosophy that its assimilation to them and the corresponding accommodation of such systems to mechanical concepts led to conflicts between mechanical knowledge and knowledge about nature as a whole. It was only after the growing body of mechanical knowledge became a vital resource of early modern societies that mechanical knowledge within its own conceptual systematization started to compete with natural philosophy by constructing its own worldviews. This finally resulted in early modern times in what has been called the "mechanization of the world picture."

The main goal of the series under the heading *The Historical Epistemology of Mechanics*, conceived in analogy to the four-volume set on *The Genesis of General Relativity*, is to explain the development and diffusion of mechanical knowledge in terms of historical-epistemological concepts. The studies presented within the series are based on a research project centered at the Max Planck Institute for the History of Science in Berlin. While the emphasis of the research has been on the period of the Scientific Revolution, the analysis also takes into account the long-term development of mechanical knowledge without which neither its emergence nor the consequences of this period can be adequately understood. Just as the reconstruction of the relativity revolution in *The Genesis of General Relativity* takes Einstein's work as the point of reference for a thorough contextualization of his achievements, the reconstruction of the transformation of mechanical knowledge during the Scientific Revolution similarly refers to Galileo's work as a point of departure for outlining a historical epistemology of mechanics.

The development of an adequate theoretical framework provides a common basis for the investigations constituting *The Historical Epistemology of Mechanics*. The longevity of mechanics makes it particularly clear that large domains of human knowledge accumulated by experience are not simply lost when theories are revised, even if this knowledge does not explicitly appear in such theories. Thus formal logic is of little use for a description of the multi-layered architecture of scientific knowledge that allows both the continuous and the discontinuous aspects of the transmission of mechanical knowledge to be accounted for. In order to explain structural transformations of systems of knowledge, it is furthermore necessary to take into account the collective character and the historical specificity of the knowledge being transmitted and transformed, as well as to employ sophisticated models for reconstructing processes of knowledge development. Concepts such as "mental model", "shared knowledge", "challenging object", and "knowledge reorganization" have turned out in our work to be pivotal for such explanations.

We conceive of mental models as knowledge representation structures based on default logic, which allow inferences to be drawn from prior experiences about complex objects and processes even when only incomplete information on them is available. Mental models relevant to the history of mechanics either belong to generally shared knowledge or to the shared knowledge of specific groups. Accordingly, they can be related either to intuitive, to practical, or to theoretical knowledge. They are, in any case, characterized by a remarkable longevity—even across historical breaks—as becomes clear when considering examples such as the mental models of an atom, of a balance, of the center of gravity, or of positional weight. Their persistence in shaping the shared knowledge documented by the historical sources becomes particularly apparent in the consistency of the terminology used, a consistency that offers one important element for an empirical control in the reconstruction of mental models and their historical development. The concept of mental model is particularly suited to study the role of practical knowledge for the transformation of mechanics in the early modern period.

We conceive of challenging objects as historically specific material objects, processes or practices entering the range of application of a system of knowledge without the system being capable of providing a canonical explanation for them. Examples run from mechanical devices challenging Aristotelian dynamics, via artillery challenging early modern theories of motion, to black body radiation challenging classical radiation theory. In reaction to such challenges, knowledge systems are typically further elaborated, occasionally to the extent that they give rise to internal tensions and even inconsistencies. Such explorations of their limits may then become starting points for their reorganization where often previously marginal insights take on a central role in an emerging new system of knowledge. Such processes of reorganization may be exemplified by the emergence of theoretical mechanics from Aristotelian natural philosophy in ancient Greece, the transformation of preclassical into classical mechanics in early modern times, or the emergence of quantum theory from classical physics at the turn of the last century.

The investigations constituting *The Historical Epistemology of Mechanics* build on this theoretical framework, centering on the role of shared knowledge, of challenging objects, and of knowledge reorganization. The first study, Matthias Schemmel's *The English Galileo: Thomas Harriot's Work on Motion as an Example of Preclassical Mechanics*, has investigated the shared knowledge of preclassical mechanics by relating the work of Thomas Harriot on motion, documented by a wealth of manuscripts, to that of Galileo and other contemporaries. While the paths Harriot traces through the shared knowledgeare different from Galileo's, the work of the two scientists displays striking similarities as regards their achievements as well as the problems they were unable to solve. The study of Harriot's parallel work has thus allowed the exploration of the structure of the shared knowledge of early modern mechanics, to perceive possible alternative histories, and to distinguish between individual peculiarities and shared structures of early modern mechanical reasoning.

This volume, *Galileo Engineer*, the second study of the series, looks more closely at the role of Galileo as a practical mathematician and engineer-scientist. It focuses on his intellectual development in the frame of the interaction between natural philosophy and the challenging objects provided by technological developments. It analyzes Galileo's contribution to the practical science of machines as well as his role as a teacher involved in the contemporary art of war. The results of this analysis highlight Galileo's profile as a military engineer. By means of two case studies this book develops a model according to which new scientific knowledge was generated on the basis of the interaction between theoretical knowledge—basically Aristotelian—and the practical knowledge Galileo shared with his contemporaries. The first case study concerns Galileo's theory of the strength of materials, namely the first of his *Two New Sciences*, and its relation to the practical knowledge of the Venetian Arsenal. The second case study concerns the emergence of Galileo's heat doctrine on the basis of the practical knowledge related to pneumatics. Galileo's work is finally reinterpreted in its entirety against the background of a historiographical investigation concerning the early modern figure of the engineer-scientist, which concludes this book.

A subsequent contribution to this series will look more closely at the reorganization of mechanical knowledge that took place in the course of Galileo's research process stimulated by contemporary challenging objects. A further study will articulate more extensively the theoretical foundations of a historical epistemology of mechanics, providing an outline of the long-term development of mechanical knowledge. The theoretical framework adopted makes it possible to analyze and make explicit the relations between diverse forms of mechanical knowledge which have hitherto been mostly treated in isolation from each other. Among these different forms is the intuitive knowledge gained through basic material activities, the practical knowledge of professionals, and the theoretical knowledge resulting from the reflection of various forms of knowledge in the context of scientific theories. On this basis it should be possible to reconstruct the long-term development of mechanical knowledge from its anthropological origins via the formation of a mechanical world view to the understanding of material interactions within the framework of quantum mechanics and of the space-time geometry of modern physics.

Introduction

Galileo Galilei (1564–1642), his life and his work have been and continue to be the subject of an enormous number of scholarly works. One of the consequences of this is the proliferation of identities bestowed on this figure of the Italian Renaissance: Galileo the great theoretician, Galileo the keen astronomer, Galileo the genius, Galileo the physicist, Galileo the mathematician, Galileo the solitary thinker, Galileo the founder of modern science, Galileo the heretic, Galileo the courtier, Galileo the early modern Archimedes, Galileo the Aristotelian, Galileo the founder of the Italian scientific language, Galileo the cosmologist, Galileo the Platonist, Galileo the artist and Galileo the democratic scientist. These may be only a few of the identities that historians of science have associated with Galileo. And now: Galileo the engineer!

That Galileo had so many faces, or even identities, seems hardly plausible. But by focusing on his activities as an engineer, historians are able to reassemble Galileo in a single persona, at least as far as his scientific work is concerned. The impression that Galileo was an ingenious and isolated theoretician derives from his scientific work being regarded outside the context in which it originated. Thanks to a series of historical research works dedicated to case studies and to a certain historiographical tradition that began in the 1920s, represented chiefly by Leonardo Olschki (Olschki 1919–1927), it has been possible to infer that Galileo's practical activities, that is, his engagement in the practical knowledge of his time, played a significant role in his scientific speculations. A relevant case study that confirms such an inference concerns Galileo's achievement of the formulation of the law of fall. In this case, it has been shown how Galileo's theoretical investigations were directly connected to the knowledge of early modern artillerists (Renn et al. 2001), and thus that the main building block of Galileo's new science of dynamics was rooted in their practical knowledge (Damerow et al. 2004). Another case study was able to show how relevant aspects of Galileo's hydrostatics, published in his *Floating Bodies* in 1612, were directly connected to metallurgy and, specifically, to the practice of bell casting (Valleriani 2008). The practical knowledge that Galileo shared thus appears to be the most suitable guide for contextualizing his theoretical speculations. In consequence of these considerations the hypothesis emerged that Galileo's science, in general, is rooted in the practical knowledge of his time.[1] Accordingly Pamela Smith argued in favor of a conception of new early modern science as first "disseminated and inculcated" in the workshops of the artisans (Smith 2004, 240).

To investigate in such a direction, both a general definition of practical knowledge and a historiographical determination of those who produced it and were active in its framework are needed. This work makes use of a definition of artist-engineer formulated on the basis of Edgar Zilsel's *The Social Origins of Modern Science* (Zilsel 2000).[2] Zilsel defined the Renaissance artist-engineer mainly on the basis of his analysis of the training of famous engineers, architects and artists. He found similarities in the training curricula for these professions and formulated the thesis that an artist, engineer or architect became such after an apprenticeship, based on the work he was commissioned to perform, and on his success in completing a project that embraced either engineering, architecture or art. Zilsel's thesis contributes to clarifying the tendency in Renaissance culture not to consider these as separate fields of activity, and therefore forces the contemporary historian to turn to a later period to seek and understand, for example, the process that led to a view of the artist as apparently and completely removed from the technological development of his era.[3]

One relevant phenomenon of the early modern period, and especially of Galileo's time, was the huge increase in textual output as result of a process of codifying practical knowledge. These texts mostly contained the knowledge of military engineers and architects, machine makers, makers of mathematical instruments and shipwrights. The texts, together with manuscripts and books on subjects such as theatrical machinery, trick fountains, automata, metallurgy, instruments, mechanical tools, and practical optics, constitute a considerable portion of the entire textual production of the age. The knowledge codified in these writings, together with the knowledge integrated directly into the results of their practical implementation, such as cathedrals, milling devices, galleys, tools and instruments, many of which were left behind by those who never wrote a treatise, constitutes this work's definition of practical knowledge.

Until now no systematic research has ever been undertaken that aims to show how Galileo's interactions with the practical knowledge of his time were more or less intensive and fruitful by taking the entire spectrum of his practical activities into consideration rather than just a selection. This work aspires to present such a panoramic view. In accomplishing this goal, however, the level of generalization implied in the leading hypothesis had to be abandoned. This work follows Galileo through the main phases of his life—his time in Florence, in Pisa, in Padova and his subsequent return to Florence, but it is also organized thematically in

---

[1] Thomas Kuhn also formulated such an encompassing hypothesis. However, he did not investigate the reasons for such a statement (Kuhn 1976, 55–56).

[2] Zilsel's book was originally published in German in 1976.

[3] For an exhaustive analysis of the historiographical consequences of Zilsel's thesis, see Valleriani (2009b, especially 116–117).

accordance with the specific activities Galileo undertook. According to this planning, the space-time area on which this research focuses narrows to the Italian peninsula in the period that includes the second half of the sixteenth and the first half of the seventeenth centuries.

Each of the practical activities undertaken by Galileo and all of the aspects of practical knowledge shared by Galileo are analyzed starting from historical evidence such as Galileo's personal and administrative documents. His correspondence, as collected and published by Antonio Favaro,[4] plays a major role. The lives and works of relevant correspondents and, in particular, their relations with Galileo, are then considered on the basis of the leading hypothesis of this work. Concerning the subjects discussed, the state of the specific practical knowledge involved is investigated in greater detail, mainly by means of treatises compiled by experts in various practical activities, as contemporary to Galileo as possible. A comparison between Galileo's arguments and those of such experts, to the extent that these can be inferred from their treatises and material works, reveals the intensity of Galileo's practical activity in each of the fields in which he was involved, and finally the "degree" to which he utilized the practical knowledge of his day. This analysis is performed for each kind of practical activity undertaken by Galileo. Galileo's major published works, finally, are related to those activities as well. Thus, the traditional historical approach to Galileo, which begins with his major publications, is reversed. All of the letters which play a decisive role in this work have therefore been appended in the author's English translation.

According to this method, the work turns out to have two main protagonists: the first is Galileo, while the other is practical knowledge, or rather those who embodied and implemented it. Although the principal aim of this book is to approach Galileo's work from the perspective of the practical knowledge, Galileo himself can be used as a lantern to elucidate the state of the art and, in general, the structure of practical knowledge between the second half of the sixteenth and the first half of the seventeenth century, as well as for other activities in which Galileo participated.

Structure of the Book

The book is divided into three main parts. The first—*War and Practice*—aims to show how Galileo followed the typical educational path of the artist-engineer in the second half of the sixteenth century and, finally, how he consequently began his career and retained the profile of a military engineer. The second part—*Practice and Science*—comprises two major case studies that are able to show how particular theoretical developments of Galileo are rooted in the practical knowledge of his

[4]Galileo's works were published several times in the form of *Opera omnia* before Antonio Favaro edited *Le opere di Galileo Galilei* in twenty volumes between 1890 and 1909. This is still the standard work used by scholars today. The second edition of Favaro's collection (1968) is used in the present work and quoted with the abbreviation *EN* (*Edizione Nazionale*). Galileo's correspondence is in *EN*, X–XVIII.

time. The first case study concerns Galileo's theory of the strength of materials, and the second his atomistic conception of heat. The third part—*The Engineer and the Scientist*—is devoted to the definition of the figure of the engineer-scientist as a historiographical key for investigating further aspects of the interaction between theoretical and practical knowledge, on which the early modern scientific revolution was based.

First part The profound changes in the art of war that took place from the end of the fifteenth century serve as the historical background for the first part. This part comprises three chapters. The first focuses on young Galileo's early training as an artist-engineer in Florence, after he abandoned the university in Pisa. In the second chapter, Galileo's activity running a smithy and as a designer and maker of instruments is taken into consideration: first his activity as a designer and producer of military mathematical instruments in Padova between 1592 and 1610 and, second, as a designer and maker of optical instruments after 1609 and until the end of his life. The third chapter approaches the topic of Galileo's private courses on fortifications, which included courses on the science of machines, technical drawing techniques, military architecture, practical astronomy and the use of artillery.

The overarching message of the first part of this work is that Galileo's activities in the realm of practical knowledge—designer, maker, producer and evaluator of instruments, and teacher—can be generally interpreted as the typical activities of most of those who received training as an artist-engineer at the end of the sixteenth century. Moreover, against the historical background according to which artist-engineers were expected to address their efforts to meet the particular needs of the art of war, and the fact that only those who did so had the chance to improve their social status (Biagioli 1989), the first part of this work is able to show how closely Galileo followed this traditional path, thus earning a strong reputation as a military engineer by 1610, up until the publication of his *Sidereus nuncius*. The title of this work—*Galileo Engineer*—is ultimately the natural output of the results of the first part.

Second part The second part of this work—*Practice and Science*—comprises two case studies that are able to show how practical knowledge and theoretical developments are related. The first case is concerned with Galileo's theory of the strength of materials, published for the first time in 1638 in the *Discorsi e dimostrazioni intorno à due nuove scienze* (*EN*, VIII:39–318), though its first developments are dated to 1592. The second case deals with Galileo's atomistic conception of heat, and how he exposed it in his *Il Saggiatore* in 1623 (*EN*, VI:197–372).

The definition of practical knowledge given above still requires further elaboration when considered within the frame of the investigations that focus on the relations between Galileo's shared practical knowledge and his theoretical developments. There was, in fact, a wide spectrum of possibilities, methods and paths for sharing practical knowledge available to people like Galileo. For example, as shown in the second chapter, Galileo produced with his own hands a good series of lenses for telescopes. In this case Galileo certainly required direct contact with lens makers to learn their craft. He also needed to work himself until he achieved a satisfactory result. In this case Galileo shared practical knowledge in that he functioned as an artisan himself. At the end of the second chapter it is also shown how Galileo was requested to act as an evaluator of machine proposals, eventually presented in the form of a machine model. In this case Galileo did not act as an artisan because he did not build anything. What he did was evaluate the potential efficiency of the machine. He analyzed the composition of motions displayed by the arrangement of the different components of the machine. He then analyzed velocities and times of the single motions and of the compound motion connecting the point where the moving force was applied to the component of the machine that moved and thus accomplished the work. From the proper historical perspective, machine evaluators of this kind were neither craftsmen nor machine makers. The latter often did not possess the reflective knowledge required to calculate the efficiency of the machine and to express it in a way understandable for those who were not experts on machines. Machine evaluators were engineers, as in people who possessed a good knowledge of practical geometry and arithmetic, as well as the fundamentals of the science of machines and experience of working with compound machines. Making something with one's own hands or evaluating something built or conceived by someone else were two ways for Galileo to be connected to practical knowledge. Both of them presuppose a process of sharing practical knowledge, but in two very different forms and with two very distinct targets.

Following these two different paths of sharing practical knowledge was a natural consequence of Galileo's apprenticeship as an artist-engineer: a person able to work with his hands and, at the same time, a person who had received enough mathematical knowledge and skill to accomplish a more reflective approach to practical knowledge as well. Galileo came into contact with practical knowledge through both of these two forms of sharing processes.

The first of the case studies shows how Galileo's science of the strength of materials is rooted in the knowledge and experience of the Venetian shipwrights employed at the Arsenal. In this case Galileo shared the knowledge of shipwrights not as a craftsman but as sort of (unsuccessful) evaluator. This case study, moreover, also shows that the theoretical paths which led Galileo through the shipyards, and especially the ones that led him to reframe such practical knowledge within the mathematical deductive structure of his theory, was specifically Aristotelian.

A similar result, though with different shadings, is achieved on the basis of the investigations directed towards understanding the research paths that led Galileo to formulate his atomistic conception of heat in 1623. In this case Galileo's research was grounded on the practical knowledge he possessed in the field of hydraulics and, specifically, pneumatics. Thanks to such a skill, Galileo was able to be one of the first to start working with the thermoscope for scientific purposes. The thermoscope was a pneumatic instrument applied from the beginning of the seventeenth century on to measure temperature, for the first time without recourse to the human senses. According to the results of this research Galileo shared practical knowledge both as a craftsman and as an evaluator. Galileo's aim to explain the functioning of such an instrument, which in modern terms worked on the basis of air's capacity to contract and expand, led him, moreover, to reconsider his practical knowledge in the light of Aristotelian doctrines, such as the doctrine of the transformation of the elements.

If the first part of this work enables the historian to define Galileo as an engineer, that is, as an artist-engineer, the second part clearly shows how Galileo generated new scientific knowledge by acting not only as an engineer, but also as an expert on Aristotelian natural philosophy.

Third part The last part of this work—*The Engineer and the Scientist*—is dedicated to unifying all of the results achieved in the previous ones. To which extent Galileo can be considered as a military engineer and to which extent Galileo was not only a military engineer are the main questions the last part aims to answer. The context of Galileo's activity as a military engineer, the method he followed as an Aristotelian engineer in generating new scientific knowledge, and all of the single results presented in the previous part, constitute the foundation of a historiographical analysis of the figure of the early modern engineer-scientist, with which the final part concludes.

Galileo is identified as belonging to the category of the engineer-scientists, the pivot around which the scientific revolution developed.

Conspicuously, this work does not make use of the word "practitioner" (aside from this paragraph). Many recent historical studies that focus on the emergence of new scientific knowledge during the early modern period already have pointed to the fact that the early modern scientific revolution is somehow connected to the work of the "practitioners." However, this term is universally used in a very vague way to denote a plethora of figures ranging from the illiterate young assistant of a machine maker to the highly educated military engineer at a sixteenth-century court. There is a qualitative difference between these figures, however, especially when the period starting from the second half of the sixteenth century is taken into consideration. The difference consists in the fact that engineers and architects, for example, from the end of the sixteenth century, already possessed relevant reflective knowledge concerned with their practical activities, and this to such an extent that many of them had already entered into the scientific discourse of that period.[5] Craftsmen and foremen, on the other side, had no such reflective knowledge. While investigating the emergence of Galileo's science, therefore, it does make a difference whether he was observing how a lens maker ground his object or whether he was speaking with a military architect educated, for example, at the *Accademia del disegno* in Florence. In conclusion, this work differentiates between craftsmen, who were those persons manually involved in practical activities such as, for example, mechanics in charge of assembling machines, and artist-engineers, who, to remain in the field of activity devoted to machine building, were in charge of conceiving, designing, and evaluating machines and acted as supervisors of their construction.

[5] Engineers had already approached theoretical investigations at the end of the sixteenth century, especially in the fields of activity related to hydraulics and pneumatics (Valleriani 2007).

How to Read this Book

The book ends with an appendix comprising a selection of letters by and to Galileo. Such correspondence is particularly relevant to evaluate Galileo's activity as an engineer; although they were published by Antonio Favaro, they have not received the attention they deserve from historians. The appended letters are published in English translation for the first time and quotations from them are marked in italics. The original text of the letters is not reprinted, as for Favaro's edition of Galileo's works is available at many libraries and also accessible via the Internet at several URLs. Appropriate references and cross-references link the book with the sources in translated form.

For the same reason, no original text is given for all of the quotations in English translation from Galileo's major works, nor for all easily accessible sources like Aristotle's major works. In all other cases, the original texts are in the footnotes.

Sources and secondary literature are given in two separate bibliographies. References to Galileo's works also published in *Le opere di Galileo Galilei* edited by Antonio Favaro make use of the bibliographic data of the latter and not of those of the original publications. These, however, can be found in a distinct bibliography containing Galileo's works that are consulted in this research.

Acknowledgments This book originated as a PhD thesis submitted to the Humboldt Universität zu Berlin. It was written in Department 1 of the Max Planck Institute for the History of Science in Berlin. Detailling the help, support and guidance that Jürgen Renn provided at many different levels while producing this work would require a separate chapter. I would like to emphasize here the crucial role he has played in this research. From May 2005 on, the work was accomplished in the framework of Project CRC 644: *Transformations of Antiquity*, funded by the Deutsche Forschungsgemeinschaft. The final structure of the book is the consequence of the research and of the exchanges conducted at the Department of History of Science of Harvard University during a 3-month stay at the beginning of 2009. I would like to thank Giunti Editore, the Biblioteca Nazionale Centrale of Florence, the Istituto e Museo di Storia della Scienza of Florence, the Biblioteca Riccardiana also of Florence, and the Staats- und Universitätsbibliothek of Hamburg for their permission to reproduce the illustrations.

The necessity to systematically investigate practical knowledge as the root of Galileo's science emerged during research group meetings of Department I of the Max Planck Institute for the History of Science in Berlin between 2000 and 2002. Those meetings, in which I had the honor to participate, were largely devoted to the discussion of long-durée visions concerning the history of mechanics and their conditions of validity, which historians had eventually to prove. Also fundamental for the emergence of the specific hypothesis concerning Galileo are several works produced against the background of the same meetings: Renn (2001), Renn et al. (2001), Renn and Valleriani (2001) and Lefèvre (2001). I would therefore like to acknowledge all of the scholars who attended those meetings: Katja Bödecker, Jochen Büttner, Peter Damerow, Marcus Popplow, Jürgen Renn, Simone Rieger, Matthias Schemmel, Markus Schnöpf and Paul Weinig. In particular, I thank Peter Damerow for his accurate and critical reading of the very first version, Marcus Popplow for his reading of Chapters 2 and 3, and Jochen Büttner for continuous exchanges up to the very last phase. For the construction of the research frame within which the investigations presented in Chapter 5 could be undertaken, I would like to thank Lorraine Daston, who kindly advised me in approaching the history of meteorology. Raymond Fredette and Massimiliano Badino, moreover, both helped with their readings of Chapter 4. The research presented in this chapter was also improved by the useful comments of Antonio Becchi and Gianni Micheli.

The preparation of the final version of the book was supported by detailed readings by Horst Bredekamp, Sven Dupré, Rivka Feldhay, Paolo Galluzzi, and Wolfgang Lefèvre. Intensive and useful exchanges have been held as well with Mark Schiefsky, Gideon Freudenthal, Alexander Marr, David McGee, Peter McLaughlin, Milena Wazeck, Dagmar Schäfer, Matthias Schemmel, Urs Schoepflin, and Ursula Klein. As mentioned, the final structure of the book was developed in consequence of the discussions and exchanges held at the meetings of the Early Modern Working Group of the Department of History of Science at Harvard University. From this perspective I am particularly indebted to Mario Biagioli, whose comments were decisive during the preparation of the last version. Ralf Hinrichsen, Tom Werner and Christian Voller helped to revise the bibliography. Oona Leganovic performed a style-sheet check. Susan Richter helped instrumentally in giving fluency to the text. Petra Schröter and Shadiye Leather-Barrow were particularly helpful in accomplishing many administrative tasks, especially during the PhD phase, and later Monika Liedtke as well. Sabine Bertram helped to obtain electronic reproductions of the illustrations and the related permissions. Lindy Divarci provided fundamental assistance at each step of this research, so that it is no exaggeration to state that without her it is doubtful that this book would have ever been completed. On the publishing end, I would like to thank Lucy Fleet, whose professionality and friendliness propelled me through the publishing process. Since this is my first monograph, I also would like to thank, finally, Alberto Artosi, now Professor for Theory of Law and Legal Logic and formerly Professor for Philosophy of Science at the University of Bologna. He helped me to realize my passion for research and gave me the solid basis on which I could begin such an intellectual adventure.

I dedicate this work to my wife Ulrike and to my sons Dante and Zeno.

Dahlem, October 9, 2009

Part I
War and Practice

Chapter 1
Artist-Engineers' Apprenticeship and Galileo

Early modern Italian artist-engineers improved their social status continuously during the sixteenth and seventeenth centuries. This cultural process was primarily a consequence of political and economic changes, like the intensifying process of urbanization and the broader distribution of wealth. Indeed, the demand for increasingly specialized artist-engineers increased to such an extent during the early modern period that the training of various specialists was actually taken over by the state. In keeping with the way the system of educating people involved in practical activities changed over time, their specializations, too, differentiated from each other more and more. Artist-engineers were in fact turning into professionals: engineers, architects, practical mathematicians, military engineers, mechanicians, artists, and shipwrights. By the time the duties of these different sorts of artist-engineers became well defined and distinguished from each other, the Italian Renaissance was over. Yet until this happened, the situation was so fluid that a single person, over the course of his life, might be considered not only as an artist but also as a mathematician or as a military engineer, depending upon the subjects to which he dedicated his time.

During the late sixteenth century there were three main educational paths for artist-engineers: the workshop, the Abaco school and the Academy established in the wake of the Florentine *Accademia del disegno* founded in 1563. By comparing the educational content of these paths and the knowledge Galileo acquired during his youth, it will be shown that he can be ascribed to the Renaissance artist-engineers, and to the engineers and practical mathematicians in particular.

The Political and Economic Context

Urbanization and the consequent increase in economic competition emerged as factors key to the blossoming of the liberal arts during the Renaissance. Thanks to a relatively broad distribution of wealth and a highly dynamic situation of this wealth's ownership, demand increased tremendously for an ever greater variety of devices related to the landscape, for manufactures based on raw materials from abroad, and for transportation (Goldthwaite 1972). The more or less absolute

M. Valleriani, *Galileo Engineer*, Boston Studies in the Philosophy of Science 269,
DOI 10.1007/978-90-481-8645-7_1, © Springer Science+Business Media B.V. 2010

control Italian merchants enjoyed over the commercial routes used to transport raw materials and manufactured products connecting Italy to the Near East and Northern Europe netted Italian businessmen a tidy profit. This wealth, added to the large total of what was inherited by the heirs to the victims of the Plague, explains why the products of the arts, including luxury goods, were in such great demand during the Renaissance, and provided the means for funding large-scale architectural enterprises. The activity of artist-engineers was obviously related to this economic situation.

Civil architecture This positive economic dynamic was accompanied by changes in the political entities and their exercise of power. That political institutions were controlled by families became apparent in the architectural developments of the age, which occupied the Italian cities increasingly from the fourteenth century on, peaking during the sixteenth. The main cities bear the vestiges of the magnificence of those who were in charge politically. Colossal building sites were developed to construct the buildings important for contemporary institutions, like cathedrals and state authorities. But the new social class of the merchants, the *noveau riche*, also stimulated this architectural change out of its need to consolidate its new, elevated social and political status. This is why fabulous buildings and towers still can be seen in cities like Florence, Rome, and Venice.

The appearance of firearms and geometrical town planning But such undertakings were not only a question of image and patronage. In the fifteenth century, gunpowder, imported from the East, was introduced to everyday life in all European cities. The consequent development of artillery and mobile heavy artillery led to a completely new kind of city planning.[1] Around or instead of the intricate and asymmetrical complexes of alleys typical of the Middle-Age city, planned areas were built, ploughed by wider roads, often crossing each other at right angles in the ancient Roman tradition. These roads were designed to allow better communication between the different sectors of the cities, and especially between the main defense points distributed along the walls, which were built around almost every city and village in response to the development of artillery.

The first example of such planning was undertaken in Ferrara as early as 1492, where the architect Biagio Rossetti (1447–1516) was commissioned by Ercole I d'Este to plan and oversee the construction of the area called the Herculean Addition by applying principles of perspective to lend the entire area a stable geometrical balance. While this was not the first case of an architect being charged with the planning and supervision of the construction of a whole urban area, it was the first time that more than merely those architectural issues important to the individual entities of the state and military needs had been taken into consideration (as reflected in the wider roads). Equally important in this case were the general functionality and overall appearance of the whole area, to which the individual aspects had to be adapted. In keeping with the example of Ferrara, geometrical town planning became

[1] Mario Biagioli (Biagioli 1989, 44) determined the starting point of this process to be the shocking invasion of Italy by the King Charles VIII of France between 1492 and 1494.

inevitable. There was a planned architectural line of continuity between the *piazze*, the squares located in the central areas of town, which served as meeting places and market grounds, and the *piazzeforti*, the strongholds along the curtains, where the artillery was positioned.

Fortresses But architects and engineers were entrusted with more than just expanding cities. It was also necessary to provide for the capabilities of responding to new potential attacks on the boundaries of the states, duchies, republics and principalities. All of the fortresses were adapted to withstand artillery fire, and many other new fortresses were built. This was one of the main priorities of any contemporary state, and the effort invested for such a need was so great that it laid the foundations for a later distinction between civil and military architects.

Around architecture Architecture was a locomotive product sector. However, no fortress, no bridge, no civil building, no church, no ship could be made without a great variety of instruments, devices and machines. Machines to lift weights were needed at every building site, shipyard and harbor. Milling devices operated by either wind or water, or by force, human or animal, had become institutions in such enterprises since the Middle Ages. Small lathes for the production of iron tools were used in the workshops, and large lathes for cannons. Water lifting machines and pumps were a necessity for the functioning of whole cities and fortresses, for irrigation and for the increasing exploitation of the subsoil. All of these devices, together with others like pile ramming machines and draining machines, allowed the realization of great architectural enterprises. For experts on machines to build a water mill, for example, they proceeded from the qualitative evaluations of the force and quantity of the water, and of the quantity of grain to be ground. Whatever the context, these assessments allowed them to build complex machinery consisting of wheels, gear wheels, sprocket wheels, axles, lanterns, and pulleys, thanks only to their experience and to what had been passed down to them during their apprenticeship. Moreover, the Italian situation, in which great quantities of goods were in the hands of business people, most of whom were already living in the cities during the Renaissance (Goldthwaite 1972), necessitated the exact quantification of their goods in the countryside. This entailed the development of many stereometrical and mathematical instruments to facilitate the performance of demanding surveying tasks. The relatively high level of wealth stimulated these same businessmen to invest in devices such as milling stones, for which there was already a centuries' long tradition, and whose productivity in economic terms was a long-established fact. All kinds of devices for milling and irrigation, especially what were known as the "water lifting machines," were the mechanical devices clearly most required from the Middle Ages up until the end of the Renaissance.

As regards mathematical instruments, the Renaissance witnessed the continued development of a tendency that already had become apparent during the late Middle Ages, to find and build instruments to measure distances in any context, from the astronomical setting to the terrestrial, for purposes including surveying and fortification, siege strategies, or geographic cartography for navigation and, finally, for what in modern terms would be called topographic survey.

Society and the new professions The architectural changes mirrored changes in the style of life. Official administrative structures emerged, whose task was to control and develop rational infrastructure such as streets, bridges, river sources and riverbanks, harbors and, where needed, actual water regulation, as in the Venetian lagoon or in the eastern part of contemporary Emilia, between Bologna and Ferrara. Good communication routes, consisting of roads and rivers, were in fact needed not only for the safety of the state, but also for commerce. When the Grand Duke of Florence ordered two engineers to change the bed of the Bisenzio river in 1630, he did so because of the needs of the farmers who held property close to the river and deplored great losses of harvests because of the continuous floods.[2] Thus architects were not the only specialized personnel increasingly in demand by a state, but also urbanists, and military and civil engineers, who often were also painters and sculptors, but also experts for roads and rivers.

Different geographical and political conditions corresponded to different needs. As concerned the seafaring republics, Venice and Genova, another important need was to improve and defend their communication routes at sea. This explains the major impulse the Venetian Senate gave toward the constitution of a great naval fleet. The Venetian Arsenal became one of the greatest pre-industrial centers in Europe, and its organizational requirements led to the emergence of other professional figures, such as shipwrights specialized in one single kind of ship. Moreover, in production centers like the Venetian Arsenal, labor was organized in such a sophisticated manner that Venice was able to launch a newly built fleet constituted of hundreds of vessels within only one year (Aymard 1980).

The dimensions of enterprises like the Venetian Arsenal, for example, or the construction site for the cathedral of Santa Maria del Fiore in Florence, where hundreds of foremen and their assistants worked, required another level of professionals able to coordinate the work on the one hand, and, on the other, to serve as a contact between the sites where these enterprises were realized and the authorities who had commissioned them.[3] In keeping with the significance of such undertakings, even the education of authorities like dukes and doges gradually changed to incorporate some of that knowledge which was normally taught to engineers.[4]

Thanks to the new systems of administration, greater wealth was accumulated not only by private families, but by cities as well. For this reason people like the Grand Duke of Tuscany, during the second half of the sixteenth century, could afford

[2]Galileo, in charge of the court engineer, remained involved in this business and was asked to evaluate the proposals by the two engineers. Galileo's proposals concerning the way to change the bed of the river Bisenzio are interesting because they could testify to the potential efficiency of Galileo's hydrodynamics (Maffioli 2008).

[3]The digital archive *The Years of the Cupola* collects all of the administrative documentation of the archive of the Opera of Santa Maria del Fiore for the period 1417–1436 when the cupola of Brunelleschi was planned and constructed (Margaret Haines and Jochen Büttner, http://duomo.mpiwg-berlin.mpg.de/Accessed October 2009).

[4]Details about the education of the highest political authorities during the sixteenth century are provided at the beginning of Chapter 4.

to commission the construction of technically complex systems like the garden of Pratolino, equipped with a myriad of pneumatic devices (trick fountains), systems of automata and two hydraulic organs, all of them powered by the most sophisticated hydraulic network of the age.[5]

The Education of Artist-Engineers

Different sorts of artist-engineers often hailed from identical educational backgrounds. At the end of the sixteenth century, for example, Genoan regulations still treated the profession of painter on the same basis as frame gilders, the craftsmen of a mechanical art (Soprani 1674; Dempsey 1980). Artists, engineers, and mathematicians were often equally expert in practical geometry, practical arithmetic, geodesy, statics, stereometry, perspective and technical drawing. The content of their training often coincided, and the focus of their adult profession was the primary factor that decided whether they were regarded as artists or as engineers. For example, architects generally came from apprenticeships as sculptors or painters, and were only considered architects after being granted their first commission for a building project. Artists, engineers and mathematicians were generally trained in either workshops or the Abaco schools. Starting from the second half of the sixteenth century, the engineers' education was also provided by academies modeled on the example of the *Accademia del disegno* of Florence.

Artist workshops Artist workshops (*botteghe*) usually offered the same apprenticeship for both future artists and certain kinds of foremen.[6] In artist workshops, in addition to the subjects needed in order to become a painter, like how to make colors and prepare canvasses,[7] the fundamentals of drawing were taught as well. This subject also included technical drawing, perspective and stereometry. Giorgio Vasari (1511–1574) stated that "for the sculptors compasses and set squares are sufficient to find and report all the proportions and measures they need; for the painters it is also necessary, besides being able to use the above-mentioned instruments well, to possess a detailed knowledge on perspective, in order to draw a thousand other things besides villages or houses, all the more because one has to have a better judgement with reference to the quantity of figures in a story, where more mistakes

[5]For the history of the garden of Pratolino, see Zangheri (1987), Valleriani (2007, 2009b, 2010, Forthcoming a).

[6]Not just any sort of foremen completed their apprenticeships at the artist workshop. In order to become a shipwright, for example, the apprenticeship could only take place in a shipyard, where no future artist was generally employed. Even in this case, however, the boundaries remained blurry, since in some cases it is possible to find civil architects who had their apprenticeships in shipyards, as happened in Venice. This is reminiscent of the similarities between naval and civil architecture stated by Leon Battista Alberti (1404–1472) back in 1454 and published in his *De re aedificatoria* of 1485.

[7]Cennino Cennini described these processes in detail (Cennini 1437).

can emerge than in a statue."[8] Finally Vasari added that "sculpture and painting are sisters in reality, because they were born from a unique father, which is drawing."[9] Precisely those topics related to learning how to draw were the ones they had in common with the education of pupils who would later become involved with the mechanical arts. The art of drawing was the common denominator of every kind of practical art, from navigation to architecture. During an apprenticeship in a workshop, young pupils thus could become familiar with drawing and mathematical instruments, which allowed them to avoid learning long and complicated geometrical theorems. According to Peter Burke, however, only some workshops in each city realized this educational mission (Burke 1972, 53). Teachers, who were not necessarily artists, went to these workshops to teach practical geometry to the young apprentices after their working day.

Craftsmen such as, for example, ironsmiths, glassmakers, and shipwrights learned their professions through an apprenticeship in workshops devoted specifically to these activities. Given the spectrum of experiences accumulated during an apprenticeship, the apprentice had the chance to try out several kinds of work before eventually specializing, which entailed opening a shop particularly oriented toward a certain kind of production. Glassmakers could therefore become lens makers; ironsmiths could become constructors of mathematical instruments. The skill of the apprentices in a particular art was assessed on the basis of the result they presented, the famous masterpiece at the end of their apprenticeship. Another kind of craftsmen, the machine maker, was a synthesis between a carpenter and a smith, in demand, for example, whenever a milling device or weight lifting device was needed.

A workshop's pupil The Florentine painter Ludovico Cardi da Cigoli (1559–1613), for example, completed an apprenticeship in decoration at the *bottega* of Bernardo Buontalenti (1536–1608),[10] where he took lessons on practical geometry from Ostilio Ricci (1535–1607). Cigoli was born to a noble family and was initially destined for literary studies. He then abandoned them in order to follow his inclination for painting, first attending the artist workshop of Alessandro Allori (1535–1607), and then Bernardo Buontalenti's. At the end of his training he even enrolled at the *Accademia del disegno*, where he studied anatomy with particular determination. At the end of the sixteenth century he was one of the favored painters at the Medicean court, although he also often received commissions for

[8]"Ali scultori bastano le seste o le squadre a ritrovare e riportare tutte le proporzioni e misure che egli hanno di bisogno; a′ pittori è necessario, oltre al sapere bene adoperare i sopradetti strumenti, una accurata cognizione di prospettiva, per avere a porre mille altre cose che paesi o casamenti; oltra che bisogna aver maggior giudicio per la quantità delle figure in una storia dove può nascer piú errori che in una sola statua" (Vasari 1568, 19).

[9]"[...] la scultura e la pittura per il vero sono sorelle, nate di un padre, che è il disegno [...]" (Vasari 1568, 26).

[10]On the life and work of Bernardo Buontalenti, see Fara (1988, 1995, 1998). For a detailed discussion about Ludovico Cardi da Cigoli, see the introduction of Cardi da Cigoli and Profumo (1992).

architectural designs, among them for the famous *Cappella de' Principi*, and as a maker of theatrical machinery. From the beginning of the seventeenth century on, he also obtained very important commissions in Rome as a painter, among them, for example, a painting in Saint Peter's Basilica (*Saint Peter Healing the Lame Man*). In addition to his artistic activities, Cigoli was deeply involved in the scientific debates of his day, first of all in those concerning sunspots and Medicean Planets. He assisted Galileo in his observations, transmitting his observational data through regular correspondence. Finally, Cigoli wrote a treatise on practical geometry, specifically on methods of drawing perspective, although he never had this published.[11]

The Abaco schools Pupils enrolled at the Abaco school at the age of ten or eleven, after having completed several years of grammar school, where they first learned reading, writing, and sometimes rudiments of Latin. The Abaco schools[12] appeared during the thirteenth century, with the aim of familiarizing merchants with the new arithmetic based on Arabic and Indian numeration, which had been imported to Italy by Leonardo Fibonacci (Fibonacci XIII century).[13] Although these schools were intended mostly for future merchants, in Galileo's day future painters, sculptors and architects could attend them as well; for example, in order to acquire the skills needed to work as a shopkeeper. According to the economic historian Richard A. Goldthwaite, in 1338 there were six Abaco schools in Florence with around 1,200 pupils in attendance (Goldthwaite 1972).

Abaco schools placed their highest educational priority on teaching practical arithmetic. Details about the schools' curriculum can be inferred from documents passed down by school owners. For example, a 1590 contract to hire a new teacher at an Abaco school in Florence, stipulates that the new teacher, namely Giuliano di Buonaguida della Valle, had to teach mathematics at seven different levels. First, basic arithmetical operations; then three courses on the operation of division, respectively with one, two and then three or more digits; the fifth level was dedicated to fractions; the sixth to the Rule of Three[14] and the seventh to the Florentine monetary system.[15] Besides working as teachers at the Abaco schools, these practical mathematicians were often official measurers, especially in order to settle accounts between patron and builders after the completion of buildings. In fact, this task was related strictly to the payment of the workers. For example, founders were normally paid for the number of cubic meters prepared. Thus, once

---

[11]Cigoli's treatise obtained the *imprimatur* twice in 1628 and 1629, at his relatives' request, but was not published until 1992.

[12]The Abaco schools were sometimes also called Abaco workshops (*botteghe dell'Abaco*), probably because of the similarities between their didactical program and that of the artist workshops.

[13]The first—unsuccessful—attempt to introduce the Arabic system was actually the work of Pope Sylvester II about 200 years earlier.

[14]$a : b = c : x$.

[15]The contract is published in its entirety in Goldthwaite (1972). An entire didactical program for mathematics at the Abaco school dating back to the first half of the fifteenth century is published in Arrighi (1965–1967).

the foundations of a building project had been completed, a *maestro* of the Abaco was called upon to measure them. The mathematicians of the Abaco schools further enhanced their status by writing treatises on practical arithmetic, which they invariably published. There is quite a large number of these treatises, written especially during the sixteenth century, but they have yet to be listed accurately. Without a doubt these treatises at least contributed to the refinement of calculation techniques and, especially, to the diffusion of the mathematical culture within urban society. Moreover, treatises by Abaco teachers often went beyond the limits of practical arithmetic to address either other practical topics like practical geometry, architecture, technical drawing, and the use and construction of mathematical instruments, or theoretical topics like algebra, Euclidean geometry, music theory, astronomy, and astrology. The first translation of Euclid's work into the vernacular, for example, was performed by Nicolò Tartaglia (1499–1557), an Abaco teacher.

Subjects like practical arithmetic and practical geometry were shared by pupils and teachers of the Abaco schools together with apprentices of the workshops and their masters. This situation certainly allowed craftsmen to shift from projects related to the production of art to those strictly related to practical mathematics, according to their talents.

An Abaco teacher Nicolò Tartaglia was a teacher of the Abaco school. He started his education at the age of 14 with the "alphabet" as his first subject. Apparently having abandoned his tutor by the time he learned the letter "k," he continued by teaching himself. A few years later he was employed as an Abaco teacher in Verona, until he moved to Venice in 1535 to become teacher of mathematics at several schools. Tartaglia is often cited for his contributions to pure mathematics, like his method of solving third-degree equations. However, Tartaglia worked almost his entire life in close contact with professionals, like artillerists, for example. The first Italian translations of Euclid's *Elements* (Euclid and Tartaglia 1543) and of relevant works on mechanics by Archimedes (Archimedes and Tartaglia 1543) and Jordanus de Nemore (Tartaglia 1565)[16] are just a few of the results of these relationships. Thanks to his practical activity, Tartaglia made important contributions to the art of warfare. He published these contributions in two works especially, the *Nova scientia* (Tartaglia 1537) and the *Quesiti et inventioni diverse* (Tartaglia 1554). Topics of these treatises include the firing of artillery, composition of cannonballs and gunpowder, the disposition of infantry, topographical surveying by means of compasses, and equilibrium in balances and statics. In ballistics, moreover, he developed a theoretical conception of the motion of a projectile and, above all, he contrived an instrument, called a "quadrant for bombardiers," with its associated "firing tables," which allowed artillerists to predetermine the elevation of artillery depending on the distance of the target and the size of the artillery used. Tartaglia was the first to theoreticize the angle of elevation for which a cannonball reaches the maximal range, namely 45°.[17]

---

[16] Published upon his death.

[17] Galileo integrated the quadrant for bombardiers into his military compass. These instruments are discussed in detail in the next chapter on pp. 27ff.

Accademia del disegno That art and practical mathematics were intimately related is also shown in the founding of the first academy for the training of artist-engineers: the *Accademia del disegno*.[18] This academy was founded in 1563 under the patronage of Cosimo I. Although the normal emphasis is placed on the academy's importance for the production of art, like painting, for example, this was nevertheless the place where architects and professionals of the mechanical art also could obtain their training and integration into Florentine professional life. The first statute reports that architects, sculptors and painters, besides their particular disciplines, had to be taught in what were known as the auxiliary arts (Adorno and Zangheri 1998, 6). These were divided into the fields of theory and practice. Theory was represented by the works of Euclid and Vitruvius, while practical knowledge was especially represented by perspective techniques, practical geometry, the art of fortification, and machine building.[19] However, these courses were far below the didactical standards set for the *accademici*. Still, in the first statute it is specified that Cosimo I's desire was that architects:

> [...] attend [...] to the matters of visiting the rivers, the sewers of the city, the deliberations for building bridges, and other things, public and private, relevant for the city and for the realm, together with the officers and the other engineers appointed to the matters of Drawing and that they report it to the College, which is obliged to discuss, to draw and to write, on the basis of plans and drawings, on what is needed and to inform His Excellence [...].[20]

The pupils of the *Accademia* therefore also had to learn by visiting, discussing and doing. In this respect their role was strictly official and dependent on the will of the Grand Duke: Their work was put under the control of the state administration. This process of change also concerned the social status of engineers and architects, and became even more evident in the statute of the *Accademia* compiled in 1585, when the *Accademia* started exerting a degree of control over the didactic activity of the independent artist workshops.[21] The Florentine *Accademia del disegno* then became a model for the training of artists and engineers in the most important cities throughout the Italian peninsula. Either by founding new academies, or by changing the character of already existing schools, or even by means of informal intellectual

[18]On the relevance of the *Accademia del disegno* as a cultural institution in Florence, and on the subjects taught there for the education of artist-engineers, see also Olschki (1919–1927, II: 187–194).

[19]The chair for mathematics, a required subject for the students of the *Accademia del disegno*, was located at the *Studio Fiorentino* until 1639, when the *Accademia* finally obtained its own chair (Adorno and Zangheri 1998, 71).

[20]"[...] intervenissero [...] alle cose delle visite de i fiumi, alle fogne della città, alle deliberazioni di fare i ponti, et l'altre cose publiche et private importanti della città, et del dominio, insieme co gl'ufficiali et gl'altri ingegneri diputati alle cose del Disegno et che riferissi tutto al Collegio, il quale fusse obbligato sopra le piante e disegni, di quel che si fusse, disputare et disegnare et scrivere sopra di ciò et informare S.E. [...]." From Chapter XXXVII of the statute: Adorno and Zangheri (1998, 14).

[21]The specific rules framing the *Accademia*'s control of the didactic activity of the workshops in Florence are published in Adorno and Zangheri (1998, 56–58).

circles, the Florentine teaching model became a standard during the second half of the sixteenth century.[22]

Galileo's Apprenticeship

Galileo was born in Pisa in 1564 and found his first position as a lecturer in mathematics, first at the University of Pisa in 1589 and then at the University of Padova in 1592.

Galileo's basic education In Pisa Galileo first attended a grammar school, where he was tutored by Jacopo Borghini (Favaro 1966, I:6). Assisted by his father Vincenzo, he also studied classical Latin and Greek literature during his early youth. Vincenzo Galilei was a musician, who also taught Galileo to play the lute. Galileo's father was active scientifically as well, engaged in studies of acoustics. His approach to this field was a practical one, entailing, for example, experimentation with several materials by means of his lute and a monochord (Galilei V. 1581). According to Thomas Settle, this influenced the intellectual vivacity of the young Galileo (Settle 1995). Vincenzo Viviani (1622–1703), a later pupil of Galileo and one of his first biographers, reported how the child Galileo was already able to construct models of mechanical devices like mills and galleys.[23] After this period, it seems that Galileo went to a monastery in Vallombrosa, where he is supposed to have enrolled as a novice.[24]

Galileo as a university student In 1581 Galileo enrolled at the University of Pisa in order to attend, in keeping with his father's will, the studies of medicine—and thus also of philosophy and of mathematics, according to the university statutes. However, there is absolutely no evidence that Galileo ever attended a single lesson in mathematics. A plausible explanation, formulated by Antonio Favaro, the editor of most of Galileo's works, is that Galileo's professor of mathematics probably limited the content of his course only to the traditional study of *The sphere*,[25] namely practical astronomy, perhaps on the basis of an expanded edition of Sacrobosco's work, a frequent practice at universities during the sixteenth century.[26] In 1585 Galileo

[22] A comparison between the local situation in the cities of Bologna and Florence was drawn in Dempsey (1980). The situation in Venice, which is more characterized by intellectual circles among patricians of the city, is best described in Olschki (1919–1927, II:195–199).

[23] Viviani wrote: "During his leisure time, he usually was trying to construct, by own hand, several sorts of instruments and small machines through imitation, and making small models of technical devices that he saw, like mills, galleys, and also every other kind of machine" (*EN*, XIX:601; Favaro 1966, I:6).

[24] The evidence on this aspect of Galileo's life is reported in Selmi (1864). Francesco Selmi is the only historian who has been able to view these sources to date. For some discussion on this point, see also Favaro (1966, I:6–7).

[25] Details on the traditional study of *The sphere* are given in Chapter 3.

[26] Vincenzo Viviani also related that, during his third year at the university, Galileo discovered, by means of observations, the isochronism of the pendulum, and that he immediately applied this discovery to the construction of an instrument to measure time while a doctor registers the heartbeat.

abandoned the university and moved to Florence, where his family had since settled. Once in Florence Galileo almost immediately began attending lessons on Euclid by Ostilio Ricci.

Ostilio Ricci Ostilio Ricci was born in Fermo in 1540 and died in Florence in 1602 (Settle 1971, 125, note no. 9). It is unknown where Ricci received his mathematical education, but Olschki suggested that he might have been one of Tartaglia's followers (Olschki 1919–1927, III:145). At the age of forty he was employed at the court of the Grand Duke of Tuscany as a teacher of mathematics and, probably, of military engineering. His first pupils were probably the pages of the court. Later in the 1590s he also instructed Don Giovanni de Medici, natural son of Cosimo I. In 1590 he was in Ferrara and Bologna to deliver an opinion on the situation of the streams of the waters in those countries, especially that of the river Reno, which was often the cause of flooding. In 1593 he was appointed to the chair of mathematics of the *Accademia del disegno*. In 1597, on the occasion of a conflict between Tuscany and France, he directed the construction of the fortress of Iff, on the islands of Marseilles. After this experience he moved to Ferrara to work with Cesare d'Este, entrusted with negotiating about the territories claimed by both the prince and Pope Clement VIII.

Galileo and Ostilio Ricci Many early biographers give different accounts of how Galileo came to attend Ricci's lessons. It is possible, for example, that Galileo already heard Ricci's lessons in Pisa, during a stay at court there in 1584, where the mathematician was working as a teacher of practical geometry and fortifications for the pages of the court (Favaro 1966, I:15; Bredekamp 2007b, 33–38). It also seems certain that Ricci was a good friend of Galileo's father and therefore well known to Galileo as a regular guest to his family home (Olschki 1919–1927, III:142). Ricci was also giving lessons on practical geometry to apprentices at Buontalenti's artist workshop (Bredekamp 2007b, 36–37). As mentioned, these pupils also included Ludovico Cardi da Cigoli, who remained Galileo's lifelong friend. According to some historians, Galileo might have attended Ricci's lessons in Buontalenti's workshop as well.[27] Although there is only circumstantial evidence as to how Galileo came in contact with Ricci, in any case it is certain that as early as 1587 Galileo's application for a chair as lecturer on mathematics at the University of Bologna included a letter of recommendation presenting Galileo as Ricci's pupil (*EN*, XIX, Doc. no. VII, 36). Thus, Galileo had certainly been Ricci's pupil between 1585 and 1587, and it is highly probable that they remained in close contact until 1589, when Galileo left Florence to take the chair at the University of Pisa (Baldinucci 1728, VII:43; Presas i Puig 2002).

Ricci's lessons Six manuscripts are preserved from Ricci's legacy. One manuscript is the report of Ricci's study on the streams of the waters in the

---

While there is no doubt that Galileo discovered the isochronism of the pendulum at a certain early point in his life, Viviani's report that he did so in 1584 and thanks to observations of a hanging lamp in the cathedral of Pisa, and that he immediately contrived and constructed that instrument, must evoke skepticism.

[27] See in particular, *La corte il mare i mercanti* 1980, 145.

area between Ferrara and Bologna, while the other five manuscripts[28] are of a didactic character, concerned with practical geometry, arithmetic, commercial arithmetic, drawing techniques, surveying techniques, the use of measuring instruments, measuring techniques and military architecture. These five manuscripts surely represented part of the teaching program normally imparted by Ricci to his pupils, including Galileo.

Ricci's *Problemi di geometria pratica* (Ricci ca. 1560–ca. 1590b) and *Problemi geometrici* (Ricci ca. 1560–ca. 1590c) deal with a selection of propositions from the first, second, third, fifth and sixth books of Euclid's *Elements*, and with some of their practical applications. Another manuscript by Ricci is a treatise on architecture (Ricci ca. 1560–ca. 1590e), a tract on military fortifications that belongs to the treatises associated with the *Accademia del disegno* in Florence.[29] Finally, a further manuscript also entitled *Problemi geometrici* (Ricci ca. 1560–ca. 1590d), but preserved among the Galilean manuscripts in Florence, contains, as T. B. Settle first discovered (Settle 1971), an integral copy of Alberti's *Ludi matematici* written between 1450 and 1452 (Alberti 1973), a set of exercises on interest calculations (with their solutions), and an inventory list of the furnishings of a villa in Fiesole, written in 1631, which probably belonged to one of Ricci's descendants, Massimiliano Ricci. Settle presumes that Galileo's training was thus a sort of synthesis between the state of practical geometry in the first half of the sixteenth century, represented by Ricci, and the summit of practical knowledge in a wider sense, represented by Alberti. The *Ludi matematici* deal, first, with methods to measure height and depth, and also water depth, by means of sight; second, with balances and other practical applications of the principle of the lever, for example, those concerned with leveling terrain and shooting cannons; third, with the measurement of real irregular surfaces using an odometer; fourth, with the measurement of time during both day and night; and, finally, with calculating the weight of heavy objects (columns, in his example) by means of the famous Archimedes' principle passed down through the legend of the crown of Geron II, the tyrant of Syracuse. Galileo certainly studied more than one of Archimedes' works with Ricci, like, for example, *On the Equilibrium of Planes* and *On Floating Bodies*.[30] In the context of these studies Galileo wrote his first treatise, *La bilancetta* (*EN*, I:210–228), in which he described a balance that applies Archimedes' principle of specific weight (Bedini 1986, 130; Bertoloni Meli 2006, 50–51).[31]

[28]Ricci (ca. 1560–ca. 1590a, b, c, d, e).

[29]For more details about the tradition of treatises on military fortifications developed at the *Accademia del disegno*, see Chapter 3 on pp. 71ff.

[30]According to a tradition started by Niccolò Gherardini (1607–1678), one of Galileo's first biographers, Ostilio Ricci gave Galileo copies of some Archimedes' works as a gift (*EN*, XIX:637).

[31]Ricci's manuscript contains the concept of specific weight and it was a subject of his lessons. On the works of Archimedes, see Archimedes (2002a, b). Galileo explicitly quoted the legend of Geron II and complained that Archimedes did not work as handed down by that legend. Galileo therefore reconstructed what he believed Archimedes' method had been.

Settle's conviction that Ricci was using Alberti's tract for his lessons seems to be corroborated by the other treatise by Ricci, which emblematically bears the same title. In fact, the other text, entitled *Problemi geometrici* (Ricci ca. 1560–ca. 1590c), is a kind of complementary treatise to the *Ludi*. For example, where Alberti described how to measure the height of a tower, Ricci described the technique to measure a whole fortress and how to find the different points where the observer must be positioned. The problem concerning how to measure the surface of a property, for example, is explained by Ricci in a much more detailed way. Ricci's writing, moreover, concludes with a similar practical example on how to weigh a column, not by introducing the concept of specific weight, however, but merely the practical way to use that concept.

According to Sven Dupré, moreover, Ricci gave lessons on perspective, which probably included the basics of catoptrics and dioptrics (Dupré 2005). In fact, there is evidence that during the 1590s, when Ricci was giving private lessons to Giovanni de' Medici, he was using a text entitled *Della prospettiva*, whose original author was most likely Giovanni Fontana (1540–1610). On the basis of the analysis of this text, Dupré concludes that Ricci was also teaching optics; therefore, there are circumstantial reasons to believe that the same content was passed on to Galileo as well. Moreover, the knowledge Galileo acquired was sufficient to understand the functioning not only of mirrors, but also of the telescope when it first appeared, especially as regards the way images are formed by convex lenses.[32]

In conclusion, according to Dupré's inference, Galileo's training, as accomplished with Ostilio Ricci and in Buontalenti's workshop, involved practical optics and was concerned with the functioning of several optical instruments or devices as well. These included mirrors of several shapes and sorts, like the parabolic burning mirror, and ordinary spectacles (Dupré 2005). Moreover, this activity was so extensive that Galileo even became familiar with the workshop practices of spectacle makers.[33]

From the Apprenticeship to the Workshop via the University

In 1587, just two years after Galileo started learning Euclid and Archimedes, he was recommended for a chair as lecturer in mathematics at the University of Bologna. This is often seen by the biographers as a sign of the Pisan scientist's ingenuity. What is normally omitted from this account is the emphasis on the fact that the University of Bologna refused to employ Galileo. In fact, as is the case still today, obtaining a chair at the university was an opportunity reserved for scholars who already had shown their capacities in the given subject. Galileo knew this, too: He started working on another subject during this period. The title of the output of this research is *Theoremata circa centrum gravitatis solidorum*, published first in

---

[32] Galileo's work on and with the telescope is discussed in Chapter 2, pp. 41ff.

[33] Galileo's activities in the frame of practical optics is discussed in Chapter 2 on pp. 41ff.

1638 as an appendix to the *Discorsi e dimostrazioni matematiche intorno à due nuove scienze* (*EN*, I:179–208). Most of the *Theoremata* were written between 1587 and July 1588.[34] It is one of Galileo's "unpublished treatises,"[35] representing a further extension of Archimedes' mechanics into the context of practical applications. They present theorems that demonstrate where the centers of gravity of certain bodies are located. In particular, Galileo found the center of gravity of ideal balances, whose weights are hung in a few different and well-determined distributions, and also the center of gravity of bodies like those determined by a cross-section of a parabolic conoid. Galileo probably did not approach this study casually. From a letter by Guidobaldo del Monte (1545–1607),[36] it is possible to infer that several mathematicians were trying to improve, or better said, to generalize the final proposition of Federico Commandino's (1509–1575) *Liber de centro gravitatis solidorum*,[37] as one of Galileo's theorems from the *Theoremata* tries to do as well (*EN*, I:196–198). Guidobaldo del Monte, who became the editor of Commandino's works after his death, had already pointed out the lack of universality of that proposition to Christophorus Clavius (1538–1612), a famous mathematician of the *Collegio romano* at that time.

---

[34]Galileo himself indicated these dates for the *Theoremata* in a letter of 1636 to Elia Diodati, Tuscan diplomat in Paris, who also acted as Galileo's agent to the Elzevir Press House in Holland, which was in charge of printing Galileo's last work. For more details, see Galileo to E. Diodati in Paris, December 6, 1636, in *EN*, XVI:523–524.

[35]Galileo often prepared sketches and outlines for publications he had in mind but ultimately never realized. An almost complete list of these works of Galileo has been denominated "Galileo's unpublished treatises." The paper which initiated this tradition is Büttner et al. (2001).

[36]For the correspondence about the *Theoremata* between Guidobaldo del Monte, Galileo and Christophorus Clavius, see especially Guidobaldo del Monte to Galileo, January 16, 1588, in *EN*, X:25–26.

[37]Commandino's proposition, which Guidobaldo del Monte did not consider to be universal enough, is the following: "The center of gravity of any frustum is on the axis in such a way that, first, from the square which is formed on the diameter of the larger base, one takes away one third, and two thirds from the square which is formed on the diameter of the smaller base. Then one takes away from the third of the square of the larger base that part, to which the rest of the square of the larger base, together with the mentioned part, has the double proportion of that of the square of the larger base to the square of the smaller base. The center [of gravity] is in that point on the axis where it [the axis] is so divided that the part touching the smaller base has the same proportion to the other part [of the axis], that the rest, once one has taken away two thirds of the square of the larger base from the square of the smaller base, together with the part from which one third of the larger square has been taken away, has to the remaining part of the same third" ("Cuiuslibet frusti à portione rectanguli conoidis abscissi, centrum grauitatis est in axe, ita ut demptis primum à quadrato, quod fit ex diametro maioris basis, tertia ipsius parte, & duabus tertiis quadrati, quod fit ex diametro basis minoris: deinde à tertia parte quadrati maioris basis rursus dempta portione, ad quam reliquum quadrati basis maioris unà cum dicta portione duplam proportionem habeat eius, quae est quadrati maioris basis ad quadratum minoris: centrum sit in eo axis puncto, quo ita diuiditur ut pars, quæ minorem basim attingit ad alteram partem eandem proportionem habeat, quam dempto quadrato minoris basis à duabus tertiis quadrati maioris, habet id, quod reliquum est unà cum portione à tertia quadrati maioris parte dempta, ad reliquam eiusdem tertiae portionem") (Commandino 1565, 46–47).

The start of Galileo's academic career Galileo used the *Theoremata* to support his applications for a chair at several Italian universities. At the same time he succeeded in involving in his work the most important mathematicians of his age, like, for example, Pietro-Antonio Cataldi (1548–1626), Giuseppe Moletti (d. 1588), Clavius and the famous scientist Guidobaldo del Monte. Thanks especially to the support of the latter and his brother,[38] who was Cardinal, Galileo finally started his academic career as a lecturer for mathematics at the University of Pisa in fall 1589.[39]

Galileo's theoretical research in Pisa Galileo remained in Pisa for three years.[40] During this time his work was characterized by the tendency to focus on theoretical developments, in the fields of mechanics and neoscholastic philosophy. He seems to have abandoned his origins and his training as an engineer. During this period Galileo produced several texts, none of which was ever published or mentioned by him in any of his later published works.[41] Some of the texts produced during this period are collected under the name *De motu antiquiora* (*EN*, I:243–419), while the rest was compiled under the name *Juvenilia* (*EN*, I:9–188) by Antonio Favaro.

De motu The *De motu antiquiora* consists of several treatises and one outline for another treatise. In these treatises Galileo, adhering to Aristotle's theory of motion, for which a motion is possible only in a medium, tries to combine it with Archimedes' account of how a medium affects a body immersed in it (Büttner et al. 2001). Galileo's target was the study of the velocity of falling bodies, which, for Galileo as for other contemporaries like Giovanni Battista Benedetti (1530–1590), Guidobaldo del Monte, and Thomas Harriot (1560–1621),[42] required a combination of the two theories. As a case study, Galileo also took projectile motion into consideration, for which he offered, in the *De motu antiquiora*, a general vision resembling that of Tartaglia, who supposed that the motion of a projectile is constituted of two different motions: the first, which follows a straight line, is the violent motion, and the second, which is bent, is the motion which naturally brings all bodies to the center of the Earth. In other words, Galileo, like Tartaglia, conceived a theory of projectile motion which involved basic Aristotelian assumptions.[43]

Early writings The so-called *Juvenilia* are remarkable texts because of their undisputed Aristotelianism. Since these manuscripts do not bear any date, only the clear content of the texts made it possible for Favaro to attribute them to the period

---

[38] Guidobaldo del Monte to Galileo, August 3, 1589, in *EN*, X:41.

[39] For a detailed description of this period of Galileo's life, which was characterized on the one hand, by the dramatic attempt to improve his demonstrations about the centers of gravity, which had been criticized in part by Clavius, and on the other hand, by the even more dramatic search for a chair, supporting his applications with reference to his *Theoremata*, see Camerota (2004, 44–55).

[40] Copies of bureaucratic acts and administrative documents concerning Galileo's employment in Pisa between 1589 and 1592 are in *EN*, XIX:37–43.

[41] For a description of the manuscripts of these early texts and their dating, see Camerota (1992, 49–101).

[42] For an extensive work on Harriot's mechanics, see Schemmel (2008).

[43] This argument is in Damerow et al. (2004, 141–147).

when Galileo was a student in Pisa. However, post-Favaro analyses showed that these texts are in fact copies, either entire or almost complete, of handbooks written by Jesuit scholars at the *Collegio romano* in Rome. According to the date of publication of these handbooks, or to the dates when their manuscripts started circulating, Galileo drafted his *Juvenilia* when he was somewhat older, in particular, during the three years of employment in Pisa (Caruso and Crombie 1983).[44]

Galileo's *Juvenilia* contain the following works: a copy of the second edition of a commentary written by Benito Pereira (1535–1610) on Aristotle's *De caelo* (Aristotle and Guthrie 1939) entitled *De communibus omnium rerum naturalium principijs et affectionibus libri quindecim* (Pereira 1576); a copy of the commentaries written by Francisco de Toledo (1532–1596) on Aristotle's *Physics* (Toledo 1574) and *De generatione et corruptione* (Toledo 1575); a partial copy of Sacrobosco's *The Sphere* written and commented by Clavius and entitled *In Sphaeram Ioannis de Sacro Bosco commentarius* (Clavius 1582).[45] Among the *Juvenilia*, finally, Favaro also placed small portions of another Galilean manuscript, which originally did not bear any title. Favaro decided to publish only a small part of that manuscript because he was convinced that it represented a simple exercise in Aristotelian logic accomplished by Galileo, at the age of fourteen during what may have been a stay at the monastery of Vallombrosa. The debate between the scholars Caruso and Crombie on the one side and Wallace on the other, however, revealed that Galileo's manuscript, which is now entitled *Disputationes de praecognitionibus et de demonstratione*,[46] might be a commented copy of notes prepared by the Jesuit professor Paolo della Valle in Rome for his course on Aristotelian logic in 1588.[47]

All of Galileo's texts which comprise the *Juvenilia*, even those not published in their entirety and considered to be works produced at a very early age, are elaborated copies, complete or partial, of commentaries on Aristotelian works produced by Jesuits at the *Collegio romano*. As a letter by Guidobaldo del Monte written in 1590 shows, Galileo often studied philosophy during these three Pisan years, sometimes together with and as a pupil of Jacopo Mazzoni (1548–1598), a scholastic professor of Philosophy also employed at the University of Pisa and Galileo's friend.[48] Thus

---

[44]The work of identifying the so-called "early writings" by Galileo carried out by Adriano Caruso and Alistar Crombie has since been shown to be wrong with regard to one text and corrected in Wallace (1984a, b, 1986b, 1990). See also Galilei and Edwards (1988). For an overview of the quarrel between Wallace on the one side and Caruso and Crombie on the other, see Camerota (1992).

[45]Clavius's work was used by Galileo to give private lessons, until Galileo himself wrote his own commentary on *The Sphere* for the same purpose. For more details, see Chapter 3, pp. 89ff.

[46]The manuscript was completely transcribed, translated and presented to the press by William F. Edwards and William A. Wallace in Galilei and Edwards (1988).

[47]Paolo della Valle did not publish his work until 1622 (Valle 1622). The debate which resulted in the identification of Galileo's manuscript is reported in great detail in Camerota (1992, 83–101).

[48]Guidobaldo del Monte to Galileo, December 8, 1590, in *EN*, X:45. For Galileo's activity as a pupil of the philosopher Jacopo Mazzoni, see Galileo to Vincenzo Galilei, November 15, 1590, in *EN*, X:44–45. In 1597 Galileo wrote a public letter, addressed to Jacopo Mazzoni, in which

Galileo certainly dedicated his three Pisan years to the study of the neoscholastic doctrines.

Galileo's intellectual production of this period does not show a single aspect of his training obtained in the fields of practical knowledge. Apart from one further theorem on the center of gravity formulated in 1590,[49] Galileo was apparently fascinated by pure deductivism. Galileo was, in fact, a young engineer, who did not even finish his university education when he was a student and who then became a university lecturer. He was trying to update his knowledge concerning the philosophy taught at the universities, that is, the high culture of his time. Concerning his *De motu antiquiora* and his work on the commentaries by Pereira on Aristotle's *De coelo* and by Toledo on Aristotle's *De physica*, since Galileo was probably genuinely interested in developing a theory of motion, he could do so only by starting from what the knowledge of his time was stating in reference to this study: He could not ignore Aristotelian physics. And in fact his early physics is constituted not only of Archimedean principles of hydrostatics, but for the most part of basic Aristotelian assumptions.[50] Finally, concerning the new social status he enjoyed upon becoming a professor at the university, he probably felt the need to become familiar with that knowledge. He was simply expected to know Aristotelian physics as well as logic. Galileo needed to make up for what he did not achieve when he was a student, and he apparently did so together with his friend, the philosopher Jacopo Mazzoni. From Galileo's perspective, the work of these three years in Pisa corresponds to what could be expected from a young engineer who has become a lecturer for mathematics at the university.

The Buzz of the Workshop

Just a few months after starting his professorship in Pisa, Galileo was already looking for another position. According to the correspondence between Galileo and Guidobaldo del Monte, the main reason for Galileo's disappointment concerning the chair of Pisa was the low salary he was receiving for his work.[51]

---

he recalls those years when they worked together in Pisa. For more details, see Galileo to Jacopo Mazzoni, May 30, 1597, in *EN*, II:193–202.

[49]Guidobaldo del Monte to Galileo, December 8, 1590, in *EN*, X:45.

[50]For a detailed analysis of the physical assumptions in Galileo's *De motu antiquiora*, see Damerow et al. (2004, 135–286). See also Fredette (1970).

[51]Galileo's complaining about his salary as well as his attempt to obtain the chair at the University of Padova are documented by the following epistolary sources: B. Zorzi to B. Valori, December 2, 1589, in *EN*, X:42; Guidobaldo del Monte to Galileo, April 10, 1590, in *EN*, X:42–43; Guidobaldo del Monte to Galileo, February 21, 1592, in *EN*, X:46–47; G. V. Pinelli to Galileo, September 3, 1592, in *EN*, X:47–48; G. V. Pinelli to Galileo, September 9, 1592, in *EN*, X:48–49; G. Uguccioni to B. Vinta, September 21, 1592, in *EN*, X:49; G. V. Pinelli to Galileo, September 25, 1592, *EN*, X:49–50; G. Uguccioni to the Grand Duke of Tuscany, September 26, 1592, in *EN*, X:50.

The meeting between Galileo and Guidobaldo In a letter dated February 21, 1592,[52] Guidobaldo del Monte invited Galileo to stay at his house in Pesaro for a time during the journey to Venice that Galileo intended to undertake in a repeated attempt to obtain the chair for mathematics at the University of Padova, which recently had become vacant upon Moletti's death. This stay, which took place between August and the first half of September of the same year, contributed to changing the orientation of Galileo's interests back to the practical challenges of engineers. In this sense, this meeting in Pesaro changed Galileo's life path.

Recent studies concerned with Galileo's discovery of the trajectory of projectiles have shown that it was on the occasion of the Pesaro meeting that the two mathematicians together performed the experiments whose result they interpreted as a proof that the trajectories of projectiles followed a line curved as a catenary, a hanging chain (Renn et al. 2001, 24–36). The need for this kind of experiment was primarily given by the fact that the actual theories, though useful, did not correspond to the everyday-life experience of any artillerist, who knew perfectly well that the trajectory of the cannonballs did not really follow a straight line. This and another experiment on the flow of water along an inclined channel, in the performance of which Guidobaldo del Monte referred to craftsmen's rules, are experiments strongly oriented towards the challenges represented by the work of the artist-engineer, for example, that of the artillerist.

Galileo's new position in Padova Galileo never again abandoned this approach in his work, and this is especially visible during the next 18 years of his life. After the three-year stay in Pisa and, again, thanks to the help of Guidobaldo del Monte, Galileo obtained the chair of mathematics at the prestigious University of Padova, a position toward which Galileo and Guidobaldo del Monte had been working since 1590, and which he held until 1610. Galileo began teaching in Padova on December 7, 1592,[53] and one of his first initiatives there was to open a workshop in his house in order to teach military fortifications and to build mathematical instruments for military officers: Galileo became a master.

[52]Guidobaldo del Monte to Galileo, February 21, 1592, in *EN*, X:46–47.

[53]On the activity of Galileo and his network to obtain the chair at the University of Padova, see Guidobaldo del Monte to Galileo, April 10, 1590. in *EN*, X:42–43; Guidobaldo del Monte to Galileo, February 21, 1592, in *EN*, X:46–47; Giovan Vincenzo Pinelli to Galileo, September 25, 1592, in *EN*, X:49–50; Guidobaldo del Monte to Galileo, January 10, 1593, in *EN*, X:53–54. See also Favaro (1966, I:37–50).

Chapter 2
Instruments and Machines

Galileo's apprenticeship allowed him to build bridges to mathematicians, philosophers and several kinds of artist-engineers and craftsmen, including mirror makers, military engineers, and machine builders. In the context of such a network, Galileo cultivated and practiced all of these activities himself. He opened his own workshop in Padova, which produced mathematical instruments for military officers; starting in 1610, upon his move to Florence in the service of the Grand Duke, he made optical instruments destined primarily for military use; and over the course of his life he became a recognized expert on machines, especially on those devices useful for living and fighting within fortresses.

Galileo held his inaugural lecture at the University of Padova on December 7, 1592. The University of Padova, although already entering its phase of decline, was still a famous European center for all those who wanted to obtain a renowned university degree in a liberal atmosphere. At the end of the sixteenth century, the University of Padova, besides the structures usual for its day, boasted a new university chair and newly founded institutions like the school for clinical instruction, the botanical garden and the anatomical theater. The University of Padova was also a famous center for its role as an important location for the cultivation, commentary and teaching of Aristotelian philosophy. According to Edward Jan Dijksterhuis, however, the orientation of the reading of Aristotle's works at the University of Padova was based heavily on Averroes' reading, who contradicted Catholic dogma by stating the eternity of the world (Dijksterhuis 1983, 261–262).

Parallel to his activity as a public lecturer, Galileo also started giving private lessons during the first years of his stay in Padova. This practice probably developed gradually according to the degree of fame that the young professor achieved during the period. In any case this activity was certainly well established by 1599, when the production of military instruments in his workshop became systematic. His house became a sort of college where pupils lodged for a fee. Besides room and board for them and their servants, they received private lessons, manuals, and the famous military and geometrical compass conceived by Galileo. It was an all-inclusive service, with differentiated payments for each task, designed to familiarize rich students with the topics of military fortifications and with siege and defence strategies, especially by means of the mathematical instruments produced by Galileo's workshop. Moreover, the workshop also produced on demand the typical

M. Valleriani, *Galileo Engineer*, Boston Studies in the Philosophy of Science 269,
DOI 10.1007/978-90-481-8645-7_2, © Springer Science+Business Media B.V. 2010

mathematical instruments of Galileo's day, like reduction and surveying compasses, and did so independently of Galileo's activity as a private teacher. As such it worked like any of the other regular workshops run by anonymous but specialized smiths.

Galileo's activity as a workshop manager nearly came to a halt when he moved from Padova back to Florence in 1610. However, the production of instruments in general did not cease at this time. In fact, during the last year of Galileo's stay in Padova, he learned of the existence of the telescope.[1] Galileo immediately started working on this instrument to improve its magnifying power. Thanks to his discovery of four of Jupiter's moons in 1610, Galileo became enormously famous all over Europe, not only as an astronomer but also as a maker of optical instruments, an activity which kept him busy well into old age. The reconstruction of Galileo's activity during this period shows very clearly not only that he was familiar with the work of the lens and mirror makers, but also how efficient his own activity as a lens maker was.

The telescope was discovered rather accidentally by manufacturers of spectacles and lenses. Just a few years after the telescope was discovered, it seemed to Galileo and to most mathematicians that the only way to improve the telescope further was to solve three problems, on whose solutions Galileo worked intensively. First, the polishing method used to achieve the curvature needed for telescope lenses was a cause of great waste. Only one third of the lenses produced were considered usable, and very few of these to be good. Second, the only possible curvature for the lenses was the spherical one, for parabolic and hyperbolic curvatures were still far from attainable, although many soon recognized the relevance of these lense shapes for the further development of optical devices. The third problem was presented by the extremely poor quality of the glass or crystal used to make lenses. In Galileo's day this was perceived to be the most relevant of the problems. Within this context Galileo never tried to change the structure of his telescope, which was originally a compound of a plano-convex lens and a plano-concave lens, in favor of other optical solutions,[2] but he did indeed re-arrange it in order to design other optical devices working on the basis of the same principle. Galileo's microscope and binoculars can be regarded as products of such efforts.

As an extension to this activity, Galileo became involved in many discussions about the construction and functioning of mirrors, and especially of curved ones. Back in Padova he had taken advantage of occasions to discuss this issue with Paolo Sarpi (1552–1623) as well, and later he played the role of mirror broker in his capacity as court engineer of the Grand Duke.

As a private teacher, Galileo's efforts focused on practical mechanics and machine building. As will be shown in this chapter and the next, his knowledge

---

[1]It seems quite certain that the first manufacturers to use the lens combination that later became the Galilean telescope were the Dutch spectacle makers Jacob Adraansz and Hans Lippershey in 1608. For more details, see Moll (1831) and van Helden (1977).

[2]In his *Dioptrice* Johannes Kepler (1671–1630) described a telescope working with two bi-convex lenses (Kepler 1611). This sort of telescope was first built by Christoph Scheiner (1575–1650) in 1630.

of practical mechanics, which he shared with the engineers of his time, is what truly constitutes the foundation of his later theoretical developments.

During the years 1592–1593 Galileo wrote a treatise called *Delle macchine* (Galilei 1592–1593a), which is generally considered to be a brief and less interesting version of his later *Le mechaniche* (*EN*, II:145–191), written in 1599. However, the earlier text, which was used by Galileo in the framework of his private lessons, is better characterized as a kind of handbook for the understanding and evaluation of those machines useful within a fortress. In this text Galileo thus not only took many construction details into consideration, but he also made the effort to visualize these machines by means of illustrations. These characteristics were then lost in the next text, which was devoted mainly to the theoretical reorganization of the science of machines.

Galileo also worked with many smiths and masters. In 1602, for example, Galileo was commissioned by his good friend, the Venetian nobleman Giovan Francesco Sagredo (1571–1620), to repair some tools for making screws. Francesco Sagredo's tools were self-made reproductions of similar ones first assembled by Master Fait, a smith who worked for Galileo fairly constantly for a while, during the central years of Galileo's stay in Padova.[3] And later, when Galileo's daughter, Maria Celeste Galilei, sought to repair a clock for the convent where she was living, an instrument that had been manufactured in Munich under commission of Michelangelo Galilei, she dispatched the clock to Galileo, who reassembled some of its components before sending it back.[4]

Moreover, the epistolary exchange among Galileo and other mathematicians is rich in details that demonstrate Galileo's interest in machines. Yet not everything that could be learned from other mathematicians was truly fruitful. This was the case when Giovanni Battista Baliani (1582–1666) told Galileo about a new invention with which one could cook without fire.[5] Doubtless to the great disappointment of Baliani's neighbors and manservants, Baliani constructed and tested a mechanical device similar to those used to produce terra cotta vases, but capable of developing much greater force while rotating on a flat circular surface of metal. Then, taking a pot, also of metal, and presumably obliging a manservant to keep it well fixed in contact with the circular surface, Baliani set the device rotating, thus causing high friction between the bottom of the pot and the circular surface of the device. Due to this friction the pot became hot and thus, without fire but with great noise, it might have been possible to cook something.

Besides Galileo's dedication to machine-building issues in the framework of his private lessons, which will be described in the next chapter, he had already displayed

[3] G. Sagredo to Galileo, January 17, 1602, in *EN*, X:86; G. Sagredo to Galileo, August 23, 1602, in *EN*, X:90. For the translations of the entire letters, see p. 221 and p. 222.

[4] Maria Celeste Galilei to Galileo, January 21, 1630, in *EN*, XIV:68; Maria Celeste Galilei to Galileo, February 19, 1630, in *EN*, XIV:81. In his *Racconto istorico*, Vincenzo Viviani (1622–1703) stated that Galileo had written a treatise on gnomonics as well (*EN*, XIX:606).

[5] G. B. Baliani to Galileo, April 4, 1614, in *EN*, XII:44. For the translation of the entire letter, see p. 238.

his talent as an expert on machines during the first years of his stay in Padova, when he conceived a water lifting machine. He was later busy evaluating machines in his role as court engineer in Florence when he was asked to analyze proposals for a machine to pound gunpowder, and for a machine to lift great weights.

In the following sections Galileo's activity as a workshop manager and his production of mathematical and optical instruments as well as his engagement as a machine maker will be analyzed with respect to the way the instruments functioned and the context of their emergence. These will be subjected to repeated comparison with similar instruments and machines made by contemporary mathematicians, engineers and craftsmen, in order to achieve a full understanding of the historical context of Galileo's activity in this field, and of the processes of exchange between Galileo and his contemporaries. The mathematical instruments considered here are represented by several kinds of proportional compasses, by the reduction compass and by the surveying compass. Optical instruments include mirrors, lenses and those compound instruments like the telescope, the microscope and the binoculars. The machines considered here will be Galileo's water lifting machine and those documented by Galileo's evaluations of other engineers' proposals. Galileo's text *Delle macchine* will be analyzed in the next chapter, as its function is mainly to be understood in the framework of his private teaching activities.

Galileo's Balance Sheet

Two manuscripts, Ms. Gal. 26 and 49 (Galilei 1598–1634 and 1609–) (Fig. 2.1), contain a multi-column register for Galileo's household, his workshop and his private lessons. The register was published with the name *Ricordi autografi* (*EN*, XIX:130–206) and is the result of a reconstruction by Antonio Favaro, whereas the real manuscripts also include many calculations concerned with the positions of the Medicean Planets[6] and some notes concerning his studies on motion disconnected to the household register, which were often ignored by Favaro. The entries are ordered in groups and generally chronologically, whereas on the original leaves these groups are scattered about, often without any order.[7] Most of the entries refer especially to the first years of the seventeenth century. However, since there are similar entries for the earlier periods and the later one after 1620, a large section of these entries probably were lost.

[6] Galileo intended to compute the ephemerides of the eclipses of the four biggest moons of Jupiter. For more details, see p. 58.

[7] The folios of the manuscript were probably bound together in random order only later for the sake of preservation.

Fig. 2.1 Folio of Galileo's multi-column household register (Galileo 1598–1634, 13r)

An entry in the *Ricordi autografi* shows that Galileo recruited the smith Master Marcantonio Mazzoleni on July 5, 1599 (*EN*, XIX:131),[8] in order to systematize the production of his military compass.

[8]Marcantonio Mazzoleni was a descendant of a family with a glorious tradition of clock-building (Martellozzo Forin 2005, 60–61).

The Production and Organization of the Workshop

In the workshop Galileo achieved quite a systematic production of three kinds of compasses, including Galileo's famous *Compasso geometrico e militare*. The workshop manufactured the military compass, which is a compass of proportion with two flat arms; the reduction compass with four points; and the compass with four curved points, a compass of reduction conceived to work as a caliber for bombardiers (*EN*, XIX:131–149; Valleriani 2001).

Workflow Usually Master Mazzoleni and Galileo purchased the construction materials themselves, but occasionally friends or relatives did so. The material was ordinarily bought in the form of plates and then sent to a foundry. Once the pieces came back in the proper shapes and sizes, they were finished by Mazzoleni, who refined and assembled them.[9]

The melting process In Galileo's workshop no process was performed to remelt the alloy of copper and zinc carbonate,[10] previously purchased in the shape of plates, in order to obtain the components that constituted a compass. This aspect of the organization of Galileo's workshop can be best understood on the basis of a relevant work on metallurgy written during the Renaissance, Vannoccio Biringuccio's (1480–1539) *Pirotechnia* (Biringuccio 1540). First, outsourcing the foundry was not the result of technical difficulties. In fact, the small sizes needed for the components of the military compass required no particularly large, complicated or costly smithy to remelt the alloy. The other materials required for such a process were not at all difficult to find. They were crushed glass or salpeter (potassium nitrate), and common oil or Greek pitch (Biringuccio 1540, 122v). The crushed glass could be found thanks to the famous glass production of Murano, whose scoriae were normally made available for many different purposes; and saltpeter, being the main component of gunpowder, was a very common product in every Italian city of the sixteenth century. Pitch was easy to find in Venice because of both Murano's glass production, which involved the use of pitch, and shipyard activities at the Arsenal and all over the city, where huge quantities of pitch were used to caulk the galleys and boats. Neither was the remelting process a serious problem, as the associated danger of falsifying the color did not arise before the brass alloy typical for the sixteenth century was remelted for the fourth or sixth time (Biringuccio 1540, 21r).

Economic organization The organization of Galileo's workshop showed strong correspondence to the standard and most economically efficient structure of any small smithy of the time. It was common for smaller workshops to send the raw materials to what were known as second-order smithies, where the material was remelted and shaped. This was and it still is the case for all small private shipyards

[9]Military compasses were normally marked by Mazzoleni. However, those instruments made of silver and destined for great personages bore Galileo's mark as well (Valleriani 2001, 285).

[10]Today brass is made of copper and zinc. In the sixteenth century, however, as is known from Biringuccio's *Pirotechnia*, brass was obtained by "painting" copper with *giallamina*, that is, zinc carbonate (Biringuccio 1540, 19v).

in Venice, too, typically organized to perform every step in the construction of a boat except the shaping of metal components, which are then decorated back in the shipyard (Penzo 1999, 134). Even the Venetian Arsenal, one of the largest and most technologically advanced industrial centers of the day, did not have its own smithy and foundry until 1526, when it was built into the new perimeter of the *Arsenale novissimo*.[11] This organization was a consequence of the fact that, since the end of the first half of the sixteenth century, the production of small brass objects began to shift toward a standardized method of production, which means that Galileo's workshop was organized according to the most profitable contemporary structure.[12] This organization was prescribed by economics, since a well-equipped smithy could afford to produce small objects shaped of brass at lower costs than could the workshops' own smithies. Galileo thus organized his workshop just as any lesser-known foreman and skilled shop manager would have. After all, as Mario Biagioli was able to show (Biagioli 2006, 7–8), Galileo's activity as a workshop keeper was far more remunerative than his work at the university.

In the *Ricordi autografi* and in Galileo's epistolary legacy, the production documented for the period when Galileo was in Padova (1592–1610) lists around eighty-five military compasses, three large military compasses, two military compasses of silver, and a total of ten reduction compasses and proportional compasses with curved points.[13]

The Military Compass

Galileo first achieved the systematic production of the military compass in 1599.[14] In Padova in 1606, he published a book describing how to use it and, especially, how to execute operations using it (*EN*, II:335–424).[15] Galileo's military compass is a

---

[11] In fact, the existence of a small smithy in the Arsenal is documented from 1464, but most of the work was performed elsewhere. Not until 1526 did the Arsenal obtain a foundry sufficient for its production (Concina 1988, 35–136).

[12] In 1530 Biringuccio visited a large manufacture in Milan that expanded so far as to include a smithy for the serial production of small objects. Positively impressed, he vividly reported how fast the production of small objects of brass could be achieved (Biringuccio 1540, 20v).

[13] Galileo's involvement with military compasses certainly continued until at least 1636: In June 1636 Ludovico Baitelli sent his thanks to Galileo for receiving a military compass from him, presumably as a gift. For more details, see Ludovico Baitelli to Galileo, June 11, 1636, in *EN*, XVI:436–437.

[14] For the date when Galileo started producing the military compass, see Cosimo Pinelli to Galileo, April 3, 1599, in *EN*, X:73.

[15] The draft of the book was already available to his pupils in 1599. One year after Galileo's publication of his treatise on the military compass, Baldassare Capra, a person previously introduced to Galileo's house, published a similar treatise in Latin (*EN*, II:425–510). Galileo immediately recognized Capra's work as a plagiarism of his own work. The case was heard by the authorities and Galileo judged to be in the right. In his defense, Galileo stated that he began the production of

Fig. 2.2 Galileo's military compass (Istituto e Museo di Storia della Scienza, Florence. Inv. 2430)

proportional compass (Fig. 2.2).[16] It is constituted of two flat arms, on which scales are marked. The two arms are joined with a central hinge at the top. Ordered from the lowest to the highest quality, the materials used by Galileo to build the compass were basin brass, Italian brass, German brass, and silver. The points were made of steel. As to the dimensions, Galileo produced two kinds of military compasses, which he referred to as normal and as large. After a compass was assembled, it was engraved. Master Mazzoleni was also trained for this purpose.[17] The two arms were perforated at equal distances from the center, ideally placed in the middle of the hinge. Into these small holes a *squadra*, quadrant, could be inserted. This was shaped like a portion of the circumference and engraved with a set of marked scales. The dimensions of each of the arms of Galileo's ordinary compass of proportion are $25\times5\times247$ mm (Righini 1980, 5). The compass was sold equipped with a ruler.

Theory of the military compass The theory embodied in Galileo's military compass is what is called the principle of proportionality. This can be reduced to the propositions VI.2 and VI.4 of Euclid's *Elements* (Euclid and Heath 1956, II:194 and

his compass back in 1597, two years earlier than is documented in the *Ricordi* and Pinelli's letter. Galileo's defense is published in *EN*, II:518. See also the introduction of Biagioli (2006).

[16]Among scholars there is a great dispute about the names assigned to instruments like Galileo's. In particular, it is "British usage" to call the Continent's "proportional compass" a "sector," whereas for British scholars the "proportional compass" is what the Continental ones call the reduction compass. This work follows the Continental usage.

[17]For more details, see Fulgenzio Micanzio to Galileo, December 1, 1635, in *EN*, XVI:355–356. For the translation of the entire letter, see p. 248. On May 17, 1609, Galileo reported the payment to Marcantonio Mazzoleni for the marking of one instrument. In 1635 Galileo's military compass was still considered to be one of the best mathematical instruments suitable for an officer, so that demand for it was still quite high. In 1635 Galileo thus suggested to Fulgenzio Micanzio (d. 1654) in Venice that he contact Mazzoleni to commission the production of compasses. However, Antonio Mazzoleni had since died of the Plague in the early 1630s.

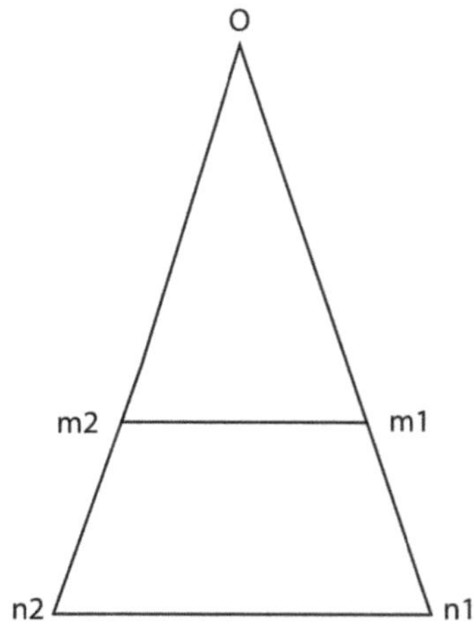

Fig. 2.3 Graphical description of the principle of proportionality

200).[18] The principle is the following: "if a straight line be drawn inside a triangle, parallel to one of the sides of the triangle, another smaller triangle is shaped. The smaller triangle is similar to the first because its sides are proportional to the sides of the first triangle" (Schneider 1970, 10) (Fig. 2.3). As far as practical application is concerned, the principle of proportionality means that given the angle at which the compass is open, and choosing two points *m1* and *n1* on a scale marked above one of the arms of the compass, and finding the corresponding points *m2* and *n2* on the same scale above the other arm, the distances *m1m2* and *n1n2* have a ratio that, given *o* as the center of the compass, can be determined by means of the distances between *m1* and *o*, and between *n1* and *o*.

The functions of the compass Galileo's military compass is conceived to accomplish quite a great number of operations and calculations. The marked scales, which included the "Arithmetic lines," the "Geometric lines," and the "Metallic lines," embodied the geometrical and practical knowledge required to accomplish those operations.[19] The execution of calculations required the user of the compass to keep a piece of paper at hand, on which measures could be recorded, and a ruler to keep track of the position at which the compass was opened.

Basic functions The first very basic operation which can be performed using the compass is the division of lines into equal parts. Actually, instruments called dividers had already been in use for this purpose since antiquity. Dividing lines

[18]On the practical character of Euclid's geometry, see Büttner et al. (2003).

[19]The entire set of operations and calculations that can be accomplished by means of Galileo's military compass is listed and described synthetically in Righini (1980). A short intruduction is also presented in Bedini (1986, 131–133).

could be useful for a vast number of tasks ranging from theoretical geometry to the work of stone cutters on construction sites.[20]

Other basic operations are, first, the one related to the rule of three and, second, the extraction of the square root. The rule of three is the method of finding the fourth term of a mathematical proportion when three terms are known. This could be required in a great variety of situations. Surveying fortresses, drawing plans, and changing currency are only a few examples of situations to which the rule of three must be applied. The extraction of the square root, apart from its immediate mathematical meaning, was certainly often requested by merchants. Plenty of instruments built before Galileo's compass were able to accomplish these operations as well. Especially in the second half of the seventeenth century, and particularly in the Venetian region, several kinds of proportional or reduction compasses were conceived and built that were able to accomplish these operations, as well many others which could executed by means of Galileo's compass. For example, the compass described by the military architect Carlo Theti (1529–1589) in his *Discorso di fortificationi*, published in Venice in 1588 (Theti 1588).

Functions for military officers Galileo's compass shows what it is really capable of when observed from the perspective of military needs on a battlefield. Surveying a fortress, reducing it to a plan, drawing new plans of fortresses or just single components of them—these are some of the main tasks for whose execution Galileo's compass is useful. For example, the military compass allows the user to produce a two-dimensional or three-dimensional drawing of an object in proportion, in larger or smaller dimensions. Second, it allows a circle to be divided into any number of portions so that any kind of polygonal figure can be drawn and, associated with this task, to square the circle or any portion of it.[21]

Surveying a fortress was quite an important task for many reasons, and special sets of instruments often were available for its accomplishment (Fig. 2.4). Military attacks generally were directed against the bastions of the fortresses, those architectural elements which were developed specifically in response to the increasing power of the new heavy and mobile artillery powered by gunpowder. The key to winning an attack against a fortress was to destroy its defences first, and this meant the bastions, because not only were they robust to artillery fire, but also because they were the elements capable of accommodating the heavy artillery that could destroy everything close to the walls and curtains outside the fortress. Thus, to enter a fortress the bastions had to be destroyed first, so that the attackers could move close to the wall and attempt to either destroy it or scale it. But since destroying a bastion was no easy task, the main strategy used was to build a sort of tower, which was also able to accommodate heavy artillery, on the field outside the fortress. Placing

[20] On the use of the compass on Medieval construction sites, see Shelby (1965).

[21] In the next chapter it will be shown that what was known as the new geometric way of designing fortresses became established in the sixteenth century, and that such designs generally assumed the shape of polygonal figures. For more details, see pp. 76ff.

the artillery at more or less the same height increased the efficiency of the shots targeting the strongholds, which were the positions in the fortress bastions where the cannons were placed.

Such towers, called cavaliers, had to be as high off the ground as the strongholds, and they had to be placed at a distance from the fortress that allowed the bombardiers to fire shots perpendicular to the strongholds (wherever the surroundings allowed, at least). Galileo's compass could be used to measure the height of the strongholds on the bastions and the user's distance from the wall of the fortress, and, more generally, to survey all of its architectural elements (Fig. 2.5).[22] However, the soldiers

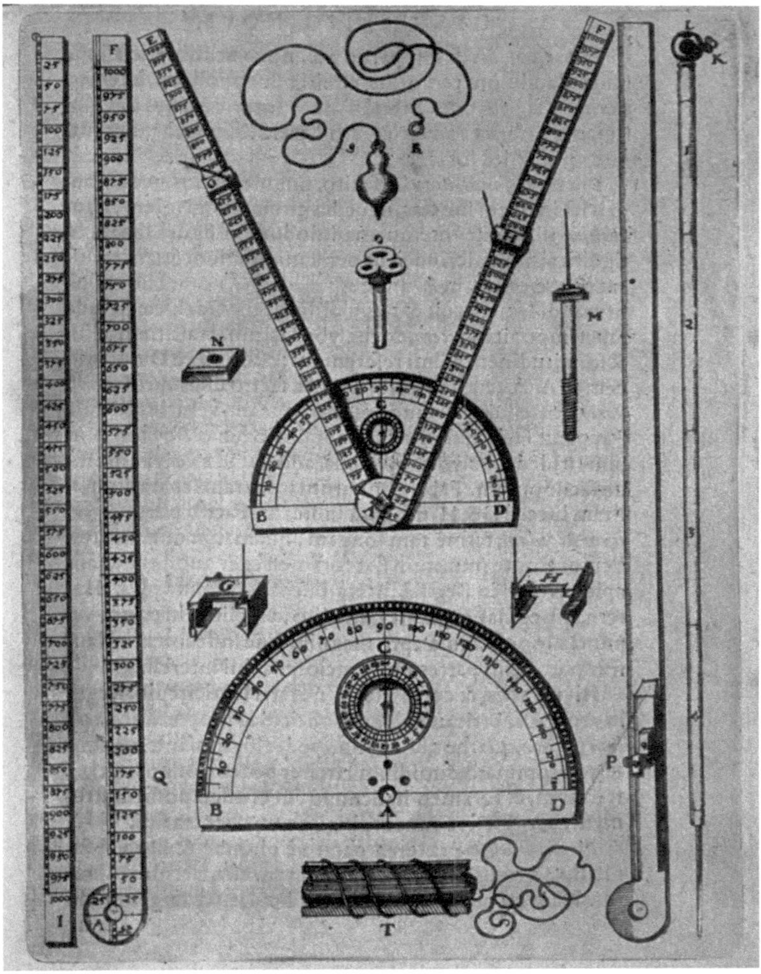

Fig. 2.4 Early modern surveying equipment (Zublero, 1670b, 5)

[22]According to the same principle, the compass could also be used to measure the height of the stars over the horizon. For more details, see *EN*, II:412.

Fig. 2.5 Description of the procedure to measure heights (Zublero 1607b, 19)

in charge of such measurements did not usually use instruments as sophisticated as Galileo's compass, but rather less sumptuous and often more efficient instruments evidently intended to accomplish only this task.[23] Moreover, even among the more sophisticated instruments, Galileo's compass was not the first to allow those operations to be performed that were typical of surveying compasses.[24]

The organization of combatants on the battle field Galileo's compass also allowed such calculations as the size of a certain shape of triangle whose area had to be equal to that of a given portion of a circle, and similar geometrical transformations. At first glance such operations seem to be little more than an interesting and challenging geometrical exercise. In fact, however, they corresponded to relevant needs on the battlefield, where many attacks were led with thousands of combatants, all of whom required commands and instructions. Depending on the field and the general surroundings of the fortress, the combatants were divided into groups and positioned at different points in the shape of figures such as triangles, semicircles, rectangles, squares or portions of circles (Fig. 2.6). In response to the development of attack strategies using artillery, prescriptions were developed as to how groups

[23]Many museums display sophisticated mathematical instruments like the proportional compass, but almost none of these was actually used on the battlefield. This is due to the fact that these "real" instruments were made of less precious materials like iron, wood or even cardboard, and thus could not be preserved for posterity.

[24]For an overview of proportional compasses conceived and built before Galileo's, see the introduction in Camerota F. (2000, 9–128).

Fig. 2.6 The Siege of Famagosta. 1571 (Stringa 1982, 309)

of combatants should approach the fortress, many of them requiring the group to change its shape, for example from a portion of circle to an equilateral triangle. For example, if the final attack on a point of the fortress required the combatants to line up in a triangular shape of one thousand square Tuscan feet, and these combatants had to wait for the final attack at a position where the most appropriate formation would be a semicircle, first the diameter and then the area of the semicircle had to be calculated in order to form the desired triangle with the same area. Once this calculation had been completed and the points of the new shape plotted on the field, the number of soldiers needed to fill the triangle were instructed to fill the semicircular formation.

Probably as a consequence of this capability of his compass, Galileo was considered to be an expert on these kinds of problems, many of which were much more complicated than the above-mentioned example. As late as 1629, for example, Matteo Carosi[25] wrote Galileo asking him for the solution to one particularly difficult problem related to the arrangement of pikemen and musketeers on the field.[26]

[25]Matteo Carosi was a Florentine doctor and member of the retinue that followed Maria de Medici to Paris when she married Henry IV of France in 1600. For extensive biographical data concerning Matteo Carosi, see Favaro and Galluzzi (1983, III:1657–1667).

[26]The problem is the following: "If we had to shape a square squadron of people with a given quantity of pikes and muskets, and we leave aside a number of musketeers sufficient to circumscribe [the squadron] on three sides, three by three, and if we wanted to place the rest of the musketry in the middle of the mentioned squadron, so that it would be defended from three sides by an equal number of lines of pikes, the question is how many musketeers we will need to leave aside for the

Galileo did not invent the geometrical procedures used to accomplish such transformations, of course, all of which found their basis in Euclid's *Elements*, and neither was he the first to find a way to perform them by using an instrument.[27] However, the broad set of geometrical transformations which can be accomplished using Galileo's compass, and the fact that their execution required no additional instruments or separate calculations, was an innovation to such compasses and, more generally, to analogous instruments conceived before Galileo's compass. Those compasses that were able to solve similar operations, albeit a more limited spectrum of them, included the proportional compass with four points conceived by Commandino in 1568; the proportional compass with two flat arms, like Galileo's, conceived by Guidobaldo del Monte in 1570 on the basis of Commandino's; and the proportional compass with eight points conceived by the practical mathematician Fabrizio Mordente (1532–1602), a description of which was published in Anversa in 1594.[28]

The quadrant One of the most relevant instruments conceived by mathematicians during the sixteenth century to provide solutions to the problems that arose on the battlefield was the quadrant for bombardiers (Fig. 2.7). In Galileo's day it was known that the maximal range of a cannon shot was obtained when the weapon was elevated to 45° over the horizon. In principle, therefore, once a piece of artillery was placed on the field, a single cannon shot at that elevation was then enough to pre-calculate the range of the trajectory of any other shot at a different elevation, so that the quadrant could then be used to adjust the level of the cannon accordingly to the desired target. This had been common knowledge since Tartaglia's publication of the *Nova scientia* in 1537 (Tartaglia 1537, Book II).[29] However, in practice this operation often turned out to be more empirical. First of all, gunpowder and cannonballs were not necessarily always the same for every shot. Second, the temperature of the cannon also influenced its efficiency. More specifically, because pieces of artillery needed to be warmed up, the bombardier had more than one chance to adjust the shot anyway. But, because of the use Tartaglia demonstrated for the first

---

mentioned circumscription, how many of them will go to the front and how many to the side, how many pikemen will there be in each line of the squadron, and, in conclusion, how many musketeers will constitute the front and the sides of the small squadron formation [...]." From Matteo Carosi to Galileo, November 2, 1629, in *EN*, XIV:49.

[27] The subject of the geometrical formations of soldiers had indeed been a relevant issue for a long time. Lengthy sections dedicated to this issue are included, for example, in Tartaglia (1554, Book IV).

[28] The transcription of Mordente's text is printed in Camerota F. (2000). For the operations concerned with regular surfaces, see Camerota F. (2000, 179–196). After Galileo, a compass capable of performing a broad spectrum of geometrical transformations was conceived by Michiel Coignet (1549–1623). For Coignet's works, see Meskens (1997).

[29] See also Henninger-Voss (2002) and Cuomo (1997). The quadrant for bombardiers was in truth already nearly a century old when Tartaglia published the *Nova Scientia*. However, since the maximal range of the shot was not yet common knowledge, such an instrument was probably used primarily to record elevations once the desired effect of the shot had been achieved empirically. The first version of the quadrant is described by Leon Battista Alberti (Alberti 1973).

Fig. 2.7 Quadrant for artillerists (Tartaglia 1554, 6v)

time, and because of the utility of such an instrument to record elevations, the quadrant became an essential tool for every artillerist. The essential role played by this instrument prompted Galileo to equip his proportional compass with an additional set of tools that allowed it to be transformed into a quadrant.

The compass as a caliber Despite the quadrant, the work of the bombardier remained very empirical for a long time. One of the reasons for this was the production of the cannonballs. A large variety of alloys was used to produce cannonballs, consisting of a whole range of different materials. Moreover, in the midst of an attack or a defence, soldiers could run out of cannonballs, and sometimes resorted to using suitably shaped stones found on the field instead. In order to maintain the target every time a different kind of ball was used, the quantity of gunpowder thus had to be changed as well. To do this, the amount of gunpowder needed to shoot a certain kind of ball was recorded, so that a procedure was needed to determine the physical relationship between the materials of which the different kinds of balls consisted. With a degree of approximation, this operation was normally accomplished using the caliber for bombardiers (Fig. 2.8), which was a kind of compass with curved arms, on which a scale was marked defining the relationships between different kinds of materials like lead, iron and stone. Theti's proportional compass of 1575, for example, succeeded in integrating the functions of the caliber by curving

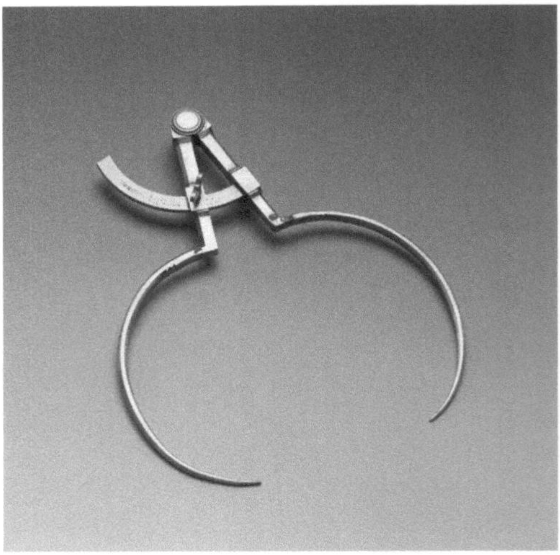

Fig. 2.8 Early modern caliber for artillerists (Istituto e Museo di Storia della Scienza, Florence. Inv. 3706)

the two arms of the instrument. Galileo's compass is also able to accomplish the same function, but dispenses with the curved arms. However, Galileo's workshop also produced a few proportional calibers with curved arms in 1600 (*EN*, XIX:148–149). No further information is preserved to help us understand the exact shape and the functions such calibers were able to perform, but the fact that Galileo's workshop produced them while it was also producing the military compass is a clear indicator that his workshop was generally producing mathematical instruments on demand, even ones that had not been conceived by Galileo himself.

The compass for the merchant Besides basic and military operations, Galileo's compass is also equipped with scales that allow commercial operations to be performed by merchants. Basically there were two of these operations. First, the compass could be useful as a tool for converting currencies, thus accomplishing an essential operation in the politically and economically fragmented Italian peninsula of the end of the sixteenth century. Second, Galileo also described a procedure according to which the compass could be used to estimate the interest on funds tied up for more than one year, earning fixed interest compounded annually. Thus Galileo was able to integrate techniques developed and known in the framework of the Abaco schools into his compass calculations as well.

The aspiration to a universal instrument Galileo's compass of proportion was an enormous success. The success of a mathematician was a function of how deeply he could enter the field of that practical knowledge of engineers and, in some cases, architects, whose work was dictated by military requirements. However, a closer analysis of those functions related to the military requirements that could be accomplished by means of Galileo's compass reveals that this instrument did not enable the user to execute any of the new operations that had required expertise on theoretical geometry back before Galileo conceived his compass. Such functions were known, and instruments that could execute them had been developed already. As to the instrument itself, separated from the context of its conception and use, as represented by Galileo's activity as a private teacher for young students training to become officers,[30] the fact that the compass succeeded nonetheless was due to the impressive spectrum of cases to which it could be applied effectively. In fact, in contrast to the fifteenth century, in the early sixteenth century the necessity emerged to develop a solitary instrument to accomplish as many operations as possible. This is what historiography refers to as the aspiration to a universal instrument (Schneider 1970, 5). Perhaps Galileo's compass was so successful because it came so close to fulfilling such an aspiration.

The precision of universal instruments According to Ivo Schneider (Schneider 1970), the effort to realize an universal instrument was intimately connected to the increasing demands placed on the precision of measurements using such an instrument. The demand for greater precision was a consequence of several factors, the most important of which was probably related to the development of cartography and thus to the introduction of new drafting methods. This development was above all a response to the increasing need for the precise surveying of the mainland, and to the expansion of navigation, which required more precise maps. But the introduction of heavy mobile artillery powered by gunpowder, and the different, geometric way of designing fortifications it provoked, also significantly influenced this development. Furthermore, when the number of operations that can be accomplished using a single instrument increases, the more sophisticated the construction of the instrument becomes. The more sophisticated the instrument, the greater the precision required in the construction of every single component of the instrument and in the process of assembly. As a matter of fact, the requirement for more universal mathematical instruments, considered as assembled devices, automatically caused an increase in the precision with which these instruments performed their functions. Galileo's compass, therefore, had the additional value of extreme precision, and his personal guarantee for quality of construction.

Universal instruments and the technological vanguard Soldiers, surveyors, and bombardiers had their own tools, which were simpler, easier to use and less elaborate than instruments like Galileo's. His universal instruments, some times made of precious metals and richly decorated, were destined for officers, noblemen or other (renowned) mathematicians. Their main use was either symbolic or didactic; in some cases they were used as an instrument to perform further research in the same field. While the materials and utilization of such instruments explain why these are the only ones that survived to enter museum inventories, the fact that they were adopted to perform further research explains why they were, and still are, so relevant from the scientific perspective. In fact, it was the aspiration to a universal instrument that created the context in which new solutions, new functions, new

---

[30]Galileo's activity as a private instructor teaching a course on fortifications will be described in the next chapter.

procedures, and new designs were conceived and realized.[31] From the sixteenth century on, universal instruments like proportional compasses were high-tech products, on the basis of which others, simpler and cheaper, were built. In conclusion, the fact that Galileo's compass was tremendously successful does not mean that every artillerist possessed one, but rather that personalities located high in the hierarchy, officers and mathematicians, were happy to receive one of them from Galileo as a gift or to be able to request one from him directly. This fame, finally, grew with every young pupil Galileo taught how to use his compass, who then purchased one to bring back to his hometown anywhere in Europe.[32]

The Reduction Compass

Another kind of compass produced in Galileo's workshop was the reduction compass with four points. This type of compass (Fig. 2.9) was well known all over Europe in Galileo's day. Three were particularly renowned: the first designed by Federico Commandino, the next made by Simone Baroccio, and the final one designed by Joos Bürgi (1552–1632). The compass consisted of four arms rotating around a mobile shaft. Arithmetic calculations, proportions and congruencies among plan figures were the problems which the compass of reduction was intended to solve. Normally made of brass like the proportional compass, Galileo did not bother to insert an explanation of how to use this compass in his treatise, even though he produced and sold it. As the evidence presented in the manuscript Ms. Gal. 26 shows (*EN*, XIX:132, 141, 147–148 and 154), it seems that Galileo was mostly

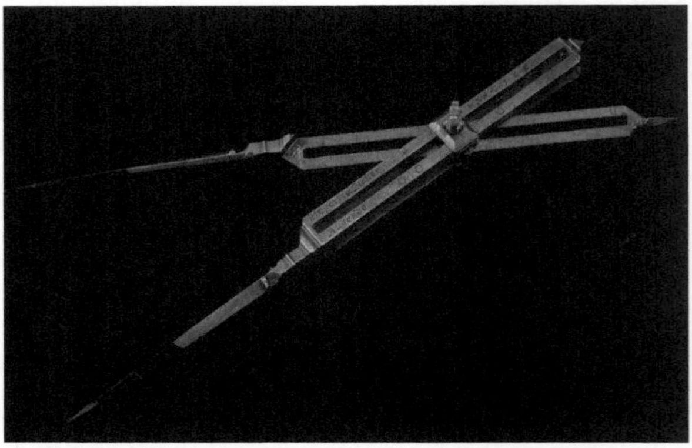

Fig. 2.9 Reduction compass with four points (Casi 1993, 29)

[31]Considering Galileo's military compass from the perspective of the aspiration to a universal compass and, therefore, as a result of and cause for the generation of new knowledge, it might become desirable to abandon James Bennett's distinction between mathematical and philosophical instruments (Bennett 1986).

[32]For an analysis of the provenience of Galileo's private students, see pp. 73ff. in the next chapter.

busy producing the reduction compass around 1599 and 1600. Other entries, only two, refer to 1603 and 1605. The reduction compass is a version of the compass whose design preceded the proportional compass. In 1570 Guidobaldo del Monte elaborated a new version of Commandino's reduction compass, transforming it into a proportional compass with only two points. Given the friendship between Guidobaldo del Monte and Galileo, it therefore cannot be ruled out that Galileo's conception for the proportional compass was related to Guidobaldo's instrument. No textual evidence supports this inference, however. As mentioned above, Galileo's proportional compass was also able to perform those operations which could be executed using the reduction compass. Thus, in this case, too, it might be inferred that Galileo's workshop produced other instruments on demand.

The Surveying Compass

During sieges and attacks, it was common practice to order soldiers to excavate tunnels as far as the curtain in order to place mines under the wall of the fortress. However, excavating tunnels could become very costly in terms of men and time, and there was absolutely no assurance that the tunnels would actually reach the points of the wall selected as the targets. The soldiers had to prepare chambers of specific shapes and sizes at these positions, which were then filled with large amounts of gunpowder. Lighting the gunpowder in the small chamber caused an explosion with strong impact waves, which wreaked the greatest damage in the direction of the chamber wall that was intentionally left weaker than the other sides. One of the instruments required to succeed in excavating tunnels was the surveying compass, and this, too, was produced in Galileo's workshop.

The origins of Galileo's surveying compass It is documented not only that Mazzoleni manufactured surveying compasses in the workshop, but also that Galileo purchased them from Master Fait in Venice, and in Florence. Entries in the *Ricordi* report the production of ten surveying compasses, and about nine sold to his students. In some cases it is certain that surveying compasses were made of carved wood and/or a perforated disc made of paper placed on their faces. The degrees marked on the surveying compasses were an early version of what became the azimuthal compass. As is known from Domenico Berti (Berti 1876, 7), Galileo had read Tartaglia's *Quesiti ed inventioni diverse*, the fifth book of which explains the construction and the use of surveying compasses. Moreover, as Sven Dupré pointed out (Dupré 2000), the version of the surveying compass presented by Tartaglia is reminiscent of the one described by Alberti in his *Ludi matematici* (Alberti 1973), which was one of the texts used by Ostilio Ricci, Galileo's teacher, in his private lessons.[33] There is thus plenty of circumstantial evidence that Galileo's surveying compass was a further development of Tartaglia's.

---

[33]On Galileo's education and the content of Ricci's private lessons to Galileo in particular, see pp. 12ff in the previous chapter.

The functions of Galileo's surveying compass Among many types of compasses, Tartaglia described one particular type made of wood (Fig. 2.10). Tartaglia's surveying compass was the first to present a circle divided into 360° and main directions defined according to the eight winds. It was equipped with two or four alidades. However, in Galileo's case no mention of alidades can be found in reference to surveying compasses, neither in his correspondence nor in the *Ricordi.* A possible reason for this could be the different purposes for which the two men's surveying compasses were intended. For Tartaglia the main purpose of the surveying compass consisted in tasks such as drawing plans of villages or sites,[34] whereas Galileo intended his military surveying compass to be used for very specific tasks, including the excavation of tunnels. In one of his two treatises on fortifications, Galileo mentioned this special use of the surveying compass, for indicating the direction in which the soldiers had to dig a tunnel between their position and the wall of the fortress they wanted to attack.[35] After having determined the straight direction the tunnel had to follow from above ground, officers had to prepare for the eventuality that obstacles underground might prevent soldiers from keeping to this path. In such cases, Galileo stated, the officer must use the surveying compass to note the degrees between North and the destination of

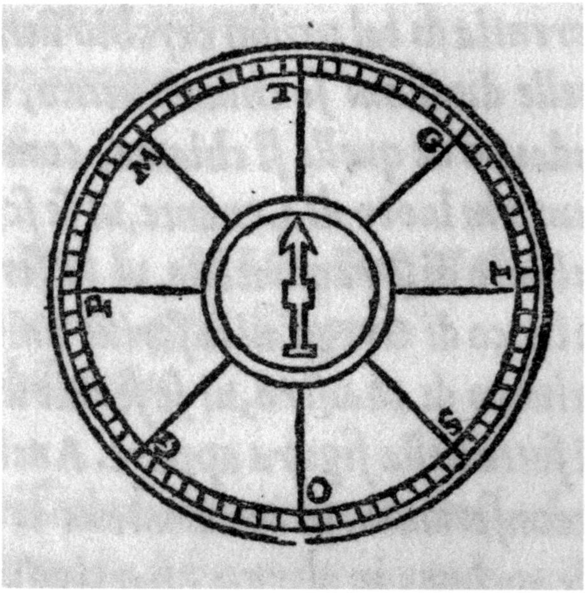

Fig. 2.10 Tartaglia's surveying compass (Tartaglia 1554, 55v)

[34]In this case the compass was often applied to one of the two arms of a proportional compass.

[35]Galileo described the use of the surveying compass in one of his two treatises on military architecture, *Breve istruzione all'architettura militare* written in ca. 1592 and published in *EN*, II:17–75. The description of the use of the surveying compass is on pp. 46–47. These treatises are analyzed in the next chapter on pp. 77ff.

the tunnel, so that, once underground, soldiers could change their direction of digging to circumvent obstacles, and then subsequently return back to the direction originally determined. This task had to be accomplished by means of a surveying compass, as it was impossible to control the direction of the tunnel visually while underground.

Other Instruments and Tools

Relatively few entries in the *Ricordi* related to production in Galileo's workshop are concerned with astrolabes, iron tools and parts of tools like screws, perpetual screws and clamps. In particular, there is only one entry related to the production of an astrolabe (*EN*, XIX:135). This probably means that the astrolabe production was not destined for the student market. The other iron tools, some registered as constructed in the workshop and others bought in Venice from Master Fait, were probably components of the devices and of machines belonging to Galileo's workshop.

Lenses

If a top ten list of mental associations with the name Galileo were to be formulated, the telescope would surely appear in second place after his trial and abjuration in front of the Holy See. Just a few years after Galileo's discovery of the Medicean Planets (Jupiter's four largest moons) in 1610, specialized telescope lens makers appeared all over Europe. Most of them were former spectacle makers, who reoriented themselves professionally, lured by the promise of higher profits now that the telescope had become such a fashionable item. After the news circulated that the Queen of France, Maria de Medici, knelt down to better view the moon through her telescope,[36] possessing such an instrument had become akin to holding a royal scepter.

Construction of the Galilean telescope Two lenses are needed to make a Galilean telescope (Fig. 2.11):[37] an objective lens, which is a plano-convex convergent lens, and an ocular lens, which is a plano-concave divergent lens.[38] Each of the lenses was placed into a tube and the two tubes were fitted together such that the one containing the ocular lens could be pushed into the other. The tubes were telescoped

[36]Matteo Botti to Galileo, August 18, 1611, in *EN*, XI:173.

[37]When the telescope first appeared, it already had the form of a Galilean telescope. Now that many sorts of telescopes are in use, this particular model is known as Galilean, because Galileo was the one who made the greatest contribution to making the instrument famous and relevant for the sciences. An introduction to Galileo's telescope is also in Bedini (1986, 138–144).

[38]From 1613 on, the Galilean telescope was equipped with bi-concave lenses as well.

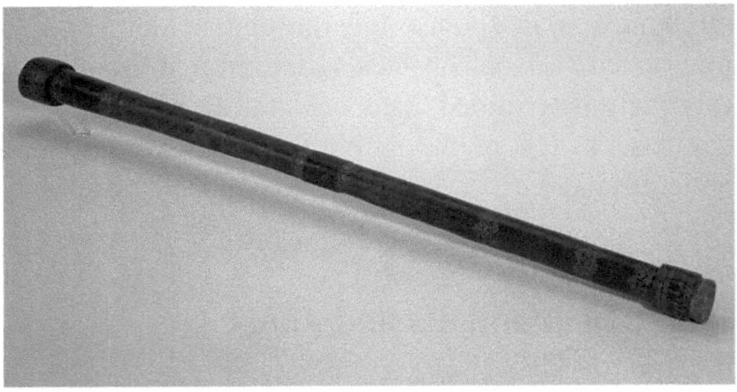

Fig. 2.11 Replica of Galileo's telescope (Istituto e Museo di Storia della Scienza, Florence. Inv. 2427)

in order to allow the focal point of the plano-convex lens to coincide with that of the plano-concave one at a time when lens production was not standardized.[39]

Improvements to the telescope's magnifying power When the telescope appeared in Venice, there were spectacle makers and mirror makers, but not a single manufacturer of telescope lenses. Although the first telescopes were equipped with spectacle lenses, it soon proved that lenses for telescopes required a more elaborate workflow and a variety of sizes that was entirely unusual at the time (Van Helden 1983). Although the method of making lenses for telescopes followed a method that did not differ fundamentally from that of making spectacle lenses, the traditional manufacture of the convex lenses, in particular, was far from sufficient to yield products that could be used in telescopes. One of Galileo's achievements in the very early life of the telescope is that he understood that the magnifying power of the telescope could be improved by equipping the instrument with plano-convex lenses that had a larger radius of curvature than was ordinarily obtained by the spectacle-makers. The greater curvature was crucial to attain a longer focal distance. As Sven Dupré pointed out (Dupré 2005),[40] Galileo was able to obtain this result only after several attempts, by combining plano-convex lenses of different curvatures with a standard concave lens. Galileo thus needed plano-convex lenses of unusual curvatures in order to achieve his results. This is the reason why Galileo ultimately decided in 1609 to become a lens maker himself, a profession closely resembling spectacle manufacture.[41]

[39]Galileo never explained the optical theories on the basis of which the telescope works. The first to do so was Kepler after Galileo's publication of the *Sidereus nuncius* (*EN*, III:53–96) in his *Dioptrice* (Kepler 1611; Dijksterhuis 1983, 423–436).

[40]For background history concerning early modern optics and especially Francesco Maurolico (1494–1575) and Giovanni Battista della Porta (1535–1615), see Lindberg (1983).

[41]According to Sven Dupré (2005), the combination of fundamentals in dioptrics and catoptrics that Galileo learned during his apprenticeship, with the reading work he undertook on Ausonio's *Theorica speculi concavi sphaerici* with Giovan Vincenzo Pinelli (1535–1601) and Paolo Sarpi, allowed Galileo to develop a method for improving the magnifying power of the telescope "on the basis of his familiarity with the workshop practices of spectacle-makers." Ausonio's work was particularly relevant for understanding how spherical mirrors function. For a description of this

Galileo was used to taking short trips to Venice to meet friends, visit the Arsenal and do some shopping. In late November 1609 he prepared a shopping list before going to Venice:[42]

[1] Shoes and hat for Vincenzo.
[2] The case of wares for Marina.
[3] Lentils, white chickpeas, rice, raisins, spelt.
[4] Sugar, pepper, cloves, cinnamon, spices, jams.
[5] Soaps, oranges.
[6] Ivory combs no. 2.
[7] Malvasia by Lords Sagredo.
[8] Artillery balls no. 2.
[9] Organ pipe of tin.
[10] German lenses, polished.
[11] Have ock crystal polished.
[12] Pieces of mirror.
[13] Tripolitan.
[14] The mirror-maker under the insignia of the Re.
[15] Small iron tuning chisels made in the *Calle delle Acque*.
[16] To haggle about iron bowls, or to make them out of stone, or like the artillery balls.
[17] Privilege for the vocabulary.
[18] Iron plane.
[19] Greek pitch.
[20] Felt, mirror to rub.
[21] [Follo].[43]
[22] Pay debts to Lord Mannucci and give him back the [Edilio].[44]

The procedure for constructing telescope lenses Apart from the needs of Galileo's household, his son Vincenzo and his girlfriend Marina di Gamba, most of the items on the shopping list are concerned with the production of lenses. The artillery balls (item no. 8) were used as dies upon which concave lenses could be ground and polished. The organ pipe (item no. 9) was supposed to work as a tube for the telescope, perhaps in combination with another of shorter diameter made of cardboard. The German lenses (item no. 10) were probably purchased so that he would have spectacle lenses as a reference for comparison with his own lenses. In fact, great quantities of lenses from southern Germany were imported to Venice and, although their quality was too poor for use in telescopes, their low cost made

work and Galileo's use of it, see pp. 64ff in this chapter. For Galileo's activity as a lens maker, see also van Helden (1983).

[42]The shopping list is written in Galileo's hand on the back of the paper bearing the letter of Ottavio Brenzoni to Galileo, November 23, 1609, in *EN*, X:269–270. The shopping list also was published by Favaro, in footnote no. 1 in *EN*, X:270. Enumeration by the author. As this work is the revision of a PhD thesis submitted to the Humboldt Universität zu Berlin in August 2007 which already fully contained this section, the analysis of Galileo's shopping list is not indebted to Strano (2009). A synthesis of this section already has been published by the author of the present work in February 2009 (Valleriani 2009c).

[43]Not identified.

[44]Not identified.

them preferable to the local products. Rock crystal (item no. 11) was a kind of crystal generally produced in Venice, though not the famous Murano crystal, and was considered to be of higher quality than ordinary glass. Rock crystal's superior clarity raised Galileo's hopes that better lenses could be produced using this material. However, because rock crystal was a much harder material than ordinary glass, Galileo needed recourse to a specialist to grind and polish it, probably a mirror maker.[45] Pieces of mirror (item no. 12) were a common purchase because they were considered raw material for the production of lenses. For example, once the surface of tin and quicksilver was taken off from a piece of curved glass mirror, further polishing its center increased its curvature and therefore potentially made it suitable for use in telescopes. A special clay was imported into Europe from Tripoli (item no. 13), very fine and mixed with stoneware, used to polish stones, glass, and crystal (Manzini 1660, 202). The Insignia of the Re in Venice is where a mirror maker (item no. 14) worked who attracted Galileo's interest sufficiently to warrant a visit to his workshop, presumably to observe the way he worked, although as the items of the shopping list show quite clearly, this could not have been the first time Galileo sought to observe lens and mirror makers. What is more, his note about chisels (item no. 15) for grinding machines make clear that Galileo either already had one of these machines at his disposal in his workshop, or had designed one. The clay from Tripoli and the felt (item no. 20) were both used for polishing by hand and by means of the grinding machine. The situation was different for the manufacture of plano-convex lenses. For this reason Galileo intended to purchase iron bowls (item no. 16), which, once filled with the clay from Tripoli and water, could be used to rub and polish the glass by hand, as no grinding machine to manipulate convex lenses existed at the beginning of the seventeenth century. For the same reason iron tools for planing (item no. 18), pieces of mirrors, and pitch (item no. 19) were also required.

Galileo's discovery of four of Jupiter's moons Galileo began using the telescope during the summer 1609. As soon as he was able to improve its magnifying power, he organized a public meeting with the Venetian authorities—the *Doge* included—in order to show them the military relevance of the instrument. According to the *forma mentis* of the early modern artist-engineer, whose career was related to his involvement in the art of war, Galileo in fact did not perceive at the beginning of his telescopic adventures the scientific relevance of the new instrument in the field of astronomy (Valleriani 2009c). This strategy worked well, indeed, for one day after the meeting, which took place on August 24, 1609, he was appointed to his previous chair permanently and with increased income. After a few months, however, in March 1610 he published the *Sidereus nuncius* (*EN*, III:53–96), the work with which he announced to the world that the sky differed from the conception of it embedded in the Aristotelian-Ptolemaic world view. As he discovered Jupiter's moons in January 1610, it could well be that the instrumentation mentioned above

[45] The possibility of producing a better material for telescopic lenses became a crucial issue for Galileo. For more details, see pp. 48ff. in this chapter.

and that Galileo purchased in Venice in November 1609 was what allowed him to produce lenses capable of increasing the magnifying of the telescope enough to do so. This accomplishment made him the most famous scientist of the era.

Galileo as a telescope lens maker Galileo continued producing lenses even after he moved from Padova to Florence in 1610. In fact, he not only had all of the tools he needed for this production brought to Florence, but even had some new tools prepared in his new house by master masons. Certainly, Galileo had in mind a grinding machine, which he designed and perhaps already built back in Padova, to make convex lenses better suited for telescopes.[46]

Galileo as a telescope lens broker Once he had settled in Florence, apart from his own production, Galileo also set about purchasing lenses from many sources thanks to his own network of friends and ex-pupils, and to the diplomatic network of the Grand Duke. In the first period after his move to Florence, he tried to obtain lenses suitable for telescopes as well as ordinary lenses that required further polishing. Along with the increasing popularity of the telescope, however, the number of mirror makers and spectacle makers shifting their professional activity increased as well, as ever more of these men became lens makers aspiring to produce lenses for telescopes. This development offered Galileo a wide selection of good lenses for purchase. In fact, Galileo did not need such a great number of lenses and telescopes himself, but his fame was so great that those lenses and telescopes that met with his personal approval were considered to be of the highest quality. He was quite overwhelmed by the demand for telescopes and lenses. For many years after the publication of the *Sidereus nuncius*, Galileo's house was literally a clearing house for lenses, most of which came from Venice and Naples, en route to every corner of Europe, serving first the most important political authorities, like the Emperor and the King of Poland; then scientists, like Pierre Gassendi (1592–1655); and, finally, significant aristocrats. The greatest care was reserved for the diplomatic corps of the Grand Duke, whose members worked de facto as ambassadors of the new Galilean discovery. Initially the most efficient lens hunter was Giovan Francesco Sagredo; after his death, his place in Venice was taken over by Fulgenzio Micanzio. The major source of lenses from outside Venice was Naples, where Benedetto Castelli (1577–1644), Galileo's ex-pupil and then professor for mathematics at the University of Pisa, apparently established efficient contacts with lens makers.[47]

---

[46]On August 19, 1610, speaking about the conceived grinding machine, Galileo wrote to Kepler: "Nobody else has built anything of equal excellence; in fact the method is very elaborate: truly I conceived a certain machine to shape and polish them, which I did not want to build, because then I could not have brought it to Florence, where my residence will be in the future." From Galileo to Kepler, August 19, 1610, in *EN*, X:421–423. On Galileo's sending his tools to Florence, see Galileo to Belisario Vinta, August 20, 1610, in *EN*, X:424–425. The necessity of engaging a master mason for his tools to make lenses was stated explicitly by Galileo in a letter he wrote to Giuliano de' Medici, the Medici Ambassador in Prague. For more details, see Galileo to Giuliano de' Medici, October 1, 1610, in *EN*, X:439–441.

[47]It is impossible to quote single documents as evidence of this lens exchange, which took place during the entire remainder of Galileo's life from 1610 on, for the simple reason that there are too many. Suffice it to say that almost every letter between Sagredo and Galileo after 1610 reports such

Sagredo's formulation of the "law of the lens maker" Over the course of his activities, Sagredo became increasingly interested in the optical characteristics of lenses for telescopes. He first started producing both bi-concave and bi-convex lenses in 1613, ultimately formulating for the first time the empirical rule known as the "law of lens makers," which, in modern terms, is the law to calculate the focal distance of bi-convex lenses.[48] Since the concave lens was placed between the convex lens and its focal point, but very close to the latter, Sagredo used his empirical formula to obtain the required length of a telescope given a certain lens. This would have provided specific information, on the basis of his own experience, about the optical characteristics of the entire telescope.[49]

Galileo as an expert lens maker at court Cosimo II de Medici and, after he died, his son Ferdinando II, proved to be particularly interested in the telescope. They continuously re-arranged and improved the ducal workshops for glass production, providing great impetus for the production of lenses for telescopes.[50] Gradually these facilities became another source of lenses for Galileo, although Ferdinando's passion for lenses also motivated Galileo and Castelli to supply the Grand Duke with the best lenses they could. Indirectly, through the mediation of the Grand Dukes, Galileo maintained his interest and attention to the procedures for manufacturing lenses into the final years of his life, even when he was held under house arrest because of the condemnation imparted by the Holy See and began losing his sight. During the late 1630s the Grand Duke had a lens maker named Ippolito Francini at his personal disposal, who often worked in direct contact with Galileo. In 1639, the seventy-five-years-old Galileo designed a new grinding machine for Francini in order to produce convex lenses (Fig. 2.12). In the preserved drawing of this machine it is possible to recognize its use of the iron bowl as a die, presumably as Galileo had conceived it back in Padova in 1609.

Galileo's lens maker apprentices There is no question that Galileo was considered to be the best lens maker and the best lens assayer, at least for the twenty years

---

an exchange. The same can be said to a slightly lesser extent about Benedetto Castelli. All of the letters are published by Favaro and the single letters can be found easily using the name index in *EN*, XX.

[48]For Sagredo's formulation of the law of lens makers, see Giovan Francesco Sagredo to Galileo, October 17, 1615, in *EN*, XII:199–200. The English translation of this letter is in Pedersen (1968).

[49]In late 1613 Sagredo achieved a degree of standardization in the production of lenses, by always using the same dies and engaging the same masters. By combining the dies to produce bi-convex lenses and observing their effects, he finally was able to formulate the law of the lens maker. For an exhaustive description of Sagredo's optical research, and for historical evidence of this work in the correspondence between Sagredo and Galileo, see Pedersen (1968), Greco and Molesini (1997).

[50]The Grand Duke of Tuscany re-organized his workshops in part as a consequence of the transfer of Venetian masters to Florence. Concerning the field of lens production, Galileo was also charged with finding a qualified lens maker in Venice through the mediation of Sagredo. See, for example, Giovan Francesco Sagredo to Galileo, August 18, 1618, in *EN*, XII:407–408. For the translation of the entire letter, see p. 255. For detailed information about Florentine glass production *a la façon de Venise*, see Taddei (1954, 49ff).

Fig. 2.12 Galileo's grinding machine designed for the court lens maker Ippolito Francini (Manzini 1660, 162)

between 1610 and 1630. During the years 1613 and 1615, epistolary communication with Galileo helped Fabio Colonna (1567–1650) in Naples to perfect a procedure for manufacturing convex lenses for telescopes. This is how Naples became another independent center for the production of this kind of lens.[51] Colonna focused especially on the polishing method for the convex lens, and also designed a grinding machine equipped with a bowl of copper; however, this last design never worked well.[52] After the Plague hit Venice around 1630, when, as Fulgenzio Micanzio complained, many craftsmen died and it thus became impossible to find good lens makers, Galileo dedicated himself to a "correspondence course" to teach the craft of producing lenses for telescopes to Sigismondo Alberghetti, the one who also tried to build Bonaventura Cavalieri's version of Archimedes' burning mirror.[53]

Parabolic lenses and technological limits Colonna's and Alberghetti's goal was to build spherical convex lenses, and for Galileo it was easy to teach them how to achieve this. Yet since 1613 Galileo had expressed the desire to produce telescope

[51]During the same period Giovanni Battista della Porta was living in Naples. However, della Porta, to whom the first optical description of the telescope was normally ascribed in Galileo's day, did not leave behind any tradition of lens making. Colonna complained bitterly about this lack of activity in Naples, which eventually flourished a few years later, until the Grand Duke of Tuscany himself began purchasing lenses from that city around 1635 through the mediation of Benedetto Castelli.

[52]For details about Colonna's efforts to improve the production of convex lenses, see especially Fabio Colonna to Galileo, July 29, 1614, in *EN*, XII:88–89, Fabio Colonna to Galileo, October 3, 1614, in *EN*, XII:102–103, Fabio Colonna to Galileo, August 14, 1615, in *EN*, XII:195–196.

[53]For more details, see Fulgenzio Micanzio to Galileo, July 8, 1634, in *EN*, XVI:108–109. For the translation of the entire letter, see p. 276. On Alberghetti's attempt to build Archimedes' burning mirror on the basis of Cavalieri's indications, see pp. 62ff in this chapter.

lenses of a parabolic shape, as he was convinced that this sort of lens would greatly improve the magnifying power of the telescope and the quality of image formation, although he was aware that parabolic lenses could not be produced for practical reasons.[54]

Glass Production

Although Galileo wished for lenses of shapes other than the spherical one, he never considered the impossibility of producing them to be the greatest limit to improving the efficiency of the telescope. His primary constraint was actually given by the quality of the material used to make lenses. During the first half of the seventeenth century modern optical glass did not yet exist, but Galileo and his contemporaries already recognized that great efforts would have to be directed to this end. Finally, Galileo and Sagredo settled on some experiments.[55]

Poor quality of glass and its causes Just a few months after Galileo produced his telescope in 1610, Giovanni Bartoli, a commissioner in Venice for the Secretary of the Grand Duke of Tuscany, remarked that there were three sorts of telescopes on the market, distinguished according to the material of which their lenses were constituted. The cheapest were provided with lenses of ordinary glass, as spectacles were. The middle category had lenses made of crystal of Murano, and the most expensive ones were equipped with lenses made of rock crystal.[56] Once Galileo succeeded in improving the magnifying power of the telescope, he presented it to the Venetian patricians and political council, especially emphasizing the instrument's military relevance. Once the Venetian Senate had awarded Galileo with a prolongation of his chair for life, interest in discovering "the secret of the telescope" experienced an immediate surge, and Bartoli was put in charge of retrieving this information. Bartoli cunningly answered that there was no secret, merely the necessities of making the lenses of good material, and of finding the proper way to fix the length of the "cannons," as the telescope was often called before 1618. Galileo himself remarked on the problem of the quality of the glass in 1610 when, between April and June, he undertook an epistolary exchange with Francesco Maria del Monte.[57] After Galileo had sent him a telescope as a present, Francesco Maria del Monte responded immediately, asking whether it would be possible to improve it, both by providing the telescope with a bi-convex lens instead of a plano-convex one, and perhaps by employing rock crystal as the material for the lenses. Galileo's lost answer

---

[54] B. Imperiali to Galileo, November 29, 1624, in *EN*, XIII:230–231.

[55] For a general overview about the quality of the lenses produced, see also Pedersen (1968).

[56] Giovanni Bartoli to Belisario Vinta, September 26, 1609, in *EN*, X:259–260. For the translation of the entire letter, see p. 225.

[57] Francesco Maria del Monte to Galileo, April 28, 1610, in *EN*, X:343–344.

apparently contained some outlines of experiments that he intended to perform using rock crystal, as the second letter by del Monte clearly indicates.[58]

Telescope lenses in 1618 In 1618 Galileo was still searching for a method to improve the quality of the material of which telescope lenses were made. In September of that year he wrote his friend Sagredo, complaining that the greatest problem in making lenses still consisted in the quality of the material.[59] Sagredo was willing to collaborate, and turned out to be a tremendous and tireless mind in the quest for a method of glass production that could furnish optical glass of higher quality. Sagredo agreed with Galileo, adding that the methods concerned with grinding and polishing were good and quite well known. Moreover, Sagredo continued, when Galileo had written him, he had already recognized the problem and begun performing experiments in Murano to improve the method of glass production. Above all, he already had formulated his own definition of the problem in the general context of the use of the telescope. First, Sagredo reported the contemporary situation to Galileo:

> Experience has shown that the more or less white color does not matter very much. The [glass] vesicles, called *puleghe* by these [workers] of Murano are not very disturbing, only the [glass] spirals, which are spiral [glass] rods that one can often see in the lenses and which originate from the mixture of different glasses. One has therefore to find a way to make homogeneous glass, very similar in all its parts, because it is conceivable that in [reference to] the variety of glasses there is a diversity of hardness that consequently causes the rays, which should run straight through the lens, to refract. Once refracted it follows paths different from the ones intended and [also] different from each other. This is the reason why one sees double and blurry images.[60]

The situation regarding telescope lenses in 1618 is well documented. Vesicles were small air bubbles, usually caused by an excessively rapid cooling process, a problem that was quite easy to solve by means of a careful cooling procedure. The presence of glass spirals, *torticci*, however, was a crucial and more difficult problem. Although glass spirals had been observed for quite some time, Sagredo's statement appears to be the first to relate this characteristic of glass to the different hardnesses exhibited by the different kinds of glass used to make a lens. In response to the demand for homogeneous glass, Sagredo had provided ash and rock crystal and had them pulverized and sifted under his own supervision in his rooms. The resulting compound, which was more homogenous because only one kind of quartz excavated in a single area was used, was then sent to Murano, first to calcinate the frit; second, to add manganese; and, finally, to perform the long and complicated

[58] Francesco Maria del Monte to Galileo, June 4, 1610, in *EN*, X:367–368.

[59] Although this letter by Galileo is not preserved, as in many other cases it is possible to infer Galileo's proposals and thoughts from the responses written by his correspondents. In this case, see Giovan Francesco Sagredo to Galileo, August 4, 1618, in *EN*, XII:403–406. For the translation of the entire letter, see pp. 252ff.

[60] From Giovan Francesco Sagredo to Galileo, August 4, 1618, in *EN*, XII:403–406. Author's italics. For the the translation of the entire letter, see pp. 252ff.

firing process.[61] In the end Sagredo reported to Galileo that the experience was a failure because the liquid glass escaped during firing and scattered over the furnace, perhaps as a result of intentional manipulation.[62]

Galileo's familiarity with glass production Sagredo's description of his experiment shows one peculiar point. He explained in detail only the addition of the manganese, a step that was performed in order to obtain a whiter material, and in such a way that it seemed he was explaining the procedure to someone thoroughly ignorant of the matter. On the other hand, however, Sagredo's description does not include any explanation about raw materials, the calcination process or firing procedure, as if Galileo were an expert on glass production. For example, he did not specify that the manganese was added after the calcination process, using what was known as a "color mill," a device normally used to mix the frit with other different compounds, the effect of which was to give the glass or crystal a certain color. Ultimately, the degree of Galileo's familiarity with glass production can be inferred on the basis of Sagredo's response of October 1618. To fully understand this document, however, it is necessary to have an idea of the state of the art among the workers in the context of this specific production in Galileo's day.

Divulging the secrets of Murano In 1612 Antonio Neri (1576–1614) published a book entitled *L'arte vetraria*, which represents the first public divulgation of the work procedures followed at Murano (Neri 1612). Antonio Neri was a "master of composition" who worked in the glass foundries of Florence, Pisa and Antwerp. Because of its crystal, Venetian glass products had been especially valued and in demand as luxury objects all over Europe since the fifteenth century. As a consequence of this trend, particularly important families like the Medici in Florence never tired of attempting to obtain permission from the head council of Venice to move artisans of Murano to their own cities in order to employ them in their glass foundries. Despite the apparent strict ban on such workers leaving the city of Venice, the Medici family turned out to be particularly successful in obtaining dispensations. At the same time Galileo and Sagredo were corresponding about the quality of crystal for telescope lenses, they arranged the move of Sagredo's crystal maker to Florence under the patronage of the Grand Duke Cosimo II in response to an official request by Galileo himself, who may have been acting in his capacity as the court's chief lens maker. Thanks to the artisans of Murano, in Florence, too, the production of glass objects *a la façon de Venise*[63] became established from the very beginning of the seventeenth century. Neri, who was not from Murano, became familiar with

---

[61]Giovan Francesco Sagredo to Galileo, August 4, 1618, in *EN*, XII:403–406. For the translation of the entire letter, see pp. 252ff.

[62]In practice, all of the experiments Sagredo related in this and other letters turned out to be failures due to similar problems. In the last letter describing such an experiment, Sagredo remarked that the workers of Murano were probably sabotaging his research for fear that he could obtain a glass which was better than what they were able to produce.

[63]*A la façon de Venise* is today's technical term to distinguish these objects from the ones actually produced at Murano. On the other hand, however, since these categories of objects resembled each other both in terms of their appearance and their methods of production, it often turns out that they

the Muranese techniques over the course of his work in the Florentine glass foundry at the Casino, which was also the residence of Don Antonio de Medici and had been built as an addition to a workshop for the intaglio of hard stones. In this sense Neri's book became the first to divulge the secrets of Murano.

Neri was a master of composition, and the entire book is concerned with this topic. With a few exceptions, all of the techniques for shaping objects are disregarded, for that was not his area of expertise. Among the seven books which constitute the *Arte vetraria*, the first is concerned with the production of ordinary glass and of crystal, while the other six deal with the production of compounds to be mixed into the glass in order to lend it color. Galileo's and Sagredo's experiments were thus an attempt to investigate the work procedures as they were described by Neri in the first book of his work (Neri 1612, 1–31).

Galileo's proposal for an experiment Toward the end of 1618 Galileo finally resolved to answer Sagredo's technical question about the use of manganese. On this occasion he also suggested a new experiment, promptly analyzed by Sagredo:

> I will speak to him[64] about what you have written to me and whether he will see fit to do some experiments in order to make clear whether, taking great care, it is possible to perfect the matter of making the desired glasses [lenses]. I placed on my to-do list an item [to make] a pan[65] of crystal cuttings, an experience I did not repeat and which could perhaps turn out well. Concerning the crushed rock crystal[66] instead of the quartz pebbles,[67] this is a thought I had myself. I even have most of the material prepared already, and I wanted to mix it with salt of tartar[68]; [...].[69]

Galileo asked Sagredo to organize many experiments to improve the crystal for making lenses, a request that prompted Sagredo to involve his crystal maker Alvise dalla Luna, who would have left Venice shortly thereafter for Florence, where he was to work at the glass foundry of the Casino under Galileo's supervision (Taddei 1954, 47–55). Moreover, Galileo proposed a specific experiment: instead of using good, but ordinary, quartz pebbles, he suggested rock crystal. This material, which is hyaline quartz, allowed the production of much higher-quality crystal because it would have not contained bubbles. The Muranese workers, who were aware of this possibility, kept on working with their own crystal rather than the rock crystal, not only because of marketing reasons, but also because

---

are in fact indistinguishable, as Rosa Badovier Mentasti showed in detail in her introduction to Neri's *Arte vetraria* (Neri 1612). The English version of her introduction starts on p. XLI.

[64]Sagredo was referring to the Muranese master of composition Alvise dalla Luna.

[65]In the terminology of glass production, a "pan" was a melting pot made of refractory material.

[66]Hyaline quartz.

[67]The siliceous material ordinarily used as glazing material for the crystal production at Murano consisted of quartz pebbles excavated near the Ticino river.

[68]Salt of tartar was used as an additional flux to make the crystal more brilliant. In Sagredo's day salt of tartar was added in form of the sediment from red wine barrels. The way to prepare the salt of tartar is explained in Neri (1612, 14).

[69]From Giovan Francesco Sagredo to Galileo, October 27, 1618, in *EN*, XII:417–418. For the translation of the entire letter, see p. 256.

the work procedures demanded by rock crystal were particularly long, expensive and complicated. Another disadvantage was that the spiral rods—the *torticci* that Sagredo wanted to avoid—were still present, though in a smaller number, a problem probably due to impurities. In reference to the lenses, moreover, the main reason for avoiding the use of rock crystal was that the grinding and the polishing procedures required very hard and long work, which was not often crowned by success. But rock crystal allowed the production of the most brilliant and transparent material, though it also required the addition of flux and other special decoloring compounds.

The workflow of Galileo's experiment From Sagredo's words and the first book of Neri's *Arte vetraria*, details of the experiment Galileo proposed can be inferred: The process of calcination entailed the use of rock crystal and vegetal ash.[70] The rock crystal had to be crushed, pulverized and sifted repeatedly, as did the ash. After at least one month, the frit was then ready for the addition of the decoloring materials and compounds. These were manganese from Piedmont[71] and *zafaro*, a kind of cobalt oxide[72] used to eliminate any green tones in the crystal. Once cobalt oxide and manganese had been added, the compound had to be pulverized and, again, sifted repeatedly, especially because of the manganese. Then the final product was ready for firing, which normally took at least one month. During firing the pan had to be placed in cold water repeatedly so that the crystal matter could separate from the alkaline salts.[73] According to the Murano workers "recruited" by Sagredo and Galileo, this step of moving the pan, and the associated requirement of changing it each time, were the cause of most experimental failures, for the simple reason that the crystal mass tended to fall on the floor. As mentioned, these falls were probably no accident.

In fact, the experiment proposed by Galileo had not yet begun in early November 1618, when Sagredo reported to Galileo[74] that he had received 200 libra of rock crystal at his home. On December 15,[75] the crystal mass had already fallen twice. The first time, part of it was recovered; the second time, the glass makers decided to take the mass out during the refiring process and work it. The Murano workers made out of it thirty *cilele*, "bottoms of the artichoke" in the Venetian dialect.

---

[70]Vegetal ash contains sodium carbonate and was the ordinary flux used at Murano.

[71]Although manganese was available in several areas of Europe, for the manufacture of particularly brilliant crystal *a la façon de Venise*, for at least a century only the extremely white kind from Piedmont was used. Neri explained how to prepare the manganese in Neri (1612, 15).

[72]This compound was normally purchased in a prepared form from Saxony. For the preparation method, see Neri (1612, 15).

[73]The firing procedure was very complicated. This and probably only a fraction of the many contrivances required for such a procedure are described in Neri (1612, 11–12).

[74]Giovan Francesco Sagredo to Galileo, November 3, 1618, in *EN*, XII:418–420. For the translation of the entire letter, see pp. 257ff.

[75]Giovan Francesco Sagredo to Galileo, December 15, 1618, in *EN*, XII:427.

Output of the experiment Sagredo tried to use two glass *cilele* to make lenses, but established that they cracked because they were not resistant enough, due to the insufficiently long refiring process. Other characteristics of the glass *cilele* were a color that was too green, probably because not enough cobalt oxide (*zafaro*) had been added, and, second, a weight that was too light in comparison with ordinary crystal. This led Sagredo to believe that rock crystal contained many air pockets, perhaps due to mistakes made during firing. If this were true, Sagredo went on, then this product would have been useless for lens production because of the obiously poor quality of the images given by the irregular refraction of the rays. One week later,[76] when Sagredo informed Galileo that he had started a new experiment, he also enclosed a lens made from another glass *cilela*. This lens still exhibited the spirals that caused improper refractions.

Not a word of Galileo's response either to this lens nor to those made from the material provided by Sagredo's last experiment[77] has survived. Sagredo, who distributed the crystal at his disposal to two lens makers and one spectacle maker, was not able to state whether or not an improvement had been achieved. Indeed, many doubts remained. For example, as soon as the material was ready, it was first immediately shaped into plain sheets. After the final experiment, Sagredo expressed his suspicion that this shaping in the form of a sheet, a process which required long polishing, could result in irregularities on the surface, which ultimately would have been one of the two sides of the lens. Research by Rolf Willach confirmed Sagredo's suspicions in 2001 (Willach 2001). In fact, all of those crystal sheets manufactured in Murano during this period, which corresponded to the flat side of the lenses and escaped processing by the lens makers, exhibit a great number of small holes clearly visible under a microscope. Sagredo died that same year, 1620, and Galileo never succeeded in setting up any further experiments.

Adapting the Telescope for other Optical Devices

Just as for the modern IPhone, soon there was demand for adaptations of the fashionable telescope (despite its inferior lenses), re-arranged into different shapes and for different functions, and equipped with different accessories suitable for other contexts. Telescopes were soon divided into those suitable for astronomical observations and those for terrestrial ones. Longer, shorter, telescopic, more or less elaborately decorated in keeping with the rank of the owner—these were typical ways of differentiating among the products on offer. Galileo did not invent any new optical instrument, but he adapted the telescope to fulfill new functions and

[76] Giovan Francesco Sagredo to Galileo, December 22, 1618, in *EN*, XII:429. For the translation of the entire letter, see p. 259.

[77] Giovan Francesco Sagredo to Galileo, March 30, 1619, in *EN*, XII:446–449. For the translation of the entire letter, see pp. 260ff.

wishes. This was the context in which Galileo undertook the designs of the binoculars and the compound microscope. After the refinements to the microscope during the first half of the seventeenth century, this instrument, especially, proved to be one of the most important technological developments in the sphere of science, bringing about dramatic changes, first and foremost in the development of the life sciences (Freedberg 2002).

The binoculars Binoculars were probably built for the first time in the Netherlands, simultaneously with the first telescopes (Abrahams 2002), while Galileo made his first version of the instrument public in 1617. In fact, between 1613 and 1614 the cartographer Ottavio Pisani, employed in Antwerp, informed Galileo that he had conceived binoculars that encapsulated two small telescopes in an oval box.[78] Galileo announced the result of his latest engineering research in 1617, however, while working at the Arsenal of Livorno, and thus in the context of his efforts to improve the military equipment of Tuscan galleys. Indeed, Galileo was commissioned with the design of new devices to facilitate the use of the telescope on the galleys of the Grand Duke; upon his announcement of the birth of the binoculars in 1617 he assumed complete paternity of the new instrument.[79] Soon thereafter Galileo had already taught his pupil Benedetto Castelli how to make binoculars so that he could test the instrument and inform Galileo of any possible defects.[80] Binoculars proved to be a particularly stable instrument for terrestrial observations, especially on ships, where oscillations prevented the efficient use of the telescope.

Functions of the binoculars The Galilean binoculars (Fig. 2.13), moreover, were conceived in such a way that the two parallel short telescopes of which they were constituted could work simultaneously as an instrument to measure distances. When two telescopes are directed in parallel toward two different objects, and when both the distance between one of the objects and the telescope directed to it and the distance between the two telescopes are known, simple geometrical formulae can be used to calculate the distance between the other telescope and the object to which it is directed.[81] Because of this function and their ease of use, binoculars initially were considered to be an instrument of military relevance; as such, its design and very existence were a guarded secret. The Tuscan head council must soon have realized that such instruments were starting to become common in town marketplaces and workshops, as they soon rescinded stipulations about preserving their secrecy. Meanwhile, however, Galileo again succeeded in activating the Grand

[78] Ottavio Pisani to Galileo, September 15, 1613, in *EN*, XI:564–565 and Ottavio Pisani to Galileo, July 18, 1614, in *EN*, XII:86–87.

[79] Galileo to Curzio Picchena, March 22, 1617, in *EN*, XII:311–312.

[80] Benedetto Castelli to Galileo, May 24, 1617, in *EN*, XII:319–320. For the translation of the entire letter, see pp. 248ff.

[81] Benedetto Castelli to Galileo, May 24, 1617, in *EN*, XII:319–320. For the translation of the entire letter, see pp. 248ff.

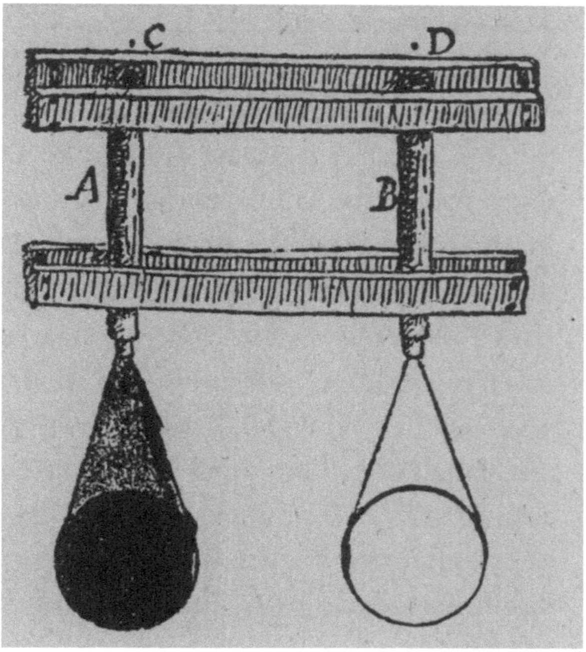

Fig. 2.13 Galilean binoculars (*EN*, XII:319)

Duke's diplomatic network to inform the ambassadors about his new invention, the best way to eliminate any doubts about its paternity.[82]

The microscope The magnifying power of single convex lenses (or bi-convex ones) had been known since the Middle Ages, of course. These lenses could be used as a microscope or as a magnifying glass. Single lenses, and devices equipped with a single lens, were apparently already in use at the beginning of the sixteenth century, in the service of the fields of botany and entomology (Wilson 1951, 72). However, not until the beginning of the seventeenth century, when Galileo adapted the principles of his telescope to invent the compound microscope, were the life sciences equipped with an instrument from which their future was to become inseparable. The compound microscope of Galilean manufacture (Fig. 2.14) was equipped with a plano-convex lens and an eyepiece consisting of either a plano-concave or a bi-concave lens.[83]

The microscope as a scientific instrument Although compound telescopes designed and built by other mathematicians and professionals were also in circulation, like Cornelius Drebbel's (1572–1633) version presented in Paris in 1621, it was Galileo who did the most to promote the instrument as a tool for acquiring knowledge. On the occasion of dispatching a microscope to the patron of the

[82]Galileo to Orso d'Elci, June, 1617, in *EN*, XII:321–328.

[83]For an introduction to Galileo's microscope, see also Bedini (1986, 145–147).

Fig. 2.14 Replica of the Galilean microscope (Turner 1991, 28)

Accademia dei Lincei,[84] Federico Cesi (1585–1630) in Rome, in 1624,[85] Galileo not only included a description of his device but also directed Cesi's attention to the scientific utility it promised:

> I send to Your Excellence a small *occhialino* [microscope] to observe the smallest things as if they were very close, and I hope you will like it and use it for no short time, for this is what happened to me. I am sending it to you quite late because I was not able to perfect it earlier, having found it difficult to produce perfect crystals [lenses]. One adheres the object to the movable circle which is at the base, and one moves it [the base] in order to watch the entire object, since what you can see at any one time is only a small part [of the object].

[84]On the relations between Galileo and the *Accademia dei Lincei*, see Westfall (1983).

[85]Galileo was probably already designing a microscope back in 1610. It is unclear when exactly he designed the apparatus in such a way that it could be presented as a new instrument, that is, small, portable and suitable for scientific use. The most relevant sources in this sense are dated from 1624. According to Horst Bredekamp, Galileo himself was not deeply interested in speculating on the subjects which could be investigated using the microscope. This situation changed during the 1620s as a consequence of his own change in attitude: The great variety and multiplicity of the microcosmos is something he initially neglected in favor of general geometrical statements about the world, and re-evaluated later in his life. For this insight I thank Horst Bredekamp. See also Bredekamp (2007b).

> And since the distance between the lens and the object must be very precise, if you observe an object with a texture [which has a height, that is, a non-negligible third dimension] one has to be able to bring the lens closer or further away, depending on the part that one is observing. Therefore the foot of the small cannon [the tube], or what we prefer to call its slide, is made movable. Moreover, one has to use it when the air is clear, and in sunlight it is even better because the object should be very much illuminated. I contemplated an infinite number of small animals with enormous admiration, among them the flea is very horrible, the mosquito and the moth are very beautiful. And with great satisfaction I have seen how flies and other small animals adhere as they walk over mirrors and also upwards. But Your Excellence will have the chance to observe thousands and thousands of details, of which I beg you to keep me informed about the most curious ones. Finally one can contemplate the greatness of nature, and how keenly nature works, and with such inexpressible diligence.[86]

In 1625 Cesi published his *Apiarium*, perhaps the first scientific work to which the microscope was applied systematically (Cesi 1625). In the same year another member of the *Accademia dei Lincei*, Giovanni Faber (1574–1629), proposed christening the newly designed instrument *microscopio*.[87]

Galileo took care to send exemplars of the microscope to important personalities throughout Europe, as usual through the diplomatic network of the Grand Duke. In particular he sent excellent telescopes and a microscope to the King of Spain in order to improve the diplomatic context within which he was trying to "sell" the secret of calculating longitude.[88]

The Quest for Longitude The context of the search for a method to calculate longitude actually served as the main impetus to develop accessories for the telescope; in particular, there was demand for variants that allowed it to be used on ships rocking at sea. A way to calculate the longitude at sea as well as on land was one of the most urgent issues, since exploratory travels had led to the emergence of relevant commercial interests on other continents, especially in the Americas. During the sixteenth and seventeenth centuries the search for such a method became increasingly intensive, with rich awards promised by all of the countries with maritime interests to anyone who could offer a solution. In Galileo's day two things were clear: First, it was "easy" to calculate longitude on a ship when both the local time and the time at the place of departure were known. Every one-hour difference between the times at the two locations designated an eastward or westward distance of 15° longitude. Second, and quite unfortunately, there was absolutely no way to keep a record on board of the time at the place of departure or any other place to use as a means of comparison. Contemporary chronometer technology was not capable of designing a clock,

---

[86] From Galileo to Federico Cesi, September 23, 1624, in *EN*, XIII:208–209. Author's italics. For the translation of the entire letter, see pp. 266ff.

[87] Giovanni Faber to Federico Cesi, April 13, 1625, in *EN*, XIII:264.

[88] For more details about the diplomatic context in which Galileo established contact with the Spanish Crown because of the "business of the longitude," see Esaù del Borgo to Galileo, September 14, 1630, in *EN*, XIV:145–147 and Francesco de' Medici to Galileo, November 26, 1631, in *EN*, XIV:309–310.

whose motion could not be influenced by the motion of ships and to other variables like changes in temperature and humidity. Not until 1760 did John Harrison (1693–1776) succeed in inventing such a clock.[89] In Galileo's day therefore some other way of measuring time (or recording it) was needed.

Galileo's method of calculating longitude Two years after discovering the satellites of Jupiter, Galileo was confident that it would be possible for him to prepare ephemerides of the motion (of the eclipses) of the four celestial bodies he was observing every night. On the basis of these tables (Fig. 2.15), the short lunar eclipses of the moons of Jupiter would have been predictable with reference to the local time of the observer who compiled the ephemerides. On the basis of the ephemerides, therefore, one could have observed from a ship the positions of the satellites, compared them with those predicted for a certain time in a certain place, and once the same configuration was found in the tables, the difference between the

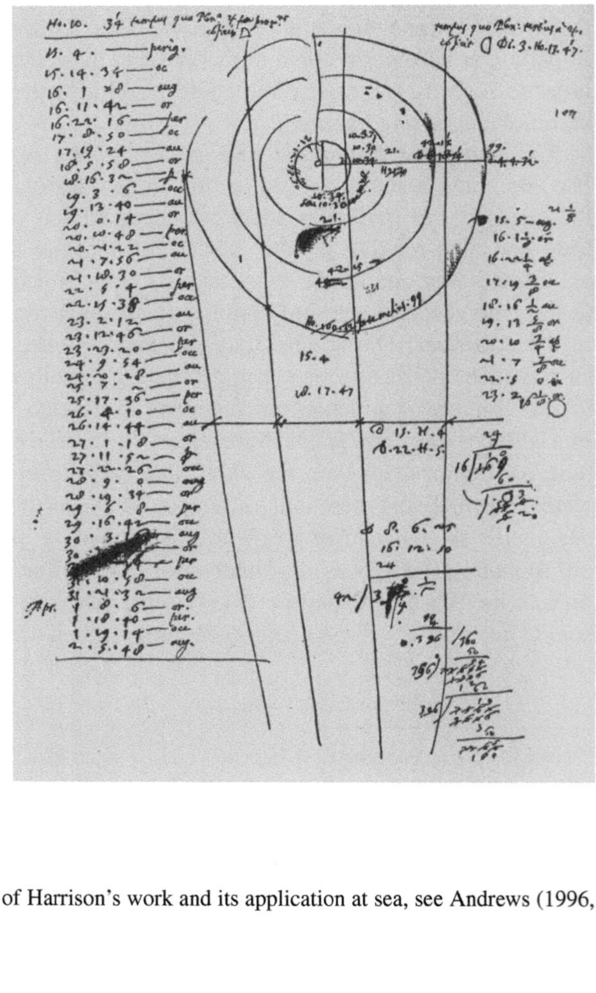

Fig. 2.15 Galileo's calculations for the computation of the ephemerides of Jupiter's satellites in 1611 (*EN*, III:820)

[89]For an exhaustive description of Harrison's work and its application at sea, see Andrews (1996, 167–280).

time predicted in the ephemerides and the local time on the ship would have provided the information required to calculate the longitude.[90] After all, the local time on board was easy to ascertain using the sun dials and lunar clocks that had been in use for centuries already. History shows that Galileo's idea was excellent, although he never succeeded compiling ephemerides with enough accuracy.[91] In fact, once such accurate ephemerides were recorded by Giovanni Domenico Cassini's hand in 1668 (1625–1712) and calculated in reference to the city of Bologna (Cassini 1668), Galileo's method was applied systematically on land, but not at sea.

The telescope on ships Although Galileo's contemporaries were quite convinced that he would have been able to deliver appropriate tables of the positions of Jupiter's moons, they immediately criticized Galileo's proposal, believing that it would be impossible to observe the moons while standing on the deck of a ship in motion. Galileo tried twice to sell his secret to calculate longitude, first to the Spanish Crown from 1613 on, and to the States General of the Netherlands starting in 1635. In both cases he encountered the same criticism. Thus in both cases Galileo came up with a new accessory to make the use of the telescope more effective on ships and for astronomic observations.

Celatone The first was called *celatone*, and consisted of a helmet of metal provided with only one short, fixed telescope to look through with one eye. The free eye was able to determine the direction in which the object could be observed, while the eye equipped with the telescope was supposed to be able to observe the magnified object (Vallerini 1992). This device was designed and built while Galileo was busy creating his binoculars at the Arsenal of Livorno.[92] It probably never produced particularly useful results because of its poor magnifying power, but the principle behind this design was intuitively correct, since applying the telescope to the head of the observer certainly made it easy for him to compensate for the movements of the ship and thus to maintain the observed object in the visual angle of the instrument. The *celatone* was tested during a galley trip from Livorno to Civitavecchia by Benedetto Castelli and Annibale Guiducci, a young pupil of Galileo. Unfortunately they met with extremely adverse weather conditions and, apart from some attempts by Guiducci, the trip's main result was to land Castelli in bed in Civitavecchia with a high fever.[93] Despite this negative outcome, in this case, too, Galileo ensured that

---

[90] For detailed research on the method of calculating longitude using Jupiter's satellites, see Helden (1996).

[91] Galileo never published any ephemerides of Jupiter's satellites. Moreover, many efforts undertaken to compile such ephemerides during Galileo's time by other astronomers, including Nicolas-Claude Fabri de Peiresc (1580–1637) and Simon Marius (1573–1624), were also failures.

[92] Galileo to Curzio Picchena, March 22, 1617, in *EN*, XII:311–312.

[93] For more details about the attempts made using the *celatone*, see Annibale Guiducci to Galileo, September 11, 1617, in *EN*, XII:344–345, Benedetto Castelli to Galileo, September 18, 1617, in *EN*, XII:346, Benedetto Castelli to Galileo, February 7, 1618, in *EN*, XII:372.

his paternity of this invention was recognized by sending a model of *celatone* to one of his patrons, Prince Leopold of Austria.[94]

The "ship over the ship" The second device was conceived by Galileo in 1637 while he was trying to sell his method for calculating longitude to the States General of Netherlands.[95] Galileo conceived an observation station to be built above deck. This station was to consist of a hemisphere, into which a second hemisphere was to be built. A seat for the observer was to be mounted over the second hemisphere, and then the two hemispheres separated from each other by a thin stratum of a liquid, preferably oil. This project was simply a further development of Galileo's first idea, as its function was to compensate for the oscillations of the ship in order to allow the observer to aim the telescope at his desired target. However, as Galileo himself admitted, it would have been able to compensate for the movements of the ship only to a certain extent. The "ship over the ship," as Galileo called it, would not have worked during particularly rough sea conditions, when all operations of the ship's screw were interrupted. Galileo's idea was never realized and the States General of Netherlands never tested his method to calculate the longitude.

Mirrors

Starting in the sixteenth century, mirror plates were normally made of glass. Behind the glass, the reflecting surface consisted of a sheet of tin coated with quicksilver. Yet as late as the early seventeenth century, curved mirrors, for instance spherical ones, often were still made of metal, especially of pewter, in keeping with the Medieval tradition (Manzini 1660, 209).

The method of constructing curved mirrors Metallic concave and convex mirrors were either hammered onto spherical dies or casted by pouring the liquid metal into a mold (Manzini 1660, 213–215).[96] Dies and molds were always spherical. If the mirror was to obtain another shape—and the most common shape other than the spherical was parabolic—first a spherical mirror was hammered, and then the shape of the parabolic line, the closest to the particular spherical one used,[97] was obtained manually by either hammering or polishing. Finally, external parabolic

[94]Galileo to Leopold of Austria, May 23, 1618, in *EN*, XII:389–392. For the translation of the entire letter, see pp. 250ff.

[95]Galileo to L. Reael, June, 1637, in *EN*, XVII:96–105. For the translation of the entire letter, see pp. 288ff. This letter also contains Galileo's description of the clock he conceived for use on ships.

[96]Carlo Antonio Manzini's work (1599–1677) is devoted primarily to the description of the method for making lenses, not mirrors. Yet as far as their shapes were concerned, lenses for telescopes and curved mirrors were made following the same procedure (Manzini 1660, 156–159).

[97]Mirror makers were not able to make particularly deep mirrors out of glass. For this reason mirrors usually had the shape of a portion of the surface of extremely large spheres. In this case, it is actually possible to draw a parabolic line that deviates only very slightly from the portion of circumference.

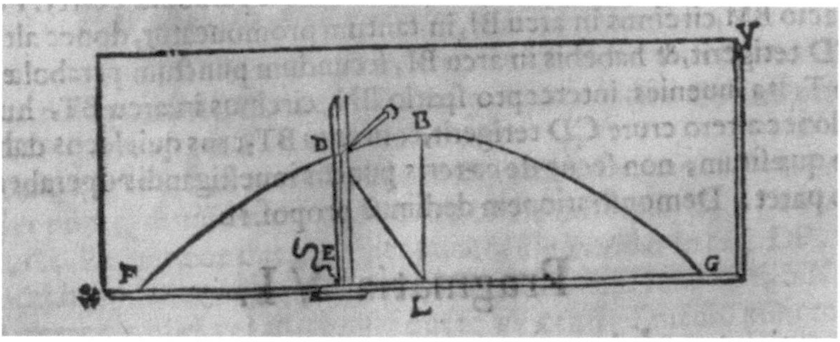

Fig. 2.16 *Parobola ope fili mechanicè describere*. Practical method of drawing a parabola using threads and nails (Kircher 1646, 308)

molds representing half of the parabola, often made of cardboard (Fig. 2.16), were used to control the precision of the mirror's shape (Manzini 1660, 218).[98]

Since the most common spherical mirrors amounted to one fourth of a sphere (or one half of a hemisphere), or occasionally one sixth, it is possible to infer circumstantially how mirror makers must have worked. The mirror maker drew a circle, ideal or not, which was supposed to represent a sphere. The horizontal diameter (or semi-diameter) was then divided into four (two), or in some cases six, equal parts. Through these points chords were then drawn. The portion of circumference obtained in this way ultimately became the profile of the bottom of the curved mirror, and the length of the chord its height. The circle was only actually drawn to construct dies and molds, which were then used to produce mirrors. Dies could be made of several materials, but those made of stone were reputed to be the best (Manzini 1660, 206–209). Of course, curved mirrors made of glass were obtained by grinding and polishing on the same kinds of dies. Although the use of grinding machines became very common during the sixteenth century, as late as the second half of the seventeenth century the best spherical mirrors were the ones made completely by hand.

The use of curved mirrors Curved mirrors were optical devices, whose functioning was the object of investigations by engineers and mathematicians. For this reason and because of the famous legend of Archimedes, curved mirrors became symbols representing the possession of knowledge; as such, the commercial demand for them was high. Finally, spherical and parabolic mirrors were devices useful for intellectual diversion, thanks to their capacities to reflect sound, create an echo, and ignite a flame. Curved mirrors like the convex ones or those shaped like a portion of a cylinder were also in particular demand because of the amusing distortion effects in their reflections. In short, curved mirrors had a plethora of characteristics that made them particularly suitable for life at court. For the last reason, especially, Galileo was concerned with the functioning of curved mirrors his whole life long.

---

[98]Constructing parabolic forms, for example of cardboard, in order to control curvature was a very old practice, probably related to the experience of trumpet makers.

The search for Archimedes' burning mirrors Galileo never made mirrors by his own hand and even probably never commissioned the construction of any. Yet he was often occupied with one of the crucial questions of those years concerning curved mirrors, that is, how to build an exactly parabolic mirror and how such a mirror works. Martin Hastal wrote from the imperial court of Prague in 1610 to Galileo that:

> [...] he [His Caesarean Majesty] ordered me to ask Your Lordship whether by chance you possess the secret of the Archimedean parabola, which kindles far away, and how far away. I answered him that I will write and also that I know with certainty that a very close friend of yours has it [the secret], but that he holds it in such value he does not want to sell it to the Grand Duke Francesco for many thousands of S[cudi]. I want to say Master Paulo.[99] I was told about this in my hometown by a great mathematician, a friend of yours.[100]

Galileo did not possess Archimedes' secret and was probably afraid that Archimedes' mirror could not be built.[101] In 1632, however, when Bonaventura Cavalieri (1598–1647), one of Galileo's pupils, published *Lo specchio ustorio* (Cavalieri 1632),[102] Galileo himself posed the same question Martin Hastal had posed to him: he asked for the geometrical description of Archimedes' burning mirror.[103] Galileo's purpose was to commission the construction of such a mirror in Florence for the Grand Duke.[104] According to the detailed description sent by Cavalieri in March of the same year,[105] the efficiency of the burning mirror was directly proportional to the curvature of its parabolic shape (Fig. 2.17).[106]

In Galileo's day the attempts to build a burning mirror like Archimedes' mirror, that is, one able to kindle strongly from a long distance, were innumerable, although none of them succeeded. In fact, parabolic mirrors were obtained by polishing spherical mirrors in such a way that their shape was altered to align with the parabolic line that was as close as possible to the circular one. However, in order to

---

[99] Paolo Sarpi. For more details, see Favaro's footnote no. 1, in *EN*, X:426–427.

[100] From Martin Hastal to Galileo, August 24, 1610, in *EN*, X:426–427. For the translation of the entire letter, see p. 226.

[101] Martin Hastal to Galileo, December 19, 1610, in *EN*, X:491–492. In 1614 and 1624 Fabio Colonna again posed Galileo the same question as Hastal. For more details, see Fabio Colonna to Galileo, July 29, 1614, in *EN*, XII:88–89 and Fabio Colonna to Galileo, August 8, 1622, in *EN*, XIII:93–94.

[102] Cavalieri's work is mainly a treatise on the geometry of conical sections, yet also a work in which the author showed the practical effects that objects like mirrors produce when they are shaped according to those sections.

[103] Bonaventura Cavalieri to Galileo, February 12, 1636, in *EN*, XVI:395–296.

[104] Giovan Paolo Casati to Giannantonio Rocca, February 13, 1636, in *EN*, XVI:396–397.

[105] Bonaventura Cavalieri to Galileo, March 11, 1636, in *EN*, XVI:401–404. For the translation of the entire letter, see pp. 285ff.

[106] Curved mirrors of elliptical and hyperbolic shapes were also contemplated, though not commonly. The Jesuit Father Christoph Grienberger (1561–1636), for example, on the occasion of a lecture at the *Collegio romano* in 1613, provided a demonstration that Archimedes's burning mirror could be nothing but a portion of an ellipse. For more details, see C. Grienberger to Galileo, February 1, 1613, in *EN*, XI:477.

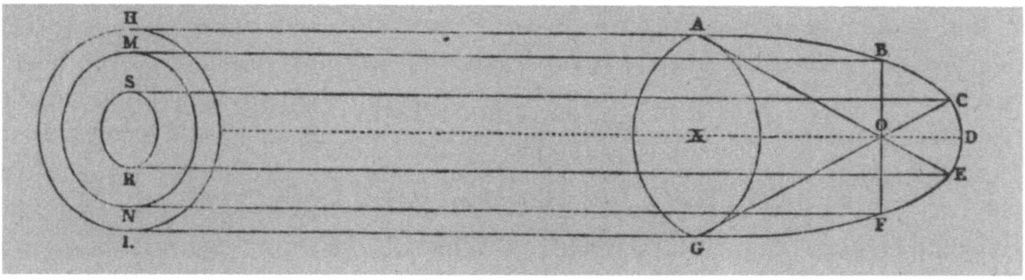

Fig. 2.17 Bonaventura Cavalieri's sketch of Archimedes' burning mirror (*EN*, XVI:402)

achieve this result, the spherical mirrors used in production had to be a small portion of a relatively large sphere, and therefore furnished with a very light curvature. This is probably the main reason why there is no trace of evidence left for Galileo attempting to commission the construction of such a mirror: As someone familiar with the procedures of manufacturing mirrors, Galileo probably decided to abandon the project immediately after learning from his pupil the shape that was supposed to constitute an efficient burning mirror.

However, the very same year, a Venetian craftsman named Sigismondo Alberghetti picked up on Galileo's idea of constructing Archimedes' mirror as conceived by Cavalieri. Alberghetti was an acquaintance of Fulgenzio Micanzio, Paolo Sarpi's former assistant and the person with whom Galileo entertained the most copious epistolary exchange in Venice in 1636.[107] Galileo probably shared with Micanzio the idea for his project and the reason why he abandoned it, and, in spite of Galileo's judgement, Micanzio probably arranged an attempt with Alberghetti, who was also building a heliocentric sphere according to the description of the world system Galileo had presented in his banned *Dialogo sopra i due massimi sistemi del mondo tolemaico e copernicano* (*EN*, VII:25–520).[108] Alberghetti's mirror project did not succeed;[109] when Micanzio informed Galileo about this negative result, Galileo's reaction was not at all surprised:

> Concerning the parabolic mirror, I always considered it very difficult, if not impossible, to reduce it [a mirror] to such a shape; but, if it is a spherical one and a portion of a very large sphere, the shape around its center differs so little from the parabolic one, that, since the spherical one can be built perfectly, as opposed to the parabolic one, the effect of burning would turn out to be stronger with the spherical one than with the parabolic, even though the latter joins the reflected rays at only one point which the other does not.[110]

---

[107] In 1636 Fulgenzio Micanzio organized a reading group for the upcoming *Discorsi* and organized the work's printing with the Elzevirs.

[108] Alberghetti notified Cavalieri of his project personally, while Galileo was kept informed by Micanzio. For more details, see Bonaventura Cavalieri to Galileo, October 21, 1636, in *EN*, XVI:508, Fulgenzio Micanzio to Galileo, February 21, 1637, in *EN*, XVII:31–32.

[109] Fulgenzio Micanzio to Galileo, October 31, 1637, in *EN*, XVII:209–210.

[110] From Galileo to Fulgenzio Micanzio, November 20, 1637, in *EN*, XVII:220–221. For the translation of the entire letter, see pp. 295ff.

The theory of curved mirrors Galileo's confidence with mirrors was probably due as much to the experience he accumulated in the context of his activity as a lens maker, as it was to what he and Paolo Sarpi learned in the cultivated circle around Giovan Vincenzo Pinelli (1535–1601) in Venice, and from his Venetian friend, Giovan Francesco Sagredo. Recent studies have shown that Galileo was familiar with medieval optics only indirectly, through the practical mathematicians of the sixteenth century (Dupré 2005). Galileo's main sources for optics were Ausonio's *Theorica speculi concavi sphaerici*, which he copied between 1592 and 1601 (*EN*, III:865–870),[111] and della Porta's *Magiae naturalis* and *De refractione* (Porta 1589 and 1593).[112] On the basis of what Galileo learned in the field of practical optics from Ostilio Ricci, he was able to copy and discuss Ausonio's *Theorica* with Giovan Vincenzo Pinelli and Fra Paolo Sarpi when he arrived in Padova and entered Pinelli's circle (Dupré 2005). Ausonio's work was reputed to be very important for the understanding of the multiple geometrical functions of curved mirrors, like image formation, kindling and causing an echo, in the case of the mirrors shaped like a portion of the surface of a sphere.

Sagredo toward a new science of mirrors From 1613 on, Giovan Francesco Sagredo again drew Galileo into theoretical speculations about curved mirrors. Sagredo supposed that curved mirrors of glass worked not only according to the principle of reflection, as common mirrors of metal did, but that their functioning also had to be explained on the basis of the principle of refraction. Sagredo opened up this field of research for the first time, today a standard chapter in every manual of geometrical optics. Sagredo unsuccessfully attempted to convince Paolo Sarpi and Agostino da Mula (1561–1621) in Venice, as well as Giovanni Antonio Magini (1555–1617) in Bologna, that the use of glass would have necessitated the creation of a new theory about the way these mirors worked. Galileo did not react to Sagredo's suggestions, either, despite the fact that the latter conceived a complete set of experiments (Fig. 2.18) to perform using curved mirrors of different shapes and requested Galileo's supervision:

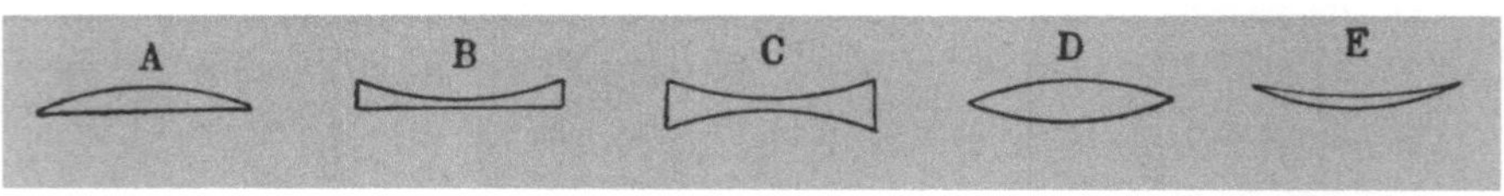

Fig. 2.18 Giovan Francesco Sagredo's set of experiments to be performed using curved mirrors (*EN*, XI:556)

---

[111] Ausonio's own copy of the *Theorica* seems to be lost. For an exhaustive analysis, see Dupré (2000, 2005). Eileen Reeves analyzes Ausonio's text in Reeves (2008, 56–58) and Galileo's work on Ausonio's text in Reeves (2008, 90–91).

[112] Other sources certainly included Kepler (1604; 1610; 1611). Concerning the explanation of the sense of sight, at least until 1612 Galileo was convinced that human beings could see thanks to visual rays emanating from the eyes. For more details, see, for example, Giovan Francesco Sagredo to Galileo, July 7, 1612, in *EN*, XI:355–356.

> Each of the drawn forms can be considered [investigated] in two ways, since one can place the leaf [stratum of quicksilver and tin] on both of the surfaces, except from C, and D, which [whose surfaces] are equals.[113]

Galileo's theory of spherical mirrors Probably in part because of Sagredo's engagement, around ten years later Galileo, too, finally revealed his confidence in theoretical explanations for the functioning of curved mirrors. A certain Giovan Battista Guazzaroni, not yet identified, reported the results of a discussion between Galileo and himself which took place some time earlier.[114] On this occasion Galileo presumably stated that della Porta's and Magini's theories about the functioning of the spherical mirrors were wrong. In particular, first he rejected the theory that the point at which the spherical mirror burns is located at the center of that sphere of which the mirror is a portion, as della Porta had stated in his *Magiae naturalis* (Della Porta 1589, Book XVII); second, he further rejected the idea that the location of this same point from the bottom of the mirror is at a distance equivalent to one fourth of the diameter of that sphere of which the mirror is a portion, as was postulated in Magini's theory.[115] Magini's and della Porta's were among the most important theories on spherical mirrors being taught and discussed in Galileo's day. Galileo opposed them with his view that the point at which a spherical mirror burns is variable, and dependent on the size of the sphere and of the portion of this sphere the mirror represents. He also added, finally, that the "point" of burning is not really a point at all, but rather a small area, certainly having noticed the effects of what nowadays is called spherical aberration.

Galileo as a mirror broker Galileo was not a mirror maker, but he was so capable of evaluating these craftsmen's products and their labor procedures that he acted as a mirror broker for the Grand Duke. For example, in 1610 he tried to get the court to acquire concave mirrors designed by the mathematician Antonio Magini.[116]

---

[113]From Giovan Francesco Sagredo to Galileo, August 24, 1613, in *EN*, XI:552–556. For the translation of the entire letter, see pp. 234ff.

[114]Giovan Battista Guazzaroni to Galileo, April 20, 1624, in *EN*, XIII:172–174. For the translation of the entire letter, see pp. 264ff.

[115]Magini published a treatise on spherical mirrors in 1611 (Magini 1611). The text was written at Galileo's request in order to facilitate the Grand Duke's purchase of Magini's curved mirrors through the mediation of Galileo. In 1602 Magini had also already resolved to publish Ausonio's *Theorica*, though revisited, corrected and, finally, significantly altered. For an analysis of Magini's changes, see Dupré (2005). Mathematicians often considered della Porta's theory to be that of the ancients and Magini's the modern theory.

[116]The specific epistolary exchange between Magini and Galileo shows that the latter was well aware of the procedure for constructing these mirrors and that he was able to understand whether a mirror was made properly or not. For more details, see Giovan Antonio Magini to Galileo, September 28, 1610, in *EN*, X:437–439, Giovan Antonio Magini to Galileo, October 2, 1610, in *EN*, X:442–443, Giovan Antonio Magini to Galileo, October 15, 1610, in *EN*, X:445–446, Giovan Antonio Magini to Galileo, October 23, 1610, in *EN*, X:450–451, Giovan Antonio Magini to Galileo, December 28, 1610, in *EN*, X:496–497, Giovan Antonio Magini to Galileo, January 11, 1611, in *EN*, XI:19–20, Giovan Antonio Magini to Galileo, January 10, 1612, in *EN*, XI:259–261.

Neither was Galileo particularly famous as a theoretician speculating on the functioning of curved mirrors, but he was familiar enough with all of the most important theoretical explanations of his time to reach the conclusion that a spherical mirror would kindle better than a parabolic one, for the simple reason that spherical mirrors had been constructed for a long time, and therefore the mirror makers were able to make mirrors of this shape that very closely approached the ideal geometrical portion of a sphere. Ultimately Galileo was able to merge these two fields of knowledge: first, in order to evaluate the practicability of building projects and, second, to create a challenge for expert mirror-makers.

Machine for Pounding Gunpowder

Within every fortress there was a machine for pounding gunpowder. Galileo was once called upon to evaluate a project for one of these machines (*EN*, VIII:585–587). In other terms Galileo was asked to write a report about another engineer's proposal in his capacity as court military engineer. The date of Galileo's document concerned with the machine for pounding gunpowder is unknown, but it can be established as 1626 thanks to a letter sent that year to Galileo by Benedetto Castelli, one of Galileo's pupils,[117] and thus sixteen years after Galileo moved from Padova back to Florence to work as the Mathematician and Philosopher of the Grand Duke.

An unknown engineer suggested a method of improving a machine to pound gunpowder (Fig. 2.19) that worked with the force of one man and had four pistons, in such a way that the same force could have been used to drive sixteen pistons. The author of this proposal also suggested that increasing the number of the pistons, while keeping the same moving force, would increase machine's efficiency. Galileo's written evaluation opens with the statement that:

> [...] being able to defraud nature, that is being able, either with less effort, or with less expenditure of time, to do those operations that could be done without the machine only with more effort or longer time, [is] something that is, absolutely speaking, completely impossible (*EN*, VIII:585).

This is a formulation of what Gianni Micheli calls Galileo's principle of conservation (Micheli 1995, 132–152).[118] It means that, if the force remains the same but

---

[117] "And as payment I beg Your Lordship to give me those writings about the perpetual motion and the motion of the pistons." From Benedetto Castelli to Galileo, August 21, 1626, in *EN*, XIII:338.

[118] When Galileo expressed his principle of conservation, he actually did so following a well-established tradition advocated by none other than Guidobaldo del Monte, one of Galileo's patrons. In fact, although Guidobaldo del Monte started the dedication of his *In duos Archimedi aequeponderantium libros paraphrasis* by discussing the *mirabilia* of the mechanics, as Micheli himself pointed out, del Monte goes on to specify on the following second page that "the art, with an admirable artifact supersedes nature by means of the same nature, just displacing the things in same way nature would, if it decided to produce for itself such effects." For Guidobaldo del Monte's paraphrase of Archimedes, see Archimedes and del Monte (1588). Micheli quotes Guidobaldo del Monte in Micheli (1995, 146). For the art-nature relationship, see pp. 91–97, and especially pp. 199ff.

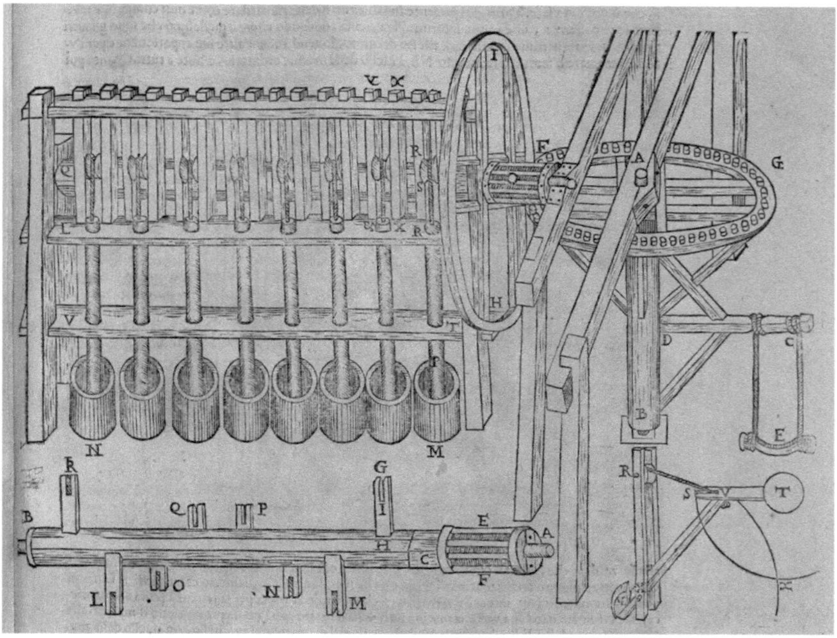

Fig. 2.19 Example of machine to pound gunpowder (Lorini 1609, 237)

the machine is bigger, there is a loss in the time and/or velocity of the machine's action. With reference to the machine for pounding gunpowder, Galileo pointed out that the greater productivity of the machine is given not by an increased number of pistons, but by the frequency of the pistons' pounding within a certain time interval. Were the force to remain constant, then the bigger machine would work more slowly and therefore the quantity of powder pounded during a given time interval would remain unchanged.

Machine for Lifting Heavy Weights

Galileo also expressed his opinion as advisor to the Grand Duke of Tuscany in the case of a machine for lifting heavy weights (*EN*, VIII:557–584) (Fig. 2.20),[119] and thus after moving back to Tuscany in 1610. The machine was a compound device consisting of a curved lever, one arm of which was five times longer than the other, with the longer arm free to move as a pendulum. Galileo criticized the machine on the basis of this design in three different texts. In the first text it is regarded as

[119]The original documents written by Galileo's hand were lost, but Favaro reproduced the copies of them, which do not have titles.

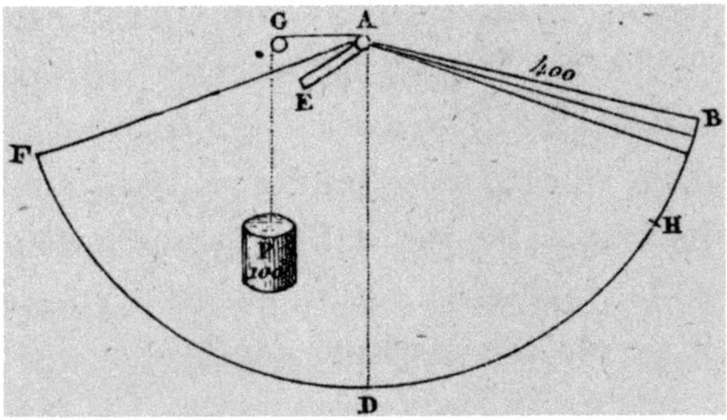

Fig. 2.20 Machine (bent arm lever) for lifting heavy weights (*EN*, VIII:575)

a winch for raising weights, in the second as part of a machine for milling grain, and in the third as a part of a machine for lifting water.[120] Fortresses typically were equipped with all of these kinds of machines. Galileo also analyzed the machine on the basis of his principle of conservation, arriving at the conclusion that the velocity of execution would have been very slow, even without considering the machine's own weight and the corresponding force required to move it.

Water Lifting Machine

Galileo not only evaluated other people's machines, he also oversaw their realization. During the first two years of his stay in Padova, Galileo was granted a patent for a new water lifting machine from the Venetian *Provveditori ai comuni*. As Galileo explained in his first specification for this patent, the machine had to be provided with twenty mouths through which the water runs and operated by the force of one horse. The *Provveditori* granted the patent for twenty years, but since nobody had seen the machine "neither in big nor in small," they gave Galileo one year's time to demonstrate such a machine in form of a model; otherwise the patent would not take effect.[121] In the *Ricordi autografi* there are two entries regarding an *edifizio*, where this word was normally used to denote water lifting machines. First, Galileo noted that Niccolò Contarini twice lent him 170 Venetian ducats for the construction of the model of such a machine. The entries in the *Ricordi* are dated September 12 and 29, and Antonio Favaro suggested that these entries were written in 1601

[120]The third text is in the form of dialog, as if it were to become part of a publication.

[121]The documents concerning the patent for the water lifting machine, granted to Galileo by the Venetian *Provveditori*, were published in *EN*, XIX:126–129, Doc. XII.

(*EN*, XIX:202).[122] Galileo, however, never realized the construction of the machine in real size.

Galileo as a Military Engineer

In perfect keeping with the the subjects learned during his apprenticeship, Galileo became an expert workshop manager and designer of mathematical and optical instruments. Like every successful artist-engineer in Galileo's day, he also performed most of these activities within the general framework constituted by the demands of warfare. The mathematical instruments he designed were destined for officers. The telescope, once he had succeeded in improving its magnifying power, he presented to the Venetian authorities as an instrument of great military relevance because it improved chances of discovering enemy ships at sea. Obviously, military security and economic success were directly connected. In this setting, Galileo went on to design the binoculars, the *celatone* and the "ship over the ship" (the gimbaled seat to observe Jupiter's moons from above deck). As an artist-engineer so involved in military culture, it was inevitable that, after moving back to Florence, Galileo attained increasing status at the Medici court. Appointed practical mathematician at court, Galileo became involved in the brokerage business, concerned primarily both with lenses for telescopes and mirrors, especially the curved ones in such fashion during his day.

With reference to the subjects discussed in this section, it is possible to perceive a slow change in the framework within which engineers like Galileo worked. The first phase, between 1592 and 1610, when Galileo lived in Padova, is characterized mainly by his commitment to the immediate technical needs of the military officer. The second phase, during his second stay in Florence where he returned in 1610, during which his activity as an engineer entered the field of economics, probably as a consequence of his role at court: the evaluations about the efficency of machines, the method to calculate longitude and the technical solutions designed to achieve

[122] Although the date of the last and definitive document concerning the approval of the patent is September 15, 1594, Favaro's suggestion is supported by two letters by Nicolas-Claude Fabri de Peiresc, who wrote to Galileo in January 1634 reminding him that they had met about thirty years earlier in Padova, more precisely, in Contarini's garden, where Galileo tested the water lifting machine. Further, in a letter to Cardinal Francesco Barberini (1597–1679) of December of the same year, the same Peiresc declared that he had met Galileo thirty-four or more years earlier. For more details, see Nicolas-Claude Fabri de Peiresc to Galileo, January 26, 1634, in *EN*, XVI:27 and Nicolas-Claude Fabri de Peiresc to Cardinal Barberini, December 5, 1634, in *EN*, XVI:169. Galileo never lost his interest in this kind of device, as two letters to him from Fulgenzio Micanzio in 1635 show. In the first letter Micanzio told Galileo about both a new water lifting machine and a hydrant made in Lyons. In the second letter, again sent by Micanzio to Galileo, he wrote that he was not able to satisfy Galileo's demand for more details about these two new machines. For more details, see F. Micanzio to Galileo, August 23, 1635, in *EN*, XVI:305; F. Micanzio to Galileo, September 15, 1635, in *EN*, XVI:310.

such a goal, and brokerage activity are the main issues which testify to such a shift of interests.

Before moving back to Florence, however, Galileo showed his skill as a military engineer, not only thanks to his smithy workshop and the mathematical instruments he was producing there. He did so also, and especially, because of his commitment to the training of the next generation of military officers. This activity is the subject of the next chapter.

Chapter 3
Galileo's Private Course on Fortifications

Galileo taught privately in Padova from 1592 to 1609. He offered lessons on a variety of subjects, which, taken together, amounted to a complete course on fortifications typical for the end of the sixteenth century. Individual topics covered by Galileo's course on fortifications included: the use of mathematical instruments; practical arithmetic and geometry; military architecture; the practical science of machines, with an emphasis on those machines needed to build and maintain fortresses; as well as elements of theoretical geometry, drawing techniques including perspective, geodesy and elements of astronomy, typically in the context of what was held to be the subject of lessons on *La sfera*, namely practical astronomy.

The balance sheet of Galileo's workshop, that is, manuscripts 26 and 29, part of which were published by Favaro as the *Ricordi autografi*, represents the main documentary evidence of Galileo's activity as a private teacher.[1] Private lessons complemented his activity as an instrument maker. The military compass, in particular, was the didactic tool which students used to complete their exercises, and lessons on the use of the compass served as the core of Galileo's overarching course on fortifications.

As a rule, the entries in the *Ricordi* concerning private lessons are labeled according to their topic. These topics are: Geodesy, Mechanics, *La sfera* and Cosmography, Perspective, Euclidean Geometry, Arithmetic, Fortifications (that is, military architecture), and Use of the Military Compass. The fact that all registered topics, taken together, correspond to a rather typical treatise on fortifications of its day, is confirmed through comparison of his courses' contents with, for example, a traditional well-known contemporary treatise on fortifications: *Le fortificationi* (Lorini 1609) by Boniauto Lorini (ca. 1540–ca. 1611).[2]

---

[1] The text of the *Ricordi autografi* was introduced in the previous chapter, for it is also the main historical evidence of Galileo's activity running a workshop. For more details, see pp. 24ff.

[2] Bonaiuto Lorini's treatise was published in Italian for the first time in 1597. In 1609 he published the second edition, which is the one considered here.

M. Valleriani, *Galileo Engineer*, Boston Studies in the Philosophy of Science 269,
DOI 10.1007/978-90-481-8645-7_3,

In the sixteenth century, it was customary for young noblemen to study and earn a university degree. After these studies, the next standard career step was service as a military officer. In preparation for this, their university curriculum generally included a "course on fortifications." The university offered a basic level of instruction, while the highest level was available through private lessons, usually given by the university lecturer and/or in private academies. The course on fortifications comprised several topics, the aim of which was to make the future officer familiar with the knowledge needed on the battlefield. These subjects were military architecture; the use of relevant mathematical instruments; practical arithmetic and geometry; drawing techniques; the use of artillery; and the use and understanding of machines like water lifting machines, machines for pounding gunpowder, and any other weight lifting machines that could be useful in such places as fortresses, battlefields, harbors, and ships. The courses generally also provided some general notions of practical astronomy: for example, what was needed to "read" the sky, and to calculate positions and times. Courses on fortifications differed from each other in terms of the details they offered on the individual topics, or in terms of the instruments the students were taught to use.

Galileo's course on fortification was distinguished primarily by the fact that it was an all-inclusive service. Galileo's pupils lodged in his house with their servants; from the workshop of their teacher, residing in the very same house, they received the mathematical instruments that functioned as the hub around which all of the lessons revolved. Furthermore, their teacher was able to provide them with treatises he had written on most of the subjects. Finally, in most cases their teacher was also their professor at the university. In fact, most of Galileo's public lectures at the University of Padova served as introductions to the topics of his private lessons. The rolls of the university (*EN*, XIX:117–119) testify that the topics Galileo taught for eight of the eighteen years he worked at the University of Padova were Fortifications, Mechanics, *La sfera*, the Theory of the Planets based on Ptolemy's *Almagest*, Euclid's *Elements*, and Aristotle's *Mechanical Questions*.

In the following, first the structure and the magnitude of Galileo's business of private instruction will be described. The contents of his lessons then will be analyzed by means of a comparison with the state of the art of each individual subject. These contents are, first, training in the fields of practical mathematics and geometry, that is, subjects like drawing techniques, perspective, stereometry, surveying, and the use of relevant mathematical instruments; second, military architecture; third, the use and function of artillery; fourth, practical astronomy; and fifth, the science of machines.

The Structure of the Business

A cursory glance at the *Ricordi autografi* reveals that Galileo's house was anything but the studio of a lone thinker, as it was inhabited by dozens of residents at any given time.

Private students who were also lodgers[a]	Other private students
Schweinitz G. (+2)	Allfeldt (von) C.
Lazocski (+1)	Filippo d'Assia
Lentowicz M.	Vinciguerra Coll'Alto
Bucau B. (+1)	Reisener B.
Buc	Luzimburg
Plesch M.	Noailles (de) F.
? Giovanni - from Lithuania	Batavilla
Ferrante (+1)	Reigesberg G
Ricques D.	Dietrichstein (de) P.
Zator G. (+8)	+3 Students whose names are unknown
Lesniowolski R.	
Soell G. C.	
Het B.	
Montalban A. (+2)	
Morelli Andrea (+1)	
Caietano Giulio Cesare (+1)	
Total: 33 persons	Total students: 28

[a] The number in parentheses is the number of servants and/or companions accompanying the student.

Galileo's private students Galileo's private students were often the offspring of distinguished, or at least rich, families who wanted to complete the curriculum at the University of Padova before starting their military career as officers or as diplomats (Valleriani 2001). For example, during the period between 1602 and 1604 alone, as demarcated by the letter to Guidobaldo del Monte[3] about the isochronism of the pendulum, and the one to Paolo Sarpi[4] which gives indications of Galileo's construction of the theory of motion, as many as twenty-eight private students are listed in the *Ricordi* (see table above), sixteen of whom also lodged at his house for either the entire period or part of it. Moreover, most of his lodgers were accompanied by at least one servant or companion, so that the total number of roomers documented during that period actually increases to thirty-three.[5]

Treatises for the lessons Four of the subjects of instruction were based on Galileo's treatises entitled *Delle macchine* (Galilei 1592–1593a), *La sfera ovvero Cosmografia* (*EN*, II:203–255), *Trattato di fortificazione* (*EN*, II:77–146), and *Le operazioni del compasso geometrico et militare* (*EN*, II:335–424). Although Galileo

[3]Galileo to G. del Monte, November 29, 1602, in *EN*, X:97.

[4]Galileo to P. Sarpi, October 16, 1604, in *EN*, X:115.

[5]The names of Galileo's private students, as they appear in the table, have not been changed according to modern convention. The way they appear is the way Galileo wrote them in the *Ricordi autografi*.

published his treatise on the military compass in 1606, a first draft of his treatise on the use of the military compass had been prepared in 1599,[6] when the production of the instrument in his workshop became systematic. After all, there is copious evidence, both in his correspondence and in the *Ricordi,* that he was selling it to his students, besides proffering it as a gift to renowned personalities. The other three treatises were prepared by Galileo during the first years of his stay in Padova.

Further entries in the *Ricordi* testify that a copyist named Messer Silvestro was working at Galileo's house in 1603. The copyist provided the private students with handwritten copies of their teacher's treatises, including the one on the use of the compass.[7] Although the only entries regarding handwritten copies of treatises concern Messer Silvestro in 1603, this seems to have been a normal procedure in Galileo's house, as indicated by the several copies written by Galileo's pupils of his *Trattato della sfera*, *Delle macchine* that are still preserved today.[8]

The organization of the entire course The sequence of subjects taught by Galileo shows some regularities (Fig. 3.1). Almost all of the students were taking lessons on the *Uso del compasso militare*, alternating these lessons on the military compass with those specifically devoted to military architecture based on the treatise *Le fortificazioni*. All of the other subjects were taught by Galileo before these lessons. Galileo's teaching of mechanics based on the treatise *Delle macchine* was in particular demand. Galileo's reading of Euclidean geometry and perspective was also propaedeutic to the lessons on military architecture and on the use of the compass. They were oriented toward the practical needs of a military man faced with challenges such as designing fortresses, aiming a cannon, and surveying terrain. Most of the students who took the class on practical astronomy did so after completing those on mechanics, and before starting instruction on military fortification and the use of the compass. Finally, there is one entry in the *Ricordi* regarding lessons in geodesy, and one for arithmetic.

---

[6]Cosimo Pinelli to Galileo, April 3, 1599, in *EN*, X:73. More than one draft of the treatise on the use of the military compass exists that was compiled before 1606, when Galileo published it. Some of these previous writings have been published by Antonio Favaro. See *EN*, II:345–361.

[7]Galileo annotated the following set of entries concerned with Messer Silvestro's work in 1603: "Note of the copies from Messer Silvestro: Fortifications, 2 copies, for Lord Giovanni Svainitz and Lord Lerbac; The same, 1 copy to Lord Bucau; The same, 1 copy to Lord Alfelt; The same, 1 copy, to Lord Staislao; The same, 1 copy, to Lord Niccolò Beatavil. For one copy of the Use of the Compass, given to Lord Staislao. For one of the Use of the Compass, given to Lord Beatavilla. For one copy of the mentioned Use, given to the Most Illustrious and Excellent Lord Langravio. For one of the mentioned copies, given to a German nobleman. For one given to Lord di Noagles" (*EN*, XIX:163).

[8]Favaro's analysis of four handwritten copies of Galileo's *Trattato della sfera* shows that Galileo's students were also permitted to copy the treatises of their teacher themselves. This is the case, for instance, for Abbot Giugni's copy of that treatise, and Galileo himself wrote that Abbot Giugni was in his house (*EN*, XIX:163).

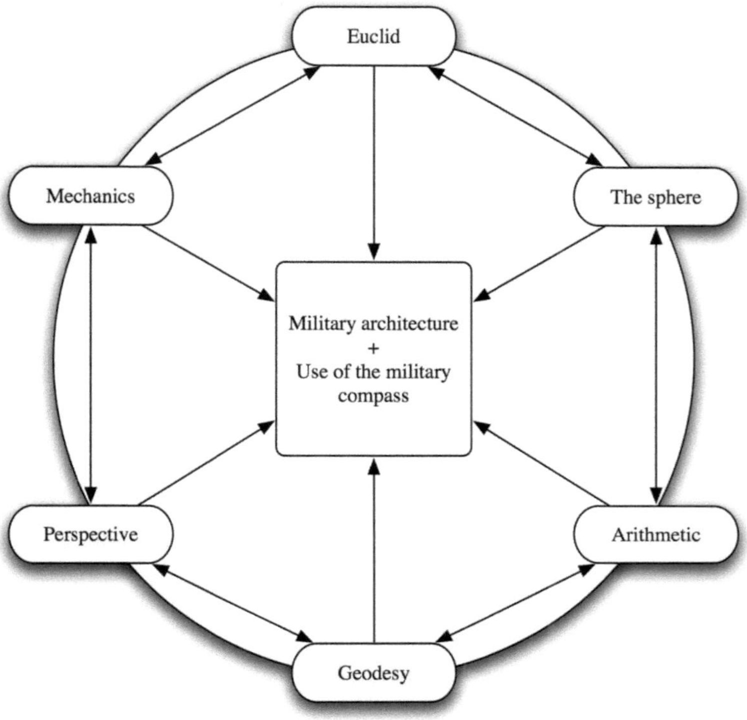

Fig. 3.1 The structure of Galileo's survey course on fortifications. The disciplines along the circumference were considered to be propaedeutic topics for those placed in the middle of the circle

Mathematics for the Military Art

The background of many of the subjects taught by Galileo consisted of notions on mathematics, practical arithmetic and practical geometry, that is, those disciplines that were often simply called mathematics.

In 1607 Galileo's friend Pietro Duodo was elected *Capitano di Padova* of the Venetian Republic, and he immediately founded an academy where the young noblemen of Padova not only were drilled in gymnastic exercises to practice for the cavalry and fencing, the usual activities offered by almost all the academies for young noblemen at the time, but, in accordance with the cultural and political context emerging in the early modern Italian cities, could also attend to learning the topics concerning the military arts.[9] This was the birth of the *Accademia Delia* of Padova, which issued a call for a lecturer of mathematics in 1610, for which position

[9]For the cultural change in favor of an education which included notions of military art, see the next chapter, pp. 117ff.

Galileo listed the required subjects of instruction during the same year. In Galileo's words:

> [1] It is primarily necessary to comprehend lower-level arithmetic, for the needs of regulating the armies and for many other eventualities.
> [2] Practice with geometry and stereometry, for measuring all planes on a surface, both regular and irregular, and for measuring all of the solid figures and bodies.
> [3] Knowledge of the mechanical sciences, not only about their reasons and common basis, but also regarding many machines and particular instruments, together with the resolutions of a great deal of questions and problems depending on that mechanical knowledge.
> [4] Practice with kinds of artillery, entailing knowledge about their differences, measures and proportions, as well as about the causes and reasons of many accidents which happen in this exercise.
> [5] Knowledge of the compass and of other instruments, for drawing all kinds of plans, both close and far away.
> [6] Use of instruments for measuring heights, distances and depths with the sight, and for leveling each site.
> [7] Some exact rules for drawing in perspective all of the things seen and imagined, by means of which the fortresses and all their parts, as well as each machine and war instrument, can be represented and visualized.
> [8] Military architecture, that is, perfect knowledge of the art of fortifying each site and area.
> [9] Instruction about the defense and attacks of fortresses.[10] (*EN*, II:607–608)

Functions of the military art Five of the points suggested by Galileo for the *Accademia Delia* course on mathematics correspond directly to what he was teaching in his lessons about the use of the military compass. The first point about the regulation of armies entails the teaching of those geometrical transformations that enable groups of soldiers to be positioned in specific, defined geometrical shapes on the battlefield, so that the army could change formation while marching, for example, starting from a semicircular formation to assume the shape of an advancing triangle of soldiers.[11] The second point describes the surveying functions an officer had to master. As mentioned in the previous chapter, measuring the dimensions of the main architectural elements of a fortress was extremely important during an attack, to ensure the proper positioning and loading of artillery.[12] On the basis of these measurements, officers were supposed to be able to draw a plan and a profile of the fortress. For all of these purposes, in points five and six Galileo also explicitly suggested teaching the use of mathematical instruments, presumably including his military compass.

Perspective as a drawing technique Drawing and surveying the plan and profile of a fortress required some experience with drawing techniques, ideally including drawing in perspective. This is not only proposed by Galileo in point seven of his notes for a course of mathematics at the *Accademia*

---

[10] Author's enumeration.

[11] The need to change the shapes of groups of soldiers on the field, and how Galileo's compass met this need, are described in the previous chapter on pp. 32ff.

[12] For fortress surveying and the related use of the military compass, see the previous chapter pp. 30ff.

Delia, but actually also was privately taught by Galileo himself as an independent discipline in the framework of the lessons concerning military architecture. Several entries in the *Ricordi autografi* indicate that Galileo gave specific private lessons on this topic. For example, on November 7, 1601 Galileo noted:

> On the seventh of November The Lord Counselor of the German Nation started Perspective (*EN*, XIX:150)[13]

where the Lord Counselor was a Mr. P. Dietrichstein.

Military Architecture

Practical mathematics and geometry found their most relevant use not merely as an intellectual background to the functions accomplished by mathematical instruments like Galileo's military compass. The new art of military architecture, which became established during the sixteenth century, was another discipline which necessarily presupposed such knowledge. Galileo wrote two treatises on military architecture; both approach the same topics, although one of them does so more extensively. They date back to the first years Galileo spent in Padova, as a handwritten copy of one of the treatises bears a date with the year 1593. One of the two treatises, the shorter, is entitled *Breve istruzione all'architettura militare* (*EN*, II:17–75), while the title of the longer is *Trattato di fortificazione*, as already mentioned. Antonio Favaro was probably right when he stated that the longer treatise was intended for private lessons, and the shorter for the public lectures held at the university (*EN*, II:9–14). Mostly, the longer treatise will be considered in this work.

Bonaiuto Lorini (1547–1611) To evaluate these of Galileo's treatises, they are compared with Lorini's treatise. According to Carlo Promis (Promis 1874, 638), Lorini was elected military engineer of the Venetian Republic in 1580. With some interruptions, he worked for the Venetian Senate for the remainder of his life. He was born in Florence, where he trained under the direct patronage of Cosimo I de' Medici, who also placed him under Buontalenti's tutelage. Lorini's treatise, which has been translated into many languages, was well known during his lifetime and for many years after his death.

Drawing techniques The treatises by both Galileo and Lorini begin with some elements of geometry, imparted in the form of definitions and basic drawing techniques, such as how to draw a line parallel to a given one. Whereas Lorini started with many definitions, Galileo opened by describing the actual techniques. In particular, he explained how to draw a line perpendicular to a given one, how to draw parallel lines, how to divide angles into equal parts, and how to draw some simple

[13] In Vincenzo Viviani's *Racconto istorico* the author affirmed that Galileo also wrote a treatise on perspective, as a text for his private lessons (*EN*, XIX:606). Unfortunately no copy of such a treatise has been ever identified.

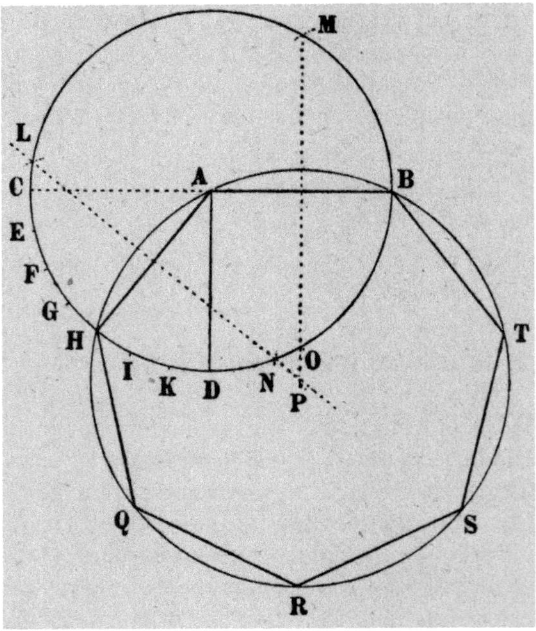

Fig. 3.2 Drawing technique to realize any kind of polygonal figure (*EN*, II:82)

polygonal figures. In fact, the main purpose of both authors was to teach how to draw any kind of polygonal figure whatsoever on the basis of only one rule (Fig. 3.2). Clearly, the objective was to enable students to draw plans of fortresses, which were generally designed in the form of some polygonal figure (*EN*, II:82; Lorini 1609, 16). Together with the propositions upon which the principle of proportion is based,[14] the Euclidean geometry involved in these drawing techniques was probably at least part of the contents of those lessons recorded under the name *Euclide* by Galileo in his *Ricordi autografi* (*EN*, XIX:155).

Attack and defense strategies After this introduction to basic geometry and the basic techniques of drawing polygonal figures, Galileo began the actual course on military architecture by explaining why fortresses are built. A fortress is built in order to allow a small number of soldiers to defend themselves against a greater number of enemies (like a force-saving machine!). Galileo elucidated the different attack and defense strategies usually used against or from a fortress.[15] In particular, he specified the five most common ways to attack a fortress: (1) opening the protective wall by means of artillery and then assault; (2) *zappa*, which means moving in close to the fortress wall and destroying it slowly with tools like picks and iron poles before assaulting; (3) scaling,

[14]The Euclidean principle of proportion is analyzed in the previous chapter on pp. 28ff.

[15]The issue of attack and defense strategies corresponds to point nine of the outline Galileo prepared for the mathematics curriculum at the *Accademia Delia*.

which means first moving in close to the fortress wall, then scaling it using ladders, and, finally, assault; (4) *mina,* which means mining underneath the fortress wall and then exploiting the advantage of surprise to assault; and (5) siege, which means simply waiting until the food and supplies in the fortress are depleted.

New defense and new design The sixteenth century experienced the diffusion of mobile heavy artillery powered by gunpowder, which immediately brought with it a new way of attacking fortresses. This subject was still considered particularly important at the end of the century, as the increasing frequency of wars made it increasingly necessary to fortify territories, villages and cities. In response to this situation a new conception of fortress was developed.[16] Galileo taught his pupils the main characteristics of such a modern fortress.

No point of the protective wall could remain concealed to the view of the soldiers inside the fortress, and therefore as few corners of the wall as possible were to be angled toward the interior. This is the reason why most of the new fortresses built in or after the second half of the sixteenth century have an external perimeter that follows a regular polygonal figure. However, it was not always possible to satisfy all of the complex contextual conditions required for the construction of a new fortress (Fig. 3.3). Thus in most cases the main task of the military architects was to renovate Medieval fortresses in order to adapt them to new military demands.

The bastion The bastion is a particularly relevant architectural element conceived to defend the fortress from cannons. This is a kind of superstructure built over the corners of the fortress and projecting toward the exterior. As this element became an established element in fortress design, a debate erupted among military

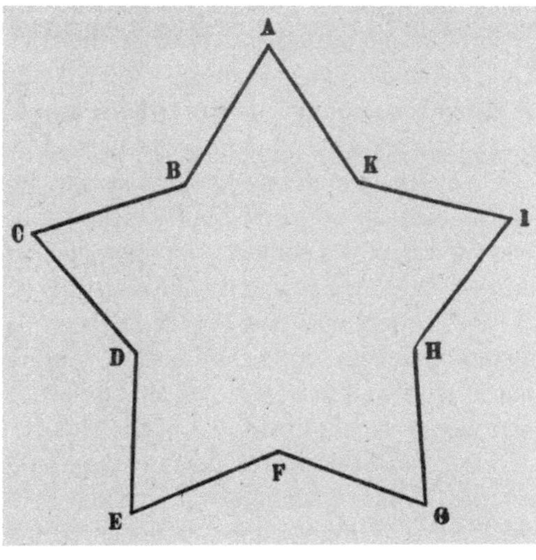

Fig. 3.3 Example of a fortress' plan (*EN*, II:85)

[16]For the development of the new early modern concept of fortress, see also Valleriani (Forthcoming b).

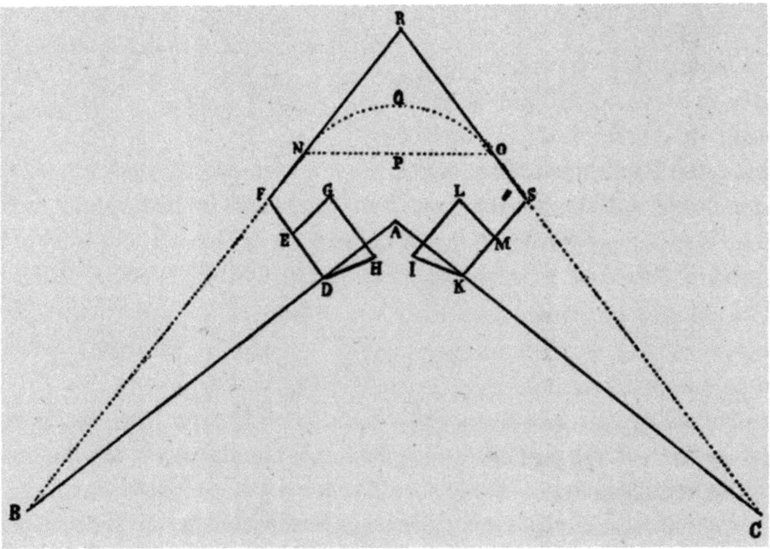

Fig. 3.4 Plant of a fortress' bastion (*EN*, II:86)

engineers about the shape for the stablest and most effective bastions. In truth, a bastion is a compound of many architectural elements (Fig. 3.4): A bastion (FDAKSR) is normally constituted of two strongholds (EDHG and IKML), where the heavy artillery was positioned; two associated shoulders, which are massive, protective walls for the strongholds (FEGN and LMSO); and the main stronghold (NOR). The two strongholds on the sides are located at a lower height than the main stronghold. A low stronghold together with its shoulder constitutes one flank of the bastion. Once the two lower strongholds had been equipped with artillery, their function was to defend the curtain (up to points B and C, for instance) in the eventuality that the enemy approached close to the curtain to scale it or attempt to destroy it manually. The function of the shoulders was to protect the two low strongholds from the enemy artillery positioned outside the fortress. From the main stronghold, which was protected by a parapet or rampart, the defenders of the fortress could use heavy artillery to launch attacks toward the fields across the moat.

The artillery positioned in the lower strongholds was the most effective defense against attackers, as the first goal of every attack was to penetrate some point of the fortress wall in order to be able to invade it, whereby the number of defenders was always smaller than that of the attackers. But in order to make a hole in the defensive wall that was sufficiently large to allow the attackers to enter the fortress, a series of other operations was required, including bridging the moat and preparing the ladders. Moreover, these time-consuming operations had to take place close to the wall. Therefore, while the attackers prepared an assault close to the wall, the artillery located in the lower strongholds executed what was called *passate*, very destructive cannonades along a line skirting and almost parallel to the wall of the fortress. This meant that an attack could not be launched until the artillery in the

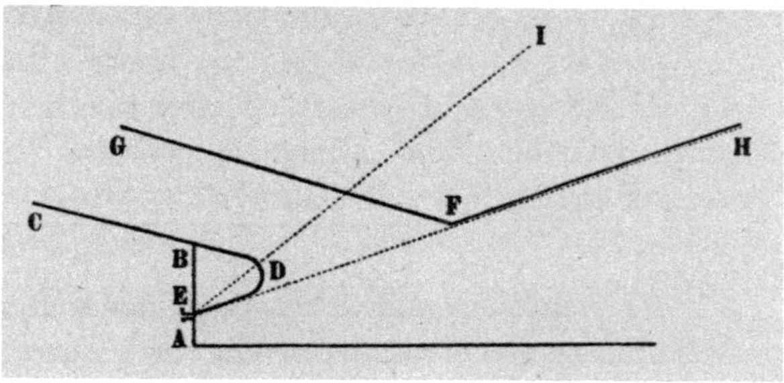

Fig. 3.5 Description of the function of the strongholds (*EN*, II:97)

strongholds had been put out of order. This is the reason why the primary target of any attacking army was always the artillery positioned in the lower strongholds.

Once the importance of the lower strongholds had been recognized, it became standard practice, as Galileo himself admitted and approved, to build additional bodies on the flanks of the bastion alongside the openings of the strongholds (Fig. 3.5). These were semicircular in shape and projected toward the exterior (BDE). They were called *orecchioni*—big ears—and their express function was either to preclude any practical possibility of shooting directly into the strongholds or, if this was impossible for architectural reasons, to force attacking artillery to shift its position further away from the strongholds but close to another point of the curtain, necessitating a dangerous maneuver like one from point I to point H.

The bastion's vulnerability Given the bastion's importance, or more precisely, its robustness, upon which the success of an attack or of a defence was directly dependent, military engineers addressed long and detailed speculations to their potential vulnerabilities. Galileo confronted this issue as well, detecting one main vulnerability that was "considered to be the main [problem] by most of the architects" (*EN*, II:94). This imperfection was related to the angle of the bastion, and therefore to the internal angle of the corner projected toward outside. According to Galileo, it was imperative that this corner not be too acute (Fig. 3.6) and this

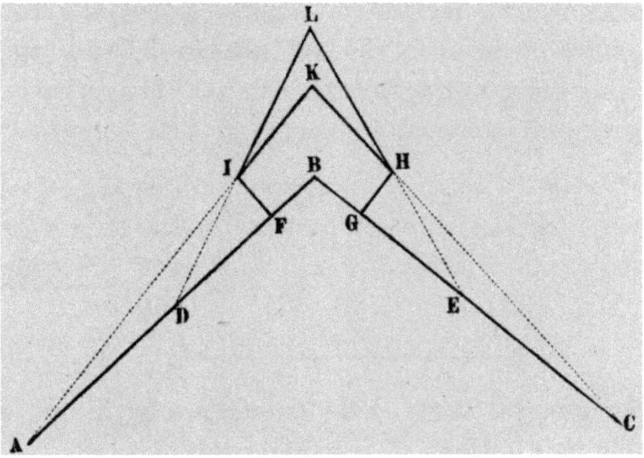

Fig. 3.6 Plan of a fortress' bastion with a corner built on an angle that is too acute (*EN*, II:95)

for two main reasons: first, because the corner would have been weak and easy for enemy artillery to cut off, and second, because the bastion would not have been wide enough for strongholds able to accommodate heavy artillery. Galileo identified three reasons for this imperfection. First, the corner is too acute when its angle inside the fortress is not obtuse; second, when the bastions are designed such that they could defend the curtain only as far as points very close to the same bastions like D and E (Fig. 3.5); third, when the flanks of the bastions were too long. As Galileo himself stated, these conclusions, especially the first and the third, were simple consequences of applying proposition 21 of the first book of Euclid's *Elements* (*EN*, II:95).

Galileo was only one of the many military architects who approached this important topic. The theme of bastion construction is discussed in almost any of this period's manuscripts or treatises on military architecture. Carlo Theti, for instance, addressed the same issue, but offered very different solutions. Since his focus was on how to change and improve fortresses built back in the Middle Ages, and therefore equipped with bastions that were either too small or had exceedingly acute angles, he did not propose any changes in the method of bastion construction, but suggested cutting them in such a way that their front would become more resistant (Fig. 3.7) (Theti 1588).

Lorini, too, devoted an entire chapter to the issue, but discussed it from the perspective of the sizes of the bastion's flanks and with reference to the kind of artillery to be deployed (Lorini 1609, 44–49). Gabriello Busca (ca. 1540–ca. 1605), finally, in his treatise of 1601, went back to Albrecht Dürer's studies on architecture and came up with the radical proposal of avoiding angular bastions altogether by making them round (Fig. 3.8) (Busca 1601, 126).

Elements of the fortress and cavaliers Galileo demonstrated his familiarity not only with the bastion, but with all of the other common architectural elements of a

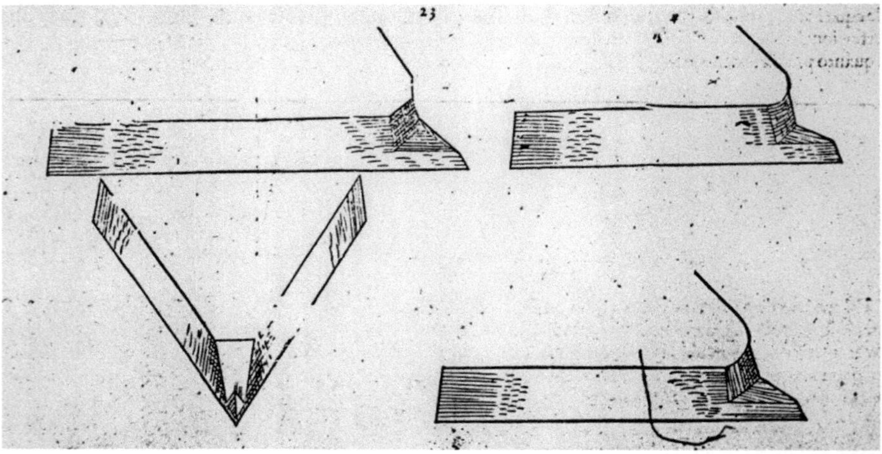

Fig. 3.7 Fortress bastions with cutted front (Theti 1588, 54)

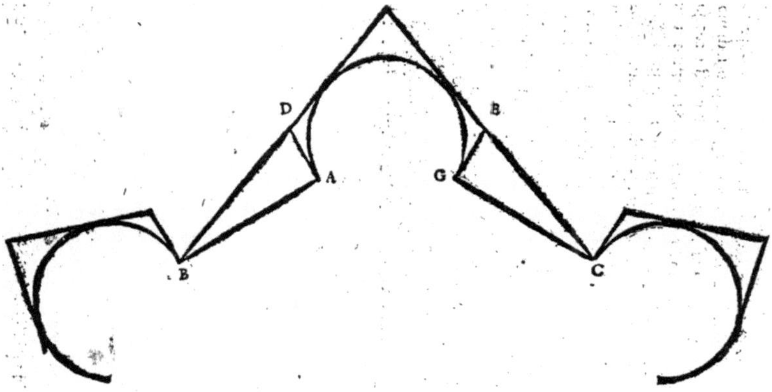

Fig. 3.8 Rounded fortress' bastions (Busca 1601, 126)

fortress as well. He diligently described the various kinds of cavaliers, strongholds, and platforms within the fortress as well as scarps, counterscarps, and covered walkways on its exterior. Moreover, he described how to build those structures outside the fortress which were needed in an attack. As mentioned in the previous chapter, the most important element that had to be built on the field to attack the fortress was the cavalier, a sort of tower as high as the lower strongholds, on which artillery was placed to shoot the cannons in the strongholds. Galileo described a method of building them using earth, reeds, and wood. In fact, this procedure, *del fortificar di terra*, was not only required by attackers launching a siege, but also turned out to be the fastest and most economical method of reinforcing old Medieval fortresses so that they could withstand the fire of the powerful new artillery developed during the sixteenth century.[17]

Quarto buono Officers had to know how to measure the inclination of the scarp (FH) and counterscarp (DEKFH) of the fortress (Fig. 3.9). This information was useful for several reasons. The inclination of the scarp was relevant primarily

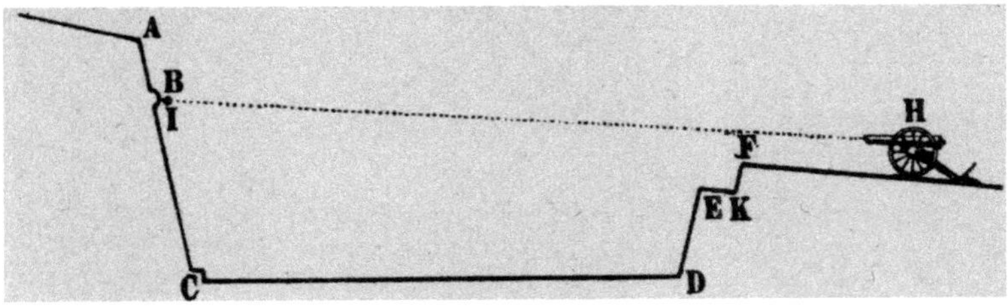

Fig. 3.9 Positioning artillery close to the scarp and the counterscarp (*EN*, II:96)

[17]The *Del fortificar di terra* procedure was used primarily during the sixteenth century in the Tuscan duchy (Lamberini 1990; 2007).

because this value could be used to determine the best intervals for positioning cannons, so that they could shoot perpendicular to the wall of the fortress at targets located as low as possible, preferably under the cordon. A second reason why it was important to know the inclination of the scarp was the practice of firing at the scarp itself to create small openings, into which the attackers closing in on the wall could lay harquebusiers to fire at the artillerists in the strongholds. This was possible only when the scarp was inclined at a relatively steep angle. Knowing the value of the inclination of the counterscarp was also useful in planning an assault at the wall, assisting the officers in determing the best method the soldiers could use to cross the moat.

For the purpose of measuring the inclination of the scarp and of the counterscarp, in the contexts of attacking a fortress, or to build them using earth, reeds and wood, Galileo presented an instrument conceived with the sole purpose of accomplishing this task. It was called *quarto buono*: the good fourth (Fig. 3.10). This instrument was not Galileo's invention.[18] Tracing the history of this instrument reveals a direct relationship between Galileo's treatises on military architecture and a particular circle of engineers and military architects. As Daniela Lamberini first pointed out (Lamberini 1990, 136–138; 2007), a tradition of treatises on fortifications became well established in the sixteenth century, starting with the manuscript

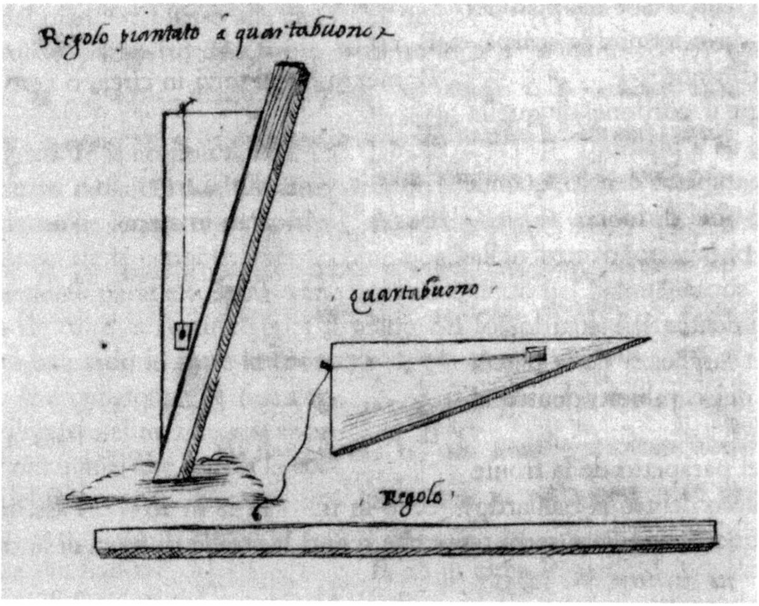

Fig. 3.10 Giovan Battista Belluzzi's illustration of the instrument called *quarto buono* (Belluzzi ca 1545, c. 27v)

[18]Galileo's compass can be configured to work as a *quarto buono* as well.

by Giovan Battista Belluzzi (1506–1554), a famous military engineer of the Grand Duke of Tuscany Cosimo I, entitled *Il Trattato delle fortificationi di terra* (Belluzzi ca. 1545).[19] Many copies of this manuscript circulated among the experts in military architecture during the entire century. In this situation the military engineers appropriated some of the contents of Belluzzi's manuscript and presented them as their own, in the form of either new manuscripts or of published treatises, according to the practice of the sixteenth century.[20]

In this context of exchange, entire paragraphs, sentences, and drawings originally by Belluzzi are found reproduced unchanged in many other sixteenth-century manuscripts and treatises on fortifications. Not only Bernardo Puccini (1521–1575), who was Belluzzi's pupil and thus officially in charge of expanding and editing Belluzzi's work, but also famous military architects like Bartolomeo Ammannati (1511–1592) and Bonaiuto Lorini were indebted to Belluzzi and Puccini. Many anonymous manuscripts still preserved at the *Biblioteca Nazionale Centrale* of Florence turn out to be more or less accurate copies of Belluzzi's and Puccini's treatises.

Galileo and Florentine military architecture There are three marks which distinguish those manuscripts on military architecture that emerged from the Florentine circle of military engineers: First, the emphasis on the fortifications made of earth, a method that achieved its widest diffusion upon the orders of Cosimo I; second, details like the description and drawing of the tools used by the sappers; third, the instructions and drawings for building the *quarto buono*.[21] Galileo's treatises on fortifications, which date back to the first years of his stay in Padova, include not only general statements similar to those in Belluzzi's manuscript, but also some of the same details, like the instructions for building the instrument *quarto buono*, clear evidence that his manuscript belongs to the Florentine tradition. In keeping with this theme, Horst Bredekamp and Daniela Lamberini accomplished a detailed comparison between Galileo's text on military architecture and Puccini's *Libro primo di fortificatione* of 1564 (Puccini 1990), reaching the convincing conclusion that Galileo's affiliation with the Florentine tradition is based mostly on what he learned from this text by Puccini (Bredekamp 2007b, 70–82; Lamberini 1990, 136–138). As a military architect, therefore, Galileo was a member of the circle of military engineers in Florence.[22]

---

[19]Reprinted in a commented edition in Lamberini (2007).

[20]For the practice among military engineers of collecting information from other engineers' manuscripts and presenting them as original proposals, see Lamberini (1987). Given the great exchange of manuscripts among military engineers during most of the sixteenth century and the practice of appropriating their contents, it becomes evident that Galileo's public charge of plagiarism against Baldassare Capra in 1607 represented quite an unusual event for their contemporaries. Indeed, this case testifies to the change in the social status of artist-engineers like Galileo, especially from the late sixteenth century on.

[21]Another mark of distinction is the way weapons are listed divided into categories defined by the weight of the balls the weapons shot. This point is considered in the next paragraph.

[22]One chapter of Galileo's treatise on fortifications is entitled: "On the *Quarto buono*. Instrument for the scarp" (*EN*, II:142). Special thanks to Thomas Settle, who, in a private discussion held in

The site of the fortress Another specific topic of sixteenth-century military architecture concerned the characteristics of the site where a fortress had to be built and the ways these characteristics could influence the design of the building. For example, a fortress on a rock was built according to principles and requirements which corresponded only in part to those valid for a fortress on flat terrain or on an island in a lake. The relevance of the setting in which the fortress had to be built became a canonical topic of treatises on military architecture thanks to the work of Florentine engineers around the mid-sixteenth century.

The status of Galileo's treatises The topics Galileo addressed in his two treatises on military architecture and fortifications were not new in Galileo's day, and he never presented them as if they were the result of new reflections. As far as their content was concerned, his lessons on military architecture must have been integrated closely with those on the use of the military compass,[23] as indicated by the high frequency of students who took classes on both of these subjects. His teachings on military architecture, along with, first, explanations of the functions of the military compass and, second, the exercises performed using the real instrument, constituted the nucleus of Galileo's activity as a private teacher.

Artillery Powered by Gunpowder

An officer at the end of the sixteenth century had to know how various pieces of artillery worked. Although Galileo never gave specific lessons devoted to this topic alone, such knowledge was transmitted in the context of the subjects of his other lessons. The analysis of the functions of Galileo's military compass back in the previous chapter showed that Galileo taught the use of the compass as a quadrant for bombardiers, the most relevant tool for artillerists. Yet the same compass could also be used as a caliber, an instrument that allowed the artillerist to calculate the quantity of gunpowder to be used, depending on the material of which the projectiles consisted.[24] During the lessons on military architecture, moreover, in the context of the question about how to position heavy artillery along the curtain of a fortress and on the bastions, Galileo furnished further relevant information about how to distinguish among different pieces of artillery.

Distinctions among sorts of artillery Galileo distinguished between artillery *reali* and *non reali*, respectively, those which shot balls heavier than eight libra, and those that shot lighter projectiles (*EN*, II:107; Lamberini 1990, 136). In fact, in his treatises on military architecture, Galileo affirmed a distinction among the various sorts of artillery that was traditional at the time and originated in the treatises of

Tenerife during the European Symposium on Galileo, first brought my attention to the relationship between Galileo's manuscripts on fortifications and the Florentine circle of engineers and military architects of the sixteenth century. See also Fara (1988, 235–236).

[23]The functions and the use of Galileo's compass were described in the previous chapter in the context of analyzing the activity of Galileo's workshop. See on pp. 29ff.

[24]The use of the compass as a caliber is explained in the previous chapter pp. 35ff.

Belluzzi and Puccini. Thus in this case, too, the close relation of Galileo's treatises with the Florentine circle of military engineers is confirmed. Many other treatises show the same distinction pointed out by Galileo, although most of them also list many sorts of weapons presented along with drawings and explanations concerning their use, whereas it seems that Galileo never composed such detailed descriptions.

On the use of artillery Although there is no further evidence directly demonstrating that Galileo occupied himself with the use and functioning of artillery powered by gunpowder in the context of his activity as a private teacher, an outline for a possible treatise (possibly another "unpublished treatise") has survived, which was completely dedicated to this topic.[25] Galileo's outline consists of 14 points (Fig. 3.11):[26]

[1] Particular advantages of artillery over other mechanical instruments.
[2] On its [the artillery's] force, and where it comes from.
[3] Whether it [the artillery] works more powerfully from a certain distance or when it is close [to the target].
[4] Whether the ball moves along a straight line, when it is not shot perpendicularly.
[5] Which line the ball describes during its motion.
[6] The cause and the time of the cannon's recoil.
[7] Hindrances that make the cannon faulty and the shot insecure.
[8] About loading them on the wagon and about taking them down.
[9] On the manufacture of the caliber.
[10] On the examination about the goodness and rightness of the cannon.
[11] Whether the longer the cannon, the further it can shoot, and why.
[12] At which elevation it shoots the furthest, and why.
[13] That, when the ball comes back down perpendicularly, it does so with the same force and velocity with which it went up.
[14] Several firing projectiles balls and lanterns,[27] and their use.

Artillery and Galileo's mechanics A closer analysis of Galileo's outline for a treatise about artillery shows that his intention was to write a text in which practical knowledge relevant for the bombardier would have been integrated with related

---

[25]The outline of this planned treatise is documented in Galilei (ca. 1602–ca. 1637, 193r). For an interpretation of this sketchbook in the context of the emergence of Galileo's mechanics, see Büttner et al. (2001).

[26]"[1] Particolari privilegii dell'artiglieria sopra gl'altri strumenti mecanici; [2] Della sua forza, et onde proceda; [3] Se operi con magior forza in una certa dist[anza] o da vicino; [4] Se la palla vadia per linea retta, non sen[do tirata] a perpendicolo; [5] Che linea descriva la palla nel suo [moto]; [6] La causa et il tempo dello stornare il pezzo; [7] Impedimenti che rendono il pezo difettoso et il tiro incerto; [8] Del metterle a cavallo e scavalcarle; [9] Della fabrica del colibro; [10] Dell'esamine circa la bontà et giustezza del pezzo; [11] Se quanto più e [è] lungo il pezzo più tira lontano, e perchè; [12] A quale elevatione tiri più da lontano, et perchè; [13] Che nel tornare la palla ingiù nel perpendicolo, torna con le medesime forze et velocità con che andò in su; [14] Diverse palle artifitiate et lanterne, et lor uso." Author's enumeration. In the manuscript the points 1, 2, 3, and 6 have been erased. For a comparison between Galileo's outline for a treatise for the artillerist and one by a military engineer, see for example Lorini (1609, 279ff), Tartaglia (1554, 37v) and Cataneo (1582), the latter being a complete treatise devoted only to this topic.

[27]Lanterns were also cylindrical gear wheels often used as components of machines. In this case, however, the term "lantern" means special projectiles able to set fire to the target, which were fired at wooden machines like the *tormenta*.

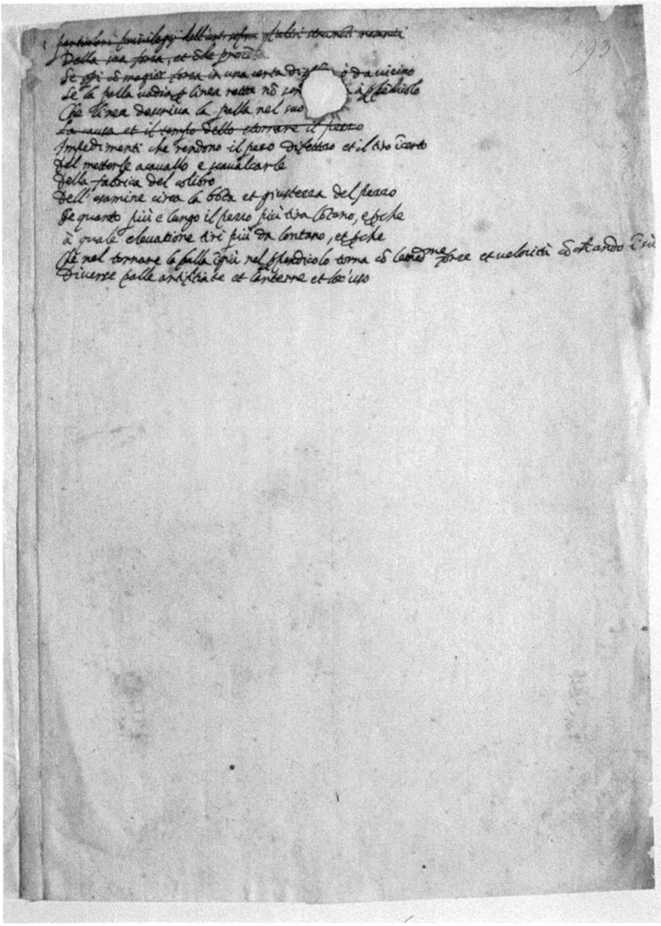

Fig. 3.11 Galileo's outline for a treatise on the use of the artillery (Galilei ca. 1602-ca. 1637, 193r)

theoretical speculations, which were particularly relevant in the structure of Galileo's theoretical mechanics. In particular, points 4, 5 and 13 are related to the development of his law of fall,[28] point 6 relates to Galileo's speculation about the force of the impact, as discussed in the Added Day of the *Discorsi* (*EN*, VIII:319–346). Points 9 and 12, finally, refer to additional operations which could be accomplished by means of Galileo's compass: the use of the caliber and the use of the quadrant for bombardiers, respectively, as described in the previous chapter.

The force of the shot The second point about the force of the artillery would probably have been a chapter dedicated to the composition of gunpowder, obviously

[28] On the relation between the artillerists' knowledge and the emergence of Galileo's law of fall, see Renn et al. (2001) and, more extensively, Damerow et al. (2004).

a relevant issue for those who had to assess its quality in order to load the cannon accordingly. The third point refers to a dispute *en vogue* in Galileo's day, as to whether cannons cause more damage when they are placed further away from the target or close to it. Galileo addressed this subject as well, in the context of his private and public lessons on military architecture.[29] In the tract used as a basis for his public lessons Galileo wrote:

> But first, before we go ahead, it seems a proper thing to me to speak a bit about a curious debate alive among some persons, who are expert bombardiers: which is, if it is better to place the artillery as close as possible to the wall that one wants to demolish; if, positioned at a certain distance away, it will have more effect than when positioned very close. And the reason for this discussion is that some believe and are completely convinced, that the artilleries will go through a thicker wall if they are positioned a certain determined distance away than if they were closer: This opinion, although it has endless proponents, can be nothing but false, and in a certain way ridiculous; as experience can make evident to anyone paying attention, and as reason can convince those who will think with right wisdom. In fact, since the motion of a ball is a violent motion, who will doubt that, once the ball is separated from its moving [force], it does not continuously lose its force? (*EN*, II:49)

Safety and cannon preparation Points 7 and 10 are about the quality of the piece of artillery and its possible imperfections. This kind of knowledge necessarily also included notions of melting and lathing procedures, and was obviously relevant for the safety of the bombardier and his assistants. Point 8 about the preparation of the cannons on the wagons concerns those machines which were required in every fortress, and which allowed the soldiers to prepare the artillery in case of attack, for they were not kept in defensive positions during peacetime. This issue will be discussed again later in this chapter.

La sfera

One piece of evidence that testifies to Galileo's lessons on *La sfera* are the entries in Manuscript 26 where he annotated the income he earned from his lessons. All of the entries are dated between 1602 and 1603, and all of the private lessons were given to Polish students.[30] The other and more important proof is the many copies still

---

[29]Galileo addressed the question of the distance between the artillery and the target in both of his treatises on fortifications. However, in the treatise identified as the one for the private lessons, he did so less extensively than in the one for lectures at the University of Padova. For the treatise for private lessons, see *EN*, II:118.

[30]Manuscript 26 documents private lessons on *La sfera* to twelve students, eleven of them between June 1602 and March 1603 and one in 1607 (Galilei 1598–1634, 149–158). University students normally attended courses on Euclid's *Elements* before starting lessons on *La sfera*. For a reconstruction of the curriculum of a university student at the end of the sixteenth century, see Favaro (1966, I:113).

preserved of a handwritten text attributed to Galileo and entitled *La sfera ovvero Cosmografia* (*EN*, II:205–209).[31]

The general topic of these lessons is specified at the beginning of this tract:

> We say thus that the subject of the cosmography is the world [...], which means nothing but *description of the world*. But we want also to point out that, of the things that could be taken into account in reference to this world, one part is the work of the cosmographer; and this is the speculation about the number and the distribution of the parts of this world, their figure, their size and their distance, and, above all, their motions; leaving the considerations about the substance and the qualities of the same parts to the natural philosopher. (*EN*, II:211)

The basic Aristotelian–Ptolemaic conception Galileo's text offers an overview of the conception of the universe, and accomplishes this task in accordance with the traditional views of his day. The universe explained in Galileo's cosmography corresponds to the Aristotelian–Ptolemaic system. The division of the universe into a sublunar and a superlunar world, the perfect circular motion of the firmament, and the central and immovable position of the Earth are the core concepts upon which this writing is founded. With this text Galileo followed the normal practice of lecturers on mathematics in his day. In fact, ever since the thirteenth century, books that exposed the Ptolemaic conception of the universe had constituted a very vital tradition that began with the *Tractatus de sphaera mundi* by Sacrobosco. Indeed, in this book written around 1240 in Paris, Sacrobosco described the motions of the celestial bodies according to the systems of Ptolemy and al-Farghânî (d. after 861 AD). The fame of Sacrobosco's book increased constantly until the end of the sixteenth century, when plenty of translated, annotated, and illustrated editions of it were in general circulation.[32] Typically, lessons on *La sfera* constituted the course of mathematics at the university during the sixteenth century.

The subjects of the *La sfera* treatises Besides the fundamental principles of the Ptolemaic conception of the universe, all of the commentaries to Sacrobosco's work address other topics as well. A later edition by Francesco Pifferi (1548–1612), a lecturer of mathematics at the University of Siena in 1604, describes how to measure the sizes of any body, how to measure distances between different places, how to recognize the winds and their relationship with the art of navigation on the sea, how to construct a solar clock, and how to find the center of gravity of a solid body. This knowledge was fundamental not only for the military arts and commerce, but

[31] William A. Wallace (1984b, 255–261) convincigly showed that Galileo's text *La sfera* was probably compiled using Clavius' homonym text (Clavius 1582) as fundamental work. For this reason, moreover, Wallace suggested that Galileo started working on his own *La sfera* during the years when he was working in Pisa as a lecturer of mathematics. This supposition, however, is not supported by any evidence.

[32] In an edition of the *Tractatum* of Sacrobosco published by Francesco Pifferi, a lecturer of mathematics in Siena in 1604, the editor had a difficult task in explaining why he published this book despite the fact that editions by famous scholars like Clavius, Maurolico, Alessandro Piccolomini (1508–1578), Magini, and Baroccio were in circulation at the same time. For more details, see his dedication to the readers in Pifferi (1604).

in general for anybody undertaking a long journey, like merchants, for example. Moreover, all of the treatises on *La sfera* contained a more or less complete outline of astrology, required by the university curriculum on medicine.

In his treatise, Galileo provided a table of the characteristics of each geographic area delimited by two parallels, though it did not consider only the seven Ptolemaic areas, but all twenty-two of the areas in early modern commentaries. The table published by Favaro, and attached to a copy of Galileo's *Trattato della sfera* preserved in Poland,[33] reports the following characteristics for each area: name, maximal duration of the solar day and a description of the changes in this duration, height of the pole of the celestial sphere, and length of the area. Finally, the data are given for each area in reference to both the northernmost parallel and the southernmost one, and with reference to the middle line between them.

Practical relevance of the lessons on Cosmography Galileo's private students copied this treatise and took lessons on it in order to be able to distinguish the celestial bodies. This then meant the acquisition of knowledge without which the use of instruments like the astrolabe, solar and lunar clocks, the armillary sphere, the astronomic quadrant and even Galileo's military compass would not make much sense. Indeed, only if the person operating such an instrument was able to recognize a celestial body was it then possible for him to determine, for example, the distance between the locations of the body observed at two different points in time. This further presupposes that the operator knows the motion of that body and, therefore, that he has a schema or conception which allows him to know the motions of all the celestial bodies. The same argument, then, is also valid for astrology and agriculture, for which Galileo taught the lunar phases.

The knowledge that Galileo taught in his private lessons on *La sfera* was generally useful to every military officer who needed to know such things as when night would fall, and also practical for merchants and even to diplomats. This was not the knowledge of philosophers, but rather that of officers and engineers like Boniauto Lorini.[34]

The Science of Machines

The last subject typical for a course on fortifications was the science of machines. During this course, students were supposed to learn enough to understand the functioning of machines relevant for the building and maintenance of a fortress, and

---

[33]Cod. 571, folios 2–45, Library of the University of Cracow.

[34]Given the aim of the lessons on *La sfera*, Favaro's hypothesis, that Galileo taught *La sfera* based on the conception of the Ptolemaic system only because he was afraid of possible repercussions from the Catholic Church, makes no sense at all (Favaro 1966, I:119–120). The only criticism that could be made against Galileo's text on cosmography is that it does not take into account all topics typical of treatises on *La sfera* at the time; for example, Galileo's treatise lacks any description of the winds, useful for navigation.

how to use them. When engineers and mathematicians approached this noble science, they rarely neglected to quote those ancient scientists upon whose reflections it was founded. Generally, Aristotle's *Mechanical Questions* (Aristotle and Hett 1980),[35] Archimedes' *Aequiponderantibus* (Archimedes 2002a) and the works of Hero of Alexandria, whose *Mechanics* was known via the homonym work of Pappus, were considered to be the works from which everything else originated. These works round out the framework of the ancient science of machines, upon which Renaissance mathematicians, engineers and machine makers founded their speculations and their imagination for new designs.[36]

The structure of the science of machines The science of machines of the Renaissance was structurally based on Hero's distinction among the five simple machines as he described them in the second book of his *Mechanica*, and as this structure was transmitted thanks to the *Mathematical Collections* of Pappus.[37] The simple machines were the lever, the pulley, the wheel, the wedge and the screw. According to this conception, the functioning of the five simple machines can be explained on the basis of the principle of the lever. All of the machines, finally, are either simple machines or compounds of two or more simple machines. The principle of the lever had been formulated first by Aristotle. This position was normally honored by Renaissance mathematicians, engineers and machine makers in the introductions to their treatises. In particular, Aristotle's introductory distinction between *Ars et Natura* (Aristotle and Hett 1980, 331), the former interpreted as products of the mechanical arts and the second as the products of nature, became a sort of *leitmotif* in publications by Renaissance artist-engineers.[38]

Early modern treatises on machines[39] Proceeding from the context of his activity as a private teacher, Galileo wrote not one, but actually two treatises about

---

[35]Up until the nineteenth century, the *Mechanical Questions* were ascribed beyond any doubt and with only one exception (Girolamo Cardano) to Aristotle. Nowadays the opinion prevails that this text was written by Aristotle's pupils. For more details see, Aristotle and Bottecchia Dehò (2000, 27–51). For a general introduction to Aristotle's *Mechanical Questions*, see Rose and Drake (1971), Gandt (1986), Laird (1986), Moletti and Laird (2000), Damerow et al. (2002), Büttner et al. (2003), and Valleriani (2009a).

[36]For an introduction to Pappus from the perspective of the history of mathematics, see Cuomo (2000). For a comprehensive work on ancient mathematics, see Cuomo (2001).

[37]The five simple machines are described in the eighth book. Pappus' work was translated and republished by Federico Commandino in Pappus Alexandrinus and Commandino (1588). In 1424 a manuscript of Pappus' *Mathematical Collections* was brought to Italy, where it is preserved at the Vatican Library today, MS. VAT. GR 218. Guidobaldo del Monte first brought Pappus to general attention during the Renaissance in Monte (1577). See in particular his dedication to Duke Francesco Maria II of Urbino. For an extensive discussion of Guidobaldo's revisitation of the works of Archimedes, Hero, and Pappus works, see Rose (1975, 230ff).

[38]During the early modern period, the Aristotelian distinction between *Ars* and *Natura* was interpreted differently by natural philosophers than by artist-engineers. This issue is discussed in Chapter 6, pp. 199ff.

[39]In the following a categorization of the early modern sorts of treatises dealing with the science of machines is proposed. This relevant topic is analyzed in this work only to the extent that is instrumentally required by the argument. For an extensive analysis of this topic, see Popplow (2004).

the science of machines, *Delle macchine* (Galilei 1592–1593a) and *Le mechaniche* (*EN*, II:145–191). In Galileo's day, moreover, the most important work on the science of machines was Guidobaldo del Monte's *Mechanicorum liber*, published first in Latin in 1577 (Monte 1577) and then in Italian in 1581 (Monte and Pigafetta 1581). Many other treatises were published as well. The most relevant were those that collected the descriptions and explanations of the relevant machines for living in a fortress, like Book 5 of Bonaiuto Lorini's *Le fortificationi*. Finally, other kinds of treatises also became very common and successful, including the "theaters of machines" and those treatises completely devoted to the analysis of one single machine. Examples of the last two sorts of treatises are *Novo teatro di machine et edificii* (Zonca 1607) of Vittorio Zonca (ca. 1568–1629) and *Tre discorsi sopra il modo d'alzar acque da' luoghi bassi* of Giuseppe Ceredi (Ceredi 1567).

Archimedes vs. Aristotle In his *Mechanicorum liber* Guidobaldo del Monte proposed a deductive system for the five simple machines such that, given some definitions and the principle of the lever, all of them could be rigorously deduced. As del Monte's aim was to furnish new theoretical foundations for the mechanical science, his work concerns only the basic definitions of the science of mechanics, the principle of the lever and the simple machines. In other words, not a single compound machine can be found in this treatise. Guidobaldo del Monte's work fully abandons Aristotle's *Mechanical Questions*, for it is based completely on Archimedes' definitions. In general, his *Mechanicorum liber* represented a rupture in the mechanical sciences because it was a sort of Archimedean attack against the Aristotelian approach (Bertoloni Meli 1992). As a consequence of Guidobaldo del Monte's publication, work on the Aristotelian *Mechanical Questions* entered a period of decline, until its role in the field of the science of machines finally became practically irrelevant during the second half of the seventeenth century.[40]

Engineers and Aristotle's *Mechanical Questions* Before and during the period when Guidobaldo del Monte published his *Mechanicorum liber*, engineers probably paid the greatest attention to the Aristotelian text.[41] Well-versed engineers like Ceredi, Zonca, Lorini, Giovanni Battista Aleotti (1546–1636), and Agostino Ramelli (ca. 1531–ca. 1600) quoted the Aristotelian text to explain their devices.[42]

[40]Guidobaldo, joining the old tradition promoted by Proclus and Plutarch, tried to defend Aristotle in his later publication, suggesting that Aristotle had offered Archimedes the idea on which he could work (Archimedes and Monte 1588). For Proclus' argument for Archimedes' mechanics compared to that of Aristotle, see Proclus and Friedlein (1873, 63). For the same topic in Plutarch, see *Vita Marcelli*, XIV:8. Del Monte displayed his familiarity with Proclus and Plutarch in his dedication to the readers in Monte and Pigafetta (1581). On the diffusion of the idea that Archimedes would be a "better" successor of Aristotle within the framework of the mechanical sciences, see Russo (2001). On the diffusion of the same idea during the early modern period, see Rose and Drake (1971).

[41]Aristotelian mechanics entered the university curriculum in Padova back in the sixteenth century. For a survey, see Bertoloni Meli (2006, 14). For an introduction to Guidobaldo's book and Aristotle's work, see also Bertoloni Meli (2006, 18–39).

[42]For a complete list of the translations of and the commentaries on the *Mechanical Questions* written by engineers during the early modern period, see Rose and Drake (1971, 97).

For example, it was an engineer who translated the *Mechanical Questions* into Italian for the first time in 1573 (Aristotle and Guarino 1573).[43] From the end of the fifteenth century on, Aristotle's work was published several times, first in Greek, then in Latin, then in Italian.[44] With each publication the text was adorned and expanded with more and more illustrations and commentaries. This process continued until the first half of the seventeenth century and more or less concluded with the commentary by Giovanni di Guevara (Aristotle and Guevara 1627).

Theaters of machines The aim of those engineers who were chiefly interested in compiling theaters of machines was in a sense opposite to that of mathematicians like Guidobaldo del Monte. In his *Novo teatro di macchine et edificii* (Zonca 1607), Zonca wrote not a single line of explanation on how simple machines work, but instead explained forty compound machines, complete with drawings and texts. As Zonca's text was for engineers, there was no need to explain again how the five simple machines work.

Didactic treatises Among these two opposite sorts of treatises on the science of machines were those compiled for a didactic purpose addressed to military officers. This is the case for Galileo's and Lorini's works, for example. This kind of treatise has a number of peculiar characteristics. First, they usually quote Aristotle's distinction between *Ars* and *Natura* as evidence of how mechanics is a science, which thus assigns men the role of dominating nature.[45] Second, they often refer to the principle of the lever as it was formulated by Archimedes. Third, they accept the theoretical structure represented by the definition of the simple machines, but then integrate this structure with issues relevant for the construction and actual uses of the machines themselves.

In fact, any of the simple machines was applied in many different ways, according to the purpose at hand and to the environment where the machine was supposed to work. As these applications were and are in principle not enumerable, most of the treatises written in the framework of officers' training open with a brief explanation of the principles of the five simple machines, and then continue by exhibiting a great number of illustrations of "real" applications of the machines.

Lorini's science of machines Bonaiuto Lorini dedicated one chapter of his treatise to the very specific topic of the machines of fortresses (Lorini 1609, 195–248). It starts with a historical introduction to the mechanical sciences. After mentioning Aristotle's distinction between *Ars* and *Natura*, the advantages that the mechanical

[43]Antonio Guarino was engineer for Duke Alfonso II d'Este of Modena. It was a tradition of the Este family to recruit engineers who also possessed a humanistic education. Another engineer of the Este family was Giovan Battista Aleotti, who translated Hero's *Pneumatics* into Italian in 1589. See also Palmieri (2003, 250).

[44]For a list of the Latin and Italian translations and commentaries of Aristotle's *Mechanical Questions* in the early modern period, excluding the late Renaissance, see the introduction to Aristotle and Bottecchia Dehò (2000, especially paragraphs two and three).

[45]The particualar view of the art-nature relationship according to which art dominates nature is normally ascribed to Francis Bacon (1562–1626). However, it has been shown that such an interpretation has its roots in the works of the artist-engineers of the end of the sixteenth century (Valleriani Forthcoming a).

sciences bring in wartime are shown first. Basing his argument on Archimedes' and Guidobaldo del Monte's works, Lorini then stated that all of the machines can be reduced to the balance and to the steelyard, and thus to the lever. Lorini then immediately provided the reader with five qualitative definitions useful for understanding how a lever works in practice—lever, force, horizon, axis, and radius—and follows them up with four propositions about the lever derived from Archimedes' *Aequiponderantibus* and Aristotle's *Mechanical Questions* (Lorini 1609, 195–201; Archimedes and Monte 1588; Aristotle and Hett 1980, 330–411). Lorini proceeded by explaining the main practical applications of the lever, that is, the simple machines, on the basis of which compound machines can be analyzed. The simple machines are the pulley, the wheel, the winch, and the screw.[46] The rest of the chapter closely resembles the kind of treatise known as a theater of machines. In this section Lorini describes a great number of compound machines and architectural elements that work on the basis of machines, including drawbridges and automatic doors.

Galileo's science of machines As mentioned, Galileo wrote two treatises on the science of machines: *Delle macchine* and *Le mechaniche*. These are two very different texts, though the second is certainly based on the first. Most of the subjects contained in Galileo's treatise *Delle macchine*, apparently an earlier version of *Le mechaniche*, are also presented in the later version. However, Galileo's method of argumentation changed significantly from one version to the next, revealing changes in Galileo's scientific interests. What is more, the first version also contains subjects that are not dealt with at all in the later one. According to Romano Gatto, the earlier treatise, *Delle macchine*, dates back to the years 1592–1593 and the later one, *Le mechaniche*, to the years 1598–1599.[47] A copy of the later version was first published by Marin Mersenne (1588–1648) in French in 1634, when Galileo was still alive, in the title of which Mersenne called Galileo *mathematicien et ingenieur* (Mersenne 1639).[48]

Galileo's *Le mechaniche* Galileo's later text on the science of machines, *Le mechaniche*, is one of the treatises written with the aim of supplying foundations for the science of mechanics, as was Guidobaldo del Monte's. This makes clear that the treatise *Le mechaniche*, usually considered only in the context of Galileo's

---

[46] Lorini did not consider the lever to be a simple machine, but a kind of fundamental tool to explain the machines.

[47] The treatise *Delle macchine* either has not been taken into account by historians, or has been regarded as a kind of poor summary of the later version *Le mechaniche* that does not show anything of interest in comparison with the second version. This interpretation was propagated by Favaro. For an outline of the discussion concerning historians' attention to the earlier version and for a detailed analysis of the differences between several copies of the two versions, see Gatto (2001, 2002). Romano Gatto transcribed, compared and finally published both *Delle macchine* and *Le mechaniche* in Gatto (2002), which offers an exhaustive analysis of these texts.

[48] Galileo's *Le mechaniche* was republished in Italian in 1649 by Luca Danesi (Galilei and Danesi 1649). No copy by Galileo's hand has been preserved. In *EN* Antonio Favaro listed eleven copies of the text in his *Avvertimento* to *Le mechaniche*. For bibliographic data, see *EN*, II:149–150.

emerging science of motion, was actually the further development of a part of a curriculum on fortifications. In the second part of *Le mechaniche*, the steelyard and the lever are explained first, followed by their applications, taking the form of axles in wheels and winches, pulleys, screws and Archimedean screws. The Archimedean screw is reduced not to the lever but to the model of an inclined plane;[49] there is also a final chapter entitled *Della forza della percossa* (On the force of percussion). No explanations of real or more complex machines is included in this treatise on mechanics by Galileo. Moreover, in this text Galileo approached the question of the relationship between products of the mechanical art and laws of nature, completely refusing to accept the view (his previous view!) that the science of machines would allow one to dominate, supersede or even defraud nature. Galileo noted that such a view was well established among artist-engineers, whom he thus criticized sharply.

Galileo's *Delle macchine* However, Galileo himself had been a strong supporter of the craftsman's view as well, which he taught to his pupils by basing his lessons on the earlier treatise on the science of mechanics, *Delle macchine*.

Four copies of the *Delle macchine* text have survived, designated by the names of the cities where they are now preserved. The one published by Favaro is in Ratisbon (Galilei and Favaro 1899); the second, discovered by Stillman Drake in 1955, is in Pasadena (Drake 1958), the third is in Hamburg (Galilei 1592–1593a) and the fourth, which probably belonged to Giovan Vincenzo Pinelli, is in the Vatican Library.[50] The Hamburg copy, which is the one considered here,[51] is contained in a folder labeled in an anonymous hand, in which the text is placed after a copy of a treatise on land surveying written in Early German.[52] After the copy of Galileo's *Delle macchine*, numerous treatises on military architecture follow. The first is a copy of Galileo's *Le fortificazioni*, used for his private lessons on military architecture. Some of the following treatises on fortifications are in Italian, some in Latin and others in Early German. A note at the bottom of almost all of the copies specifies that the copy was written in Padova. All of this taken together constitutes solid circumstantial evidence for the assumption that this folder was created by a German student, who took classes at Galileo's house for a certain period after 1592. The student probably also took other classes and used the opportunity of his stay at the University of Padova to collect as much material as possible relevant for the military and mechanical arts.

---

[49]For an exhaustive analysis of the relations between Galileo's treatment of the Archimedean screw, as expressed in *Le mechaniche,* and Galileo's theory of motion, see Galluzzi (1979, 199–227).

[50]For more details about the four preserved copies of Galileo's *Delle macchine*, see Gatto (2002, especially pp. CXLV–CLVIII).

[51]The following analysis of the Hamburg copy of Galileo's *Delle macchine* as well as the translated citations included in the present text, are performed on the basis of a transcription made by the author of the present work.

[52]The term Early German is not really exhaustive. The text is written in *Frühneuhochdeutsch*.

Art and nature in the *Delle macchine* In the very brief introduction to Galileo's *Delle macchine*, the text reads:

> The science of machines is that faculty that teaches us the reasons and gives us the causes of the *miraculous* effects that we see happen [...] about the movement of very great weights by means of a very small force.[53]

Galileo pointed out the contrary relation between *Ars* and *Natura*, as was usual for all the engineers or mechanicians of his time.[54]

After the introduction Galileo described the plan of the work:

> [...] we will start speculating from the first and simplest instruments, to which all the others reduce, or of which they are compounded, and they are called first instruments. First of all, there are the lever, the winch, the pulley, the screw and the wedge. All of which also reduce themselves, in some way, to only one, that is, the scales, or, the balance [...].[55]

The balance as the foundation of mechanics Galileo went on to introduce the case of a balance AB (Fig. 3.12), with a central support placed below the beam, and whose ends support two equal weights. Because of the equal distances between the ends of arm AB to the support, and because the weights are equal, the balance remains in equilibrium. This is clearly what Archimedes said in the first postulate of his *Aequiponderantibus*, and what is also written in many other introductions to treatises on machines of that time.

Still following Archimedes and, in particular, the second and the third postulates, Galileo then explained, first, how the balance loses its state of equilibrium, if either the weights or the distances from the support are unequal, or how it mantains the equilibrium if both are unequal in inverse proportion (Fig. 3.13).

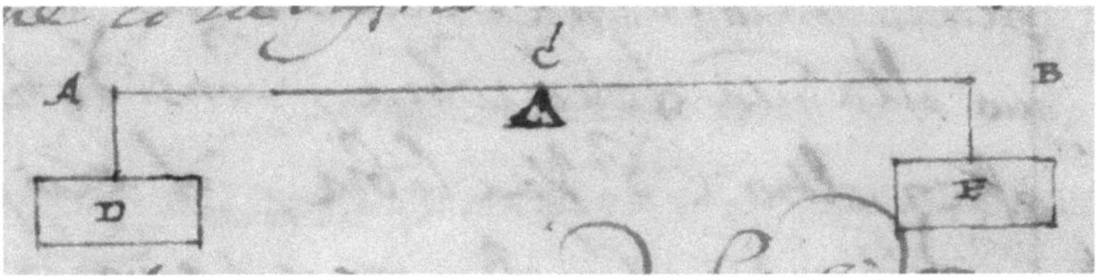

Fig. 3.12 Scale with equal arms supporting two equal weights (Galilei 1592–1593a, 15)

53"La scienza delle Machine è quella facoltà la quale c'insegna le ragioni e ci rende le cause, degli effetti miracolosi, che vediamo farsi [...] circa gli instromenti, circa il muover e alzar pesi grandissimi con pochissima forza" Galilei (1592–1593a, 15). Author's italics.

54Galileo's introductions to his *Delle macchine* and *Le mechaniche* are discussed at length on pp. 199ff.

55"[...] cominciaremo a specolare la natura dei primi e più semplici instromenti, ai quali gli altri si riducono, o di essi si compongono, e sono detti primi instromenti. In primero luogo c'è la Lieva. L'Argano. La Taglia. La Vite e il Conio. I quali tutti si riducono anchora, in certo modo, ad un solo, cioé, alla libra, ovvero bilancia [...]" Galilei (1592–1593a, 15).

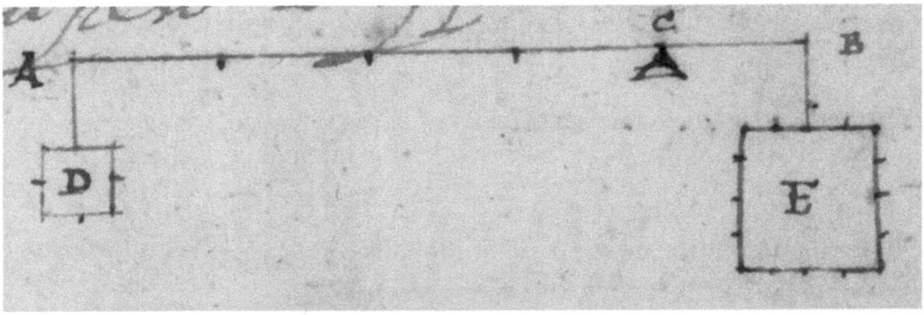

Fig. 3.13 Scale with unequal arms and weights (Galilei 1592–1593a, 15)

After having clearly enunciated that such a balance stays in equilibrium when the relation between the weights is exactly the opposite of the relationship of the distances between the fulcrum and the ends on which the weights are suspended, Galileo stated that "this is the fundament of the whole mechanics, as will become evident as we proceed with the study of particular instruments" (Galilei 1592–1593a, 16).

The lever The second chapter approaches the topic of the lever in the strictest sense. Galileo began by indicating that the lever is an instrument everybody can see, and one that is used by all the masons. He then suggested considering the lever as if it were a balance, with the fulcrum the support on which the "stake of the mason lies." With the help of a drawing (Fig. 3.14), Galileo enunciated the principle for the lever in equilibrium: a lever is in equilibrium if the relation between the weights in B and D is the same of that between the distances between the weights and the support. In other words, there is an inverse relationship between the weights and the distances between these weights and the support.

The actual use of the lever A lever is used by the mason to move weights, not to keep them in equilibrium. Galileo followed this argument and therefore specified that it is sufficient to add a "minimal momentum" (Galilei 1592–1593a, 18) to weight B, meaning that, once the lever is in equilibrium, in order to move it, it is sufficient to add a weight so small that it must not be taken into consideration in the lever's construction. Any mason who would like to make a lever for himself needs

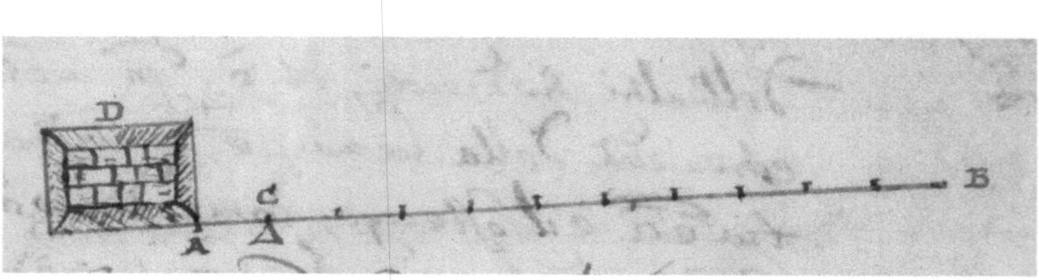

Fig. 3.14 Galileo's illustration of the principle of the lever (Galilei 1592–1593a, 17)

to know only the proportion between the weights and distances that will allow him to make a lever that remains in equilibrium.[56]

Galileo's principle of conservation Galileo proceeds by explaining the proportion between the movements of the moving weight, without that minimal momentum, and the moved weight. This inverse proportion is an important reminder that a lever with a moving weight, or force,[57] applied to the end of an arm, say, ten times greater than the distance between the support and the weight to be moved, also has to be moved a distance that is ten times greater than the space along which the moved weight would be moved. This means, finally, that a lever that uses a very small moving weight and, therefore, places this weight at the end of a very long arm, will need a longer time to accomplish its work than would a lever with a shorter arm and a correspondingly heavier moving weight.[58]

The winch and the axle in the wheel In the next chapter Galileo approached the winch and the axle in the wheel. These two machines are usually described in the opposite order. In fact, even if it is true that both the winch and the axle in the wheel are explained by reducing them to the lever, authors like Guidobaldo del Monte nevertheless preferred to explain the axle in the wheel first, and then the winch as an application of the former. Galileo started with the winch, and provided no hierarchically ordered explanation for the two instruments.[59] This could be seen as an evidence that the goal of *Delle macchine* was not the theoretical foundation of the mechanical sciences, but as simple as possible an explanation of each simple machine to which all the machines can be reduced.

In his discussion of the winch, Galileo took the vertical one into consideration. He presented a drawing reproducing the instrument as it could be normally seen, for example, on any building site (Fig. 3.15). Taking G as the middle point of the block around which the rope winds, the winch is reduced to the lever FGH, where the moving force is applied to F and the moved weight is in H. The proportion between force, weight and time of work, that is, the proportion between the force and the weight on the one hand, and the two parts of the arm on the other, is explained in

---

[56] Abandoning pure static reasoning in favor of the argument of minimal momentum would later bring Galileo into theoretical opposition with Guidobaldo del Monte. In fact, in his *Mechanicorum liber* Guidobaldo rejected treating minimal momenta, *insensibilia*, because they are not mathematically definable. However, at this step of *Delle macchine* the conflict did not arise, since this writing was not intended to provide any theoretical foundation for the lever, but only to explain what was important for the production, evaluation, and use of such machines. For Galileo's early reasoning on *insensibilia*, see Galilei (1592–1593a, 18); for Guidobaldo del Monte's, see Monte (1577, 43–44) and Monte and Pigafetta (1581, 39–40). For an extensive study on the concept of momentum in Galileo, and especially as concerns its development, see Galluzzi (1979, 199–206). For the relations between Galileo and Guidobaldo del Monte as regards the topic of the *insensibilia*, see Rose (1975, 233).

[57] Galileo interpolated the two terms "moving weight" and "moving force" for the same denotation.

[58] This argument is what Gianni Micheli called Galileo's principle of conservation. For more details, see in the previous chapter pp. 66ff.

[59] In the later *Le mechaniche* Galileo, too, inverted the order of the explanations.

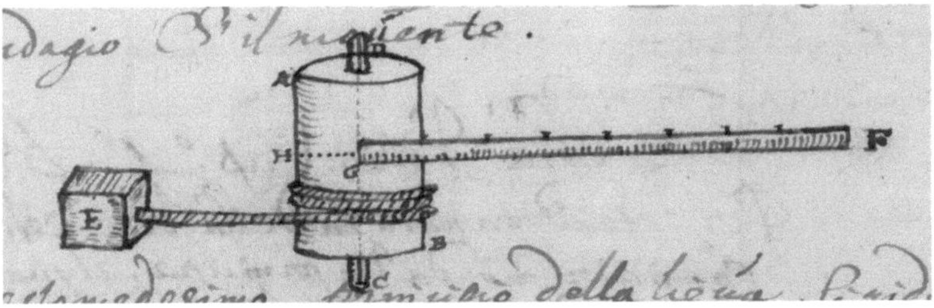

Fig. 3.15 Vertical winch (Galilei 1592–1593a, 19)

terms of "*facility* acquired by the force put in F" and "loss in time and velocity" to move the weight.[60]

The axle in the wheel is the same as the winch, where the block is an axle placed horizontally. In practice, since the block of the winch is vertical, the lever is moved by a horizontal arm, and the winch is thus more efficient when the force is applied by an animal walking around it. When the axle is horizontal and fixed in a wheel, necessarily vertical, this instrument is more easily operated by the force of a man. Of course this is true only if the machines are simple, that is, if no further transmission of the movements takes place. In his *Delle macchine* Galileo explained the axle in the wheel by means of a practical example represented by a drawing of a machine quite typical in Galileo's day (Fig. 3.16, left).[61] In particular, machines with this shape were used within fortresses, when it was time to prepare the cannons. Once the piece of artillery had been fastened and pulled up by operating the wheel, the wagon of the cannon was positioned under the axle and the cannon was lowered down onto it. For large and thus heavy pieces of artillery, there were two ways to make the machine capable of lifting that weight. One, also explained by Galileo, consisted in applying the axle to a wheel whose semidiameter is greater than that of the axle. Like the winch, the axle in the wheel can be considered as a lever, such that the moving force is applied at one point of the circumference of the wheel, the support is the line passing through the middle of the axle, and the weight to be moved is ideally located on the surface of the axle at a point opposite the point where the moving force is applied. The second way of making this simple machine more powerful is to apply a gear wheel instead of the wheel, so that the movement required by the moving force can be transmitted to such a position that the arm of the lever can be operated by an animal and thus with greater moving force.[62]

---

[60] Author's italics. In *Le mechaniche* Galileo gave the same explanation, stating that "the force acquires momentum equal to the resistance" (*EN*, II:170).

[61] Guidobaldo del Monte opened his chapter concerned with the axle in the wheel with exactly the same drawing (Monte and Pigafetta 1581, 102).

[62] In *Le mechaniche* Galileo abandoned this practical explanation for a geometrically more precise description of the change of the effects by changing the proportion between the semidiameter of the wheel and that of the axle (Fig. 3.16, right).

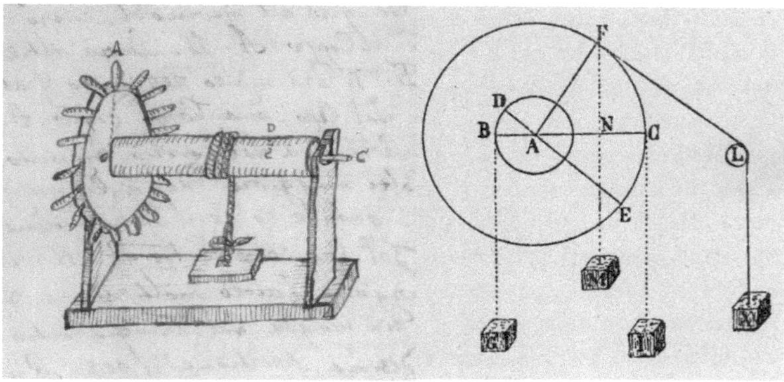

Fig. 3.16 Practical description of axle in the wheel on the left (Galilei 1592–1593a, 20) and geometrical illustration of the same instrument on the right (*EN*, II:167)

The screw As to the screw, Galileo took no pains to depict this machine in terms of the lever and balance as did, for example, Guidobaldo del Monte (Monte and Pigafetta 1581, 115–117). Galileo considered the helices of the screw as the surface of an inclined plane turned around a cylinder, i.e. the axle of the screw. The functioning of the screw is then explained in full by appealing to the behavior of a weight on an inclined plane. Although Galileo introduced his famous theorem of the inclined plane (Fig. 3.17) to explain how a weight moves along it, he did not provide any demonstration for this theorem, as he did later in *Le mechaniche* (*EN*, II:180–186).[63] Galileo pointed out that, on a plane parallel to the horizon, a ball would be moved by a minimal force and that this force has to be increased in proportion to the plane's elevation relative to the horizon. At this point he suggested skipping the "speculation" required to understand how much the force should be increased for any given elevation of the plane, and proceeded instead to the conclusion:

> [...] the weight has the same proportion to the force as the length of the elevated plane to the perpendicular height and thus, if we draw from point C the line CB, perpendicular to plane AB, the number of times that the length AC is greater than the height CB is the same number of times that the force sufficient to move the weight above the plane AC is smaller than the force that would be necessary to lift it [the weight] over the perpendicular AG [...].[64]

Galileo then quickly proceeded to approach the screw as an instrument to lift weights and, in greater detail, described a very famous, and useful, practical application of the screw: the "not only marvelous, but also miraculous" Archimedean

[63] Because of this theorem, Galileo entered in conflict with Guidobaldo del Monte, who provided for it a rigorous static demonstration. For more details, see Bertoloni Meli (2006, 35–39).

[64] "[...] il peso alla forza ha la medesima proportione che la lunghezza del piano elevato all'altezza perpendicolare, e cosi se dal punto C. tiraremo la linea C.B. perpendicolare sopra il piano A.B., quante volte tutta la lunghezza A.C. sarà maggiore dell'altezza C.B. tanto minor forza basterà per mover il peso sopra il piano A.C. di quella che sarebbe necessaria ad inalzarlo su per il perpendicolo A.G. [...]" Galilei (1592–1593a, 21). Galileo considered this explanation sufficient and did not even bother to remind the reader that the screw could also be conceived as a lever.

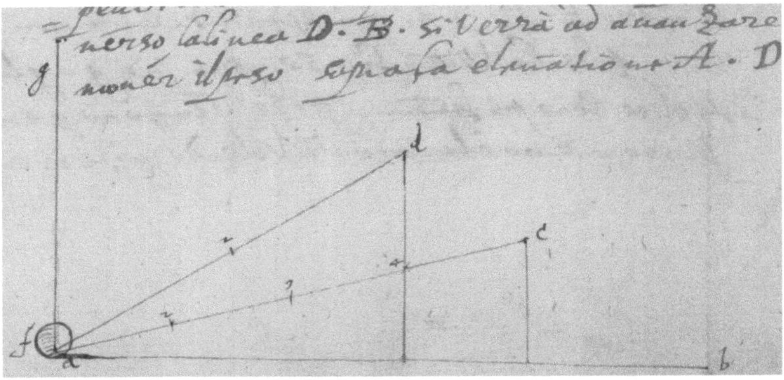

Fig. 3.17 Illustration of Galileo's theorem of the inclined plane (Galilei 1592–1593a, 21)

screw, that is, the machine to lift water. Galileo's clear, synthetic explanation considers every helix to be an inclined plane, whose angle of inclination to the horizon is always the same. If, in order to lift up water from one place to another, the screw is supposed to have an inclination of one fourth of a right angle, then the helices should have the same inclination in the opposite direction. This way the plane of the water would no longer be inclined, or even descending, so that the water which is lifted would actually follow a plane parallel to the horizon and thus the smallest of forces would be enough to move the water. Moreover, folio 21 exhibits marginalia concerning the threads of the screw and, in particular, the extent to which these threads should be hollowed and covered. These contrivances, which could be the results of further oral explanations, are not related to the theory of the instrument, but only to its efficiency when applied in practice.[65]

The pulley Finally Galileo proceeded to introduce the machine consisting of a system of pulleys. Corresponding with Guidobaldo del Monte, Galileo, too, explained at the beginning that a simple pulley works as a lever (Fig. 3.18). In particular, if it is placed over the moving force and over the weight to be pulled upwards, the arms of the lever are equal, thus establishing a perfect correspondence between the force and the "gravity" of the weight. Only a pulley placed under the moving force[66] is a lever with unequal arms, and therefore, given a certain force, able to lift weights greater than those lifted by simple force. In fact, the lower pulley corresponds to a lever, one of whose arms is the semidiameter of the pulley, and the other the whole diameter, i.e. one is double the length of the other. If a given force is able to sustain a given weight by means of a pulley placed over it, then the same force is able to sustain double the weight thanks to a pulley placed below it.[67]

---

[65] Practical contrivances related to building issues disappear in *Le mechaniche*.

[66] For more details, see the next section in this chapter.

[67] If the pulley is placed under the moving force, the weight cannot hang from the rope which turns around the pulley, but must be connected directly to the center of the same pulley, while the end of the rope where there is no moving force should be fixed somewhere else, like to a nail or a hook.

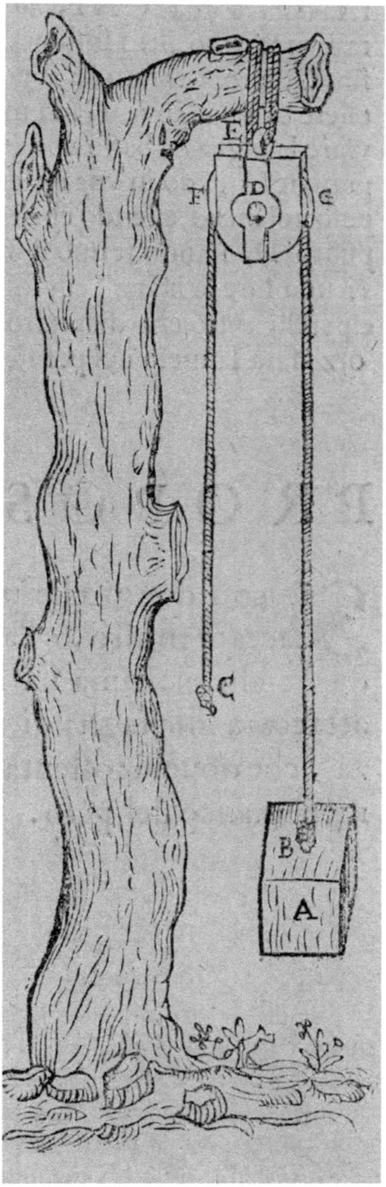

Fig. 3.18 Simple pulley explained on the basis of the principle of the lever (Lorini 1609, 201)

The third section of Galileo's *Delle macchine* The first part of the text *Delle macchine* is constituted of basic definitions and the explanation of the simple machines, just like Guidobaldo del Monte's work, for example. However, the character of this text is evidently more practical and not devoted to the foundations of mechanics. This conclusion is based on five issues. First, the Aristotelian introduction interpreted in such a way to remark on the dominance of mechanics over nature. Second, the simple machines are described not in the framework of a deductive system, but each as an independent issue. Third, the famous Galilean theorem of the inclined plane is mentioned only as a conclusion, with Galileo asserting explicitly that, for the purpose of the text, there is no need for its demonstration. Fourth, practical details, like those about the concavity of the helices of the screw to lift up water, clearly address the reader's attention to building issues. Finally, the text is followed by a series of examples of compound machines, as in Lorini's treatise. The last two parts of the manuscript, in fact, are, first, a series of exemplary combinations of simple machines in order to show how the force executed by a machine can be multiplied; and second, a series of examples of particular machines useful in the fortresses and their building details. The next two parts of this chapter will discuss these issues in detail.

Compounds of Simple Machines to Multiply Force

In his manuscript Galileo expatiated upon the many, in principle countless, possibilities of multiplying the effect of a given moving force, first in the single machines and then with reference to some typical and particularly efficient combinations of simple machines.[68]

Compound pulleys The system of two pulleys is introduced because the simple lower pulley would oblige anyone wanting to lift a weight with it to apply the moving force from the bottom up, that is, in a quite uncomfortable way. If a second upper pulley is introduced, the rope from the lower one goes around the upper one so that it hangs down where the hands of the workers could handle it easily (Fig. 3.19). For this reason alone, Galileo doubled the number of pulleys each time he showed increasingly efficient systems of pulleys. The principle is very simple: according to the relations between the weight, the moving force and the time to move the weight, every lower pulley allows the same moving force to lift double the weight, and every lower pulley should be accompanied by an upper one in order to allow the worker to work more easily. Together with this principle Galileo also presented a quick method to calculate how many times greater the "gravity" of the lifted weight can be than the gravity of the weight that could be lifted by the given moving force without any machine. One had only to count the "threads" of which the system is constituted, decrease this number by one unit, and the result is the number of times by which the given force is "multiplied" thanks to the given system of pulleys. As a matter of fact, this chapter of *Delle macchine* is better characterized as a collection of practical hints about the machine than as an explanation of its function.[69]

Uneven system of pulley In the later *Le mechaniche*, however, whereas all the practical hints disappear, Galileo extended the theoretical explanation of the

[68]Although Galileo promised to explain all the five simple machines, in neither *Delle macchine* nor *Le mechaniche* did he provide any sort of description of the wedge, which is one of them.

[69]On the basis of the explanation of the system of pulleys, Galileo also furnished a concrete example to show how the force is multiplied.

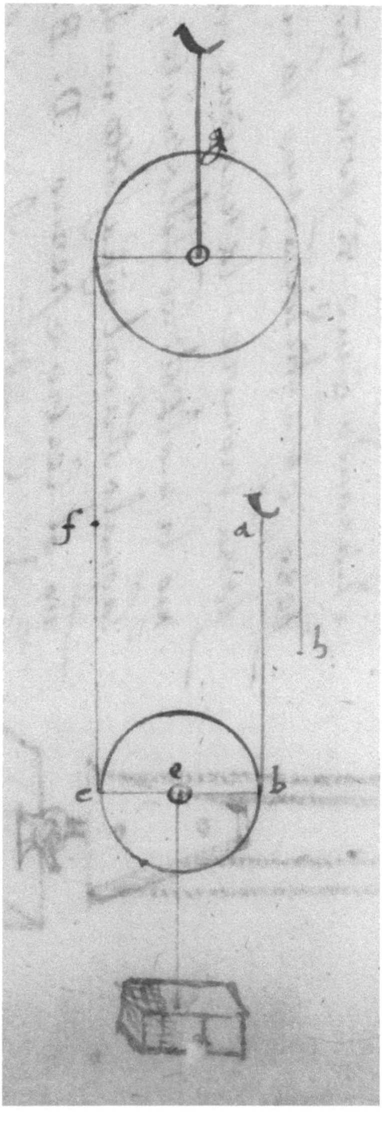

Fig. 3.19 Compound of two pulleys (Galilei 1592–1593a, 27)

machine constituted of pulleys by also taking into consideration what he called the "uneven" systems of pulleys. These systems differentiate themselves from the ones presented above because one end of the rope, rather than exiting from the lower pulley to be attached at some fixed position, exits from the upper pulley and is fixed to the lower one. In this manner the weight is sustained at three points: the two usual extremities of the diameter of the lower pulley, plus the point where the rope is fixed. In such a system a given force actually can sustain a weight three times heavier than in a system without auxiliary machinery.

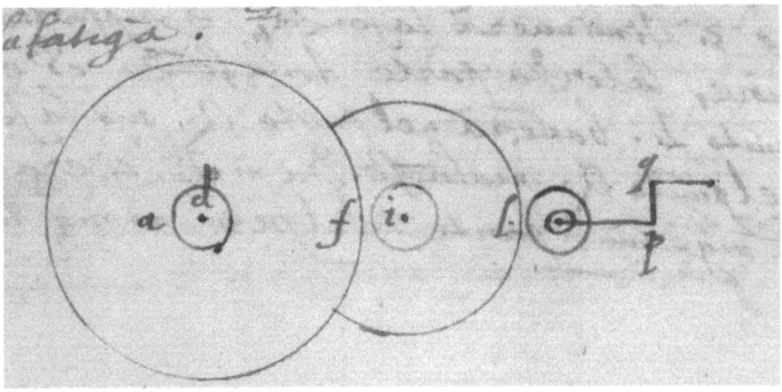

Fig. 3.20 System of axles in the wheels (Galilei 1592–1593a, 30)

Compound system of axles in the wheel As to the axle in the wheel, Galileo first described how a compound system combining several of them can be constructed (Fig. 3.20):

> First there will be the axle [AB] around the center and pivot [B] and turned around this axle there is the rope, which sustains the weight. To turn this axle with less effort we fix the wheel [F] to it [...]. If we want to add another wheel, we will fix another axle, which, turned around, will make the wheel [L] turn and this will be done by meshing the gears to this [axle] and to the mentioned wheel [F]. Then, in order to make the [second] axle G turn easily, we will fix it to the second wheel [L], which, turned around, will make the axle turn [...], and finally, in order to move the last wheel [L] we will add another gear axle, or we could call it sprocket wheel [...], whose gears correspond to those of the wheel [L] and in order to turn this sprocket wheel [...] around, we will add to it a crank handle of iron [...] so that, when the hand is put on its handle, we can turn the sprocket wheel around with effort reduced to the same degree as the [semidiameter of the crank handle] is longer than the semidiameter of the sprocket wheel, because the [semidiameter of the crank handle], turned around, becomes the semidiameter of a circle or wheel described around [its middle] point, which, together with the sprocket wheel [...] constitutes the same instrument as the axle in the wheel.[70]

[70]The following description (Galilei 1592–1593a, 28–29) refers to a drawing in which space has been left free alongside the text, but was never filled. However, in folio 30 a drawing of exactly the same machine is given, using slightly different letters for the references on the figure. Reference letters therefore have been adapted. The original text reads: "Prima sarà l'asse A.B.C. intorno al centro e perno D.B. intorno a questo asse c'è avvolta la corda, che sostiene il peso, e per girare e rivolgere questo asse con minor faticha, vi innestiamo la ruota G.H.F.A. [...]. ma volendo noi agiungere un altra ruota, accomodaremo un altro asse; il quale girato atorno farà volger la ruota C.H.I. il che si farà col far i denti ad esso e alla ruota detta. Dipoi per poter volger l'asse G. con facilità, lo metteremo nella seconda ruota Q.H.L. la quale girata, menera in volta l'asse G. [...], e finalmente per mover l'ultima ruota H.I. l'aggiungeremo un'altro asse dentato, o vogliamo chiamar rocchetto M.N. [...] e per menar atorno questo ultimo rocchetto M.N ci vi aggiungeremo il mangano di ferro O.P.Q.R. di [...], che posta la mano nel manico Q.R. volgeremo con tanta minor fatiga il rocchetto M.N. quanto la linea P.Q. sarà maggiore del semidiametro del rocchetto, perche la linea P.Q. girata intorno, diventa semidiametro d'un cerchio o ruota descritta dal punto Q. la quale con il rocchetto M.N. fa il medesimo instromento, che l'asse nella ruota."

Given the instructions for how to make such a machine, Galileo proceeded with a concrete example of a machine constituted of three axles and three wheels (Fig. 3.20). If the first wheel F, to which the axle around which the rope winds is attached, has a semidiameter five times greater than that of the axle, and if, in reference to the second wheel and its sprocket wheel, the semidiameter of one is one four times greater than the other, and, finally, if the semidiameter of the crank handle is three times longer than that of the next smaller sprocket wheel, then a given force will be able to move a weight which is $5 \times 4 \times 3 = 60$ times heavier than the weight which the same force would be able to move without the help of the system of wheels (Galilei 1592–1593a, 29). However, at this point Galileo did not neglect to introduce a calculation which demonstrates accurately that, if the increase in efficiency of the machine corresponds to a multiplication of the given moving force by 60 times, this means that the space along which the force is transmitted is 60 times longer, that is, that the time to move the weight would be 60 times longer than if the weight were moved by a force 60 times greater but without the auxiliary machine.

Compound of axle in the wheel, screw and winch Galileo then presents a similar way of proceeding to a compound of three different simple machines, a machine to lift weights: axle in the wheel, screw and winch (Fig. 3.21). This typical machine was called a perpetual screw, where perpetual refers not to perpetual motion, but means that the motion of the machine is mechanically not restricted to a range between two end points. In fact, the axle in the wheel corresponds to the axle D, which is fixed to the wheel ABC. The given moving force is multiplied in this machine according to the relationship between the semidiameters of the axle and the wheel. By means of gears around the wheel, then, the axle in the wheel can be assembled to the screw. For example, if the length of the distance along which the weight has to be moved is divided into ten helices, then the given moving force is multiplied by ten (Galilei 1592–1593a, 31). This means that the degree to which a

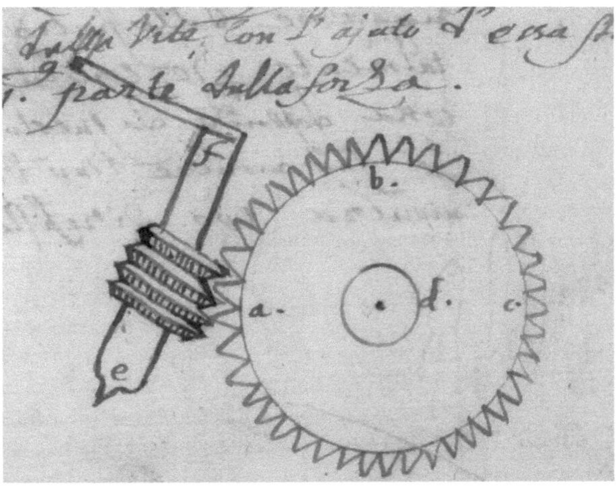

Fig. 3.21 Compound of axle in the wheel, screw and winch (Galilei 1592–1593a, 31)

composition of screw and wheel helps the worker depends on the number of helices on the screw in a given space along its axle. The more helices there are, the more the force is multiplied.[71] Finally, there is the winch, which works as an axle in the wheel and corresponds to the crank handle of the machine. Galileo concluded that, if the semidiameter of the wheel ABC is five times longer than that of its axle, and if the screw multiplies ten times, and, finally, if the semidiameter of the crank handle is four times longer than the semidiameter of the axle of the screw, then the given moving force is multiplied $5 \times 10 \times 4 = 200$ times.

The actual treatise *Delle macchine* closes with a discussion about the force of percussion.[72] To determine how much force a stroke can generate was important for engineers, in constructing machines like pile rams, for example. Such machines were required on building sites, in particular, and especially on those where a fortification of earth had to be built. Galileo explained that the stroke follows the usual mechanical principle which relates the force, the distance along which the force is moved, and the weight, which, in this case, corresponds to the resistance of the object that receives the stroke to be moved.

Compound Machines Useful in the Fortress

Galileo himself stated in his outline for the course on mathematics for the *Accademia Delia* that students have to gain "Knowledge of the mechanical sciences, not only about their reasons and common basis, but also regarding many machines and particular instruments." The third section of the manuscript fills this gap. In the copy preserved at the Staat-Universitätsbibliothek of Hamburg, the treatise *Delle macchine* is followed by thirteen folios on which different machines and their details are drawn. Some of these drawings also have captions written by the same hand that copied Galileo's treatise and wrote the notes on its margins.[73] It can therefore be supposed that either the earlier version *Delle macchine* generally was provided with these drawings, or that further discussions on particular machines took place during private lessons at Galileo's home. This second hypothesis seems more probable as these drawings were hardly new or unfamiliar, but rather familiar drawings, which had been studied and analyzed by a great many engineers and machine makers in Galileo's day. They are drawings of machines designed by the famous engineer Francesco di Giorgio Martini (1439–1501).[74]

---

[71]The more helices there are along the same section of the cylinder, the smaller the gears of the wheel have to be.

[72]The last folio of *Delle macchine* is entitled *Della forza della percossa*. This short text exhibits only minor differences from the one published by Favaro at the end of the later version *Le mechaniche* (*EN*, II:188). For a comparison between the structures of the texts of the four preserved copies of Galileo's *Delle macchine*, see Gatto (2002, CLI).

[73]Most of the notes and all of the captions are written in Early German. The collection of drawings is followed by a copy of Galileo's treatise on military architecture.

[74]Thanks to Marcus Popplow for helping me with the first approach to these drawings of machines.

Francesco di Giorgio Martini "Master Giorgio" was born in Siena in 1439. He was a painter, sculptor, architect and engineer, who worked mostly in Siena and in Urbino. Author of many architectural works in these cities, he also compiled notebooks with drawings of machines for many purposes throughout his lifetime. He finally wrote his famous *Trattato di architettura civile e militare* in Urbino, probably in 1481 (Giorgio Martini 1967), one year before his death on the field near Ferrara. Martini's *Trattato* and notebooks were quite well known throughout the Renaissance. Copies of his drawings can be found in a great many treatises by other authors, and many anonymous manuscripts of that period. Not only in Siena and Urbino, but also in Naples, Rome and Venice, Florence, Martini's drawings of machines constituted the first point of reference for all engineers. Reminiscences or even exact copies of Martini's drawings can be found, for example, in the treatises of Bonaiuto Lorini, Vittorio Zonca, Oreste Vannocci Biringucci (1558–1585, nephew of the famous Vannoccio Biringuccio), Agostino Ramelli and Jacques Besson (1540–1576).[75] At the Biblioteca Marciana of Venice, an annotated copy of Martini's treatise is preserved, bearing a bookplate with the inscription "Organa Mechanica Gui. Ub. ex Mar. Mon," that is, which belonged to Guidobaldo del Monte,[76] Galileo's patron. There are many other indications suggesting that Galileo knew Martini's work as well, not only because the engineer Zonca was active in Padova during exactly the same years as Galileo, or because Guidobaldo del Monte owned Martini's treatise. As a matter of fact, Martini's drawings were so well known that one could even say that they belonged to the common imagery, as demonstrated in the ludic competition that took place in the waters of San Marco on October 23, 1530. On that occasion the masters of the Venetian Arsenal actually erected a wooden castle above two platform boats, clearly evoking Martini's typical machinery (Sanuto and Fulin 1969–1970, LIV:col. 79–80).

The machine drawings in the manuscript All of the drawings in the Hamburg manuscript are of machines whose mechanical parts are set into a box, which is a typical characteristic of Martini's drawings of machines. In general they are compositions of winches, wheels and screws. Most of the twenty drawings of machines in the Hamburg copy can be easily identified in Martini's *Trattato di architettura*, some of them reproduced exactly and others with some variations. The rest of the drawings of machines were copied from Martini's *Opusculum de architectura.* This notebook, dated by Paolo Galluzzi to the years between 1475 and 1478,[77] is considered to be Martini's second notebook of drawings and was catalogued at the Biblioteca Ducale of Urbino until 1722, when it was brought to England (Giorgio Martini 1475–1478). Ultimately there was only one drawing that could not be identified with any other drawing by Martini or by other engineers.

---

[75]Lorini (1609), Zonca (1607), Vannocci Biringucci (after 1562), Ramelli (1588) and Besson (1578). For a reconstruction of Martini's influence during the Renaissance, see Reti (1963).

[76]Ms. Lat. VIII 87 (3048), Biblioteca Marciana, Venice.

[77]Paolo Galluzzi considers the problem of dating this notebook to be objectively insoluble. For a complete overview about this notebook, see Galluzzi (1991, 203).

The function of the drawn machines Apart from one pile driver, one millstone and one water-lifting pump, all of the remaining drawings represent devices to lift weights. No machine requires running water or wind as a moving force; all of them are conceived to be operated by either men or animals. This selection thus shows a particular interest in machines that were especially useful on construction sites. And since the approximately 500 folios of the manuscripts remaining are copies of treatises on fortifications and military architecture, one can presume that the goal of this collection of drawings is to report on the machines which could be useful both on the site where a fortification was being built, and to complete the various tasks typically needed by the residents of such a place.

Compound machine to lift weights (1) The drawings represented in (Fig. 3.22) are exact copies of Martini's drawings of machines to lift weights.[78] Both machines operate by means of a lever or crank handles that can be inserted at the top of an axle, along which there is a screw. In the machine on the left, the screw operates on a gear wheel, the lower part of which is connected to a lantern positioned horizontally along an axle. Finally, a rope, which is tied to the machine at one end, winds around the latter axle, with the weight suspended from its other end. The lifting machine on the right is a variation of the first one. The wheel operated by the screw in this machine, which can lift two weights simultaneously, has an axle along which there

Fig. 3.22 Machine to lift weights constituted of a crank handle, a screw, a wheel, an axle, a pulley and a lantern (Galilei 1592–1593a, 51)

[78]Ms. Cod. Math. 200b, Hamburg, folio 51. These drawings are copies of Giorgio Martini (1475–1478, 4r).

is a further screw that is connected with a double axle in the wheel. As the drawings clearly show, these machines could be expanded by connecting a pulley or pulley system. Connecting such additional machines not only improved the efficiency of these devices, but also made it easy to adapt them to the practical context of each given construction site and to the particular requirements of the given workflow.

Compound machine to lift weights (2) Some of the drawings of machines of the manuscript preserved in Hamburg are drawn first in large format on a whole folio and then, some folios later, in a smaller size but accompanied by the drawings of the simple machines of which the first is constituted. This is the case, for example, for the lifting machine illustrated in Fig. 3.23, which is also an exact copy of a drawing by Giorgio Martini.[79] The prime mover of this device is a winch that is operated by a multiple crank handle. Depending on the dimensions of the device, such a machine can be operated either by men or by animals. The winch, a simple version of which is drawn in the middle, is connected to an axle in the wheel by means of two gear wheels. The axle in the wheel, drawn on the right, is represented here by an axle, around which the rope winds, and by a crank handle, whose vertical component corresponds to the wheel.

Positioning artillery onto wagons One of the chapters listed in the outline of the treatise mentioned earlier in this chapter, entitled "Particular Advantages of the Artillery in Comparison with other Mechanical Instruments," (Galilei ca. 1602–ca. 1637, 193r) deals with the operation of positioning artillery on wagons inside a fortification. This task was indeed a very important one. As Lorini told, great effort was spent to organize the fortifications so that this operation could be accomplished as quickly as possible. The soldiers also had to be drilled for it and, obviously, extremely efficient lifting machines were required. Galileo's potential chapter thus touched a nerve center for everything concerned with a fortification. This topic involved questions about the defense strategy of the fortification, such as where the artillery room should be located and how long the whole operation should take,

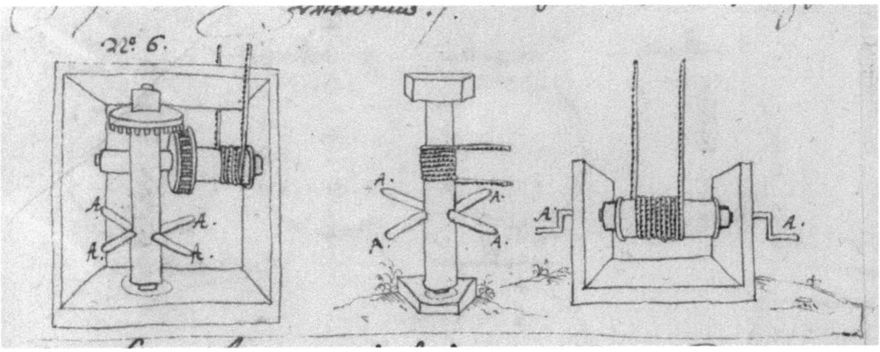

Fig. 3.23 Machine to lift weights constituted of a multiple crank handle, a winch and an axle in the wheel (Galilei 1592–1593a, 56)

[79]Ms. Cod. math. 200b. Hamburg. folio 56. These drawings are copies of Giorgio Martini (1967, f. 51, Table 93).

and, again, with how many machines and which ones the fortification should be equipped.

Compound machine to place cannons onto wagons For this purpose a complex machine is depicted among the drawings of the Hamburg manuscript (Fig. 3.24). What is peculiar about this drawing is that it is not a copy from Martini's treatise or notebook. In fact, this machine and its representation seem more modern that those of Martini. The device is constituted of a winch as a prime mover, which can be operated either at the top or at the bottom, and of a first lantern along the winch, further connected with a gear wheel in the middle. This gear wheel drives another lantern that has internal threads for a perpetual screw, which is not supposed to turn, but only to ascend or descend. The solution of a perpetual screw that does not turn is already present in Martini's treatises, but there it is always a mother screw positioned above the machine, in direct contact with the surface through which the screw ascends. The constitution of the machine drawn in the Hamburg manuscript, therefore, appears to be an attempt to decrease the effect of friction. The peculiarity of this solution is also shown by the attention paid to it by the author of the drawing. The three drawings on the left, indeed, illustrate how this part of the machine works. From right to left, the first cutaway shows the longitudinal section of the second lantern with its internal threads, the second demonstrates that two hollows have to be made along the screw and that the screw has to go over another small component illustrated in the third horizontal section. This last small component is a small pivot, which fits with the hollow of the screw and prevents it from turning, allowing only vertical motion.

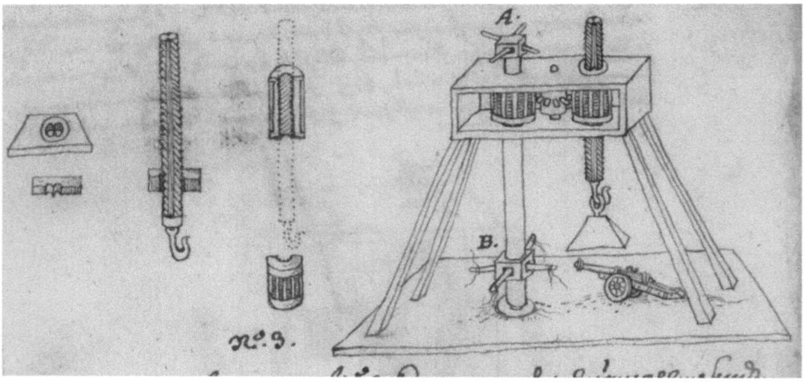

Fig. 3.24 Compound machine to prepare heavy artillery (Galilei 1592–1593, 55)

The Art of War and the Materiality of Machines

To ensure the success of his enterprise, Galileo had to make contacts outside of university life. His house, crowded with people, and housing a workshop for mathematical instruments, was not merely a meeting place for craftsmen like smiths, with whom Galileo performed systematic work, but also a sort of training camp for future military officers.

Galileo integrated the practical knowledge he acquired from the Florentine circles of military engineers and architects together with his experience as a designer and producer of mathematical (military) instruments and, finally, also with the practical arithmetic, geometry and mechanics that he learned from Ostilio Ricci during his youth. The result of such an integration of knowledge was a complete overarching course on fortifications typical of the military engineers of the second half of the sixteenth century.

Galileo's apprenticeship, his activity as an instrument designer and workshop manager and, finally, his copious teaching activity, demonstrate strongly how Galileo's profile was that of the artist-engineer, who, in keeping with the trend of the sixteenth century, tried to place his work within the framework of the early modern art of war. Until the end of 1609, when Galileo suddenly and unexpedectly turned the telescope toward the sky, he worked and was perceived primarily as a military engineer.

Engineers and military architects like Galileo were not the ones who built machines, but they were often the ones who oversaw their construction or evaluated them. However, Galileo's knowledge of mechanics, as can be inferred from the content of his private course on fortifications, does not seem sufficient for those tasks: To oversee the construction of a machine or to evaluate it, the materiality of the machine itself had to be taken into account. Issues like thickness, robustness, and the enlargement of devices from the scale of a model up to a real machine were part of this knowledge. As Lorini added:

> And since the demonstrations, and the propositions among the superficial lines, and the imagined bodies, and separated from the matter, do not properly work, when they are applied to the material things, that is, that the mental concepts of the mathematician do not receive and are not affected by those hindrances, that the matter always entails by its nature, and with which the mechanician works. [...]. Therefore, for the mentioned things, I will remind all of those, who would like to engage themselves in such enterprises such as judging or leading the execution of whichever machine, that it is necessary not only to have knowledge of mathematics, but also to be a shrewd and experienced mechanician [...].[80]

The insights presented in Galileo's private course on fortifications, related to his activity as an instrument maker, show that he shared manifold aspects of the practical knowledge, but did not reflect much on the materiality of the machines. Nevertheless Galileo, too, completed Lorini's training program to become what Lorini called a "mathematician–mechanician." He did so at the Venetian Arsenal, whose knowledge became the fundament of his First New Science, namely the science of the strength of materials. This is the topic of the next chapter.

---

[80] "E perche le dimostrationi, e proportioni, che si ritrouano tra le linee superficie, e corpi imaginarij, e separati dalla materia, non rispondono cosi esquisitamente, quando alle cose materiali si applicano, cioè che i concetti mentali del Matematico non riceuono nè sono sottoposti a quegli impedimenti, che di sua natura sempre porta seco congiunti la materia, con che opera il Mecanico; [...] Adunque per le cose dette ricorderò a quelli, che si vorranno porre a così fatte imprese nel giudicare, ouero comandare l'essecutione, di qual si voglia machina, esserli necessario non solo hauer cognitione delle Matematiche, ma ancora essere aueduto, e pratico Mecanico [...]" Lorini (1606, 196).

Part II
Practice and Science

Chapter 4
The Knowledge of the Venetian Arsenal

Galileo crossed the threshold of the Venetian Arsenal in 1593, where, thanks to the mediation of its executive body, he came into contact with shipwrights and oar makers. Later in his career Galileo opened up a new field of modern science, one concerned with the strength of materials, thanks to the publication of the first of his two new sciences in the *Discorsi e dimostrazioni matematiche intorno à due nuove scienze* (*EN*, VIII:39–318).[1] These two events in Galileo's life are intimately connected: Galileo's first new science is rooted in the practical knowledge of the shipwrights of the Venetian Arsenal, the high-tech center of the Republic of Venice.

Politically, sixteenth-century Venice was characterized by the development and accession of a party known by the moniker "Of the Youth." Its ambition to occupy prestigious offices in the Venetian political system was related directly to its members' perceived need for cultural emancipation through studies, first, of ancient knowledge according to the humanistic spirit, and, second, of practical knowledge related to the needs of contemporary Venetian society. Necessary were (a) the knowledge of a military architect, because the fortifications on the mainland required a complete reorganization. This was a consequence both of the increasing tendency of the Venetians to address their life to the mainland, and of the development of heavy artillery; (b) the knowledge of a *Proto dei marangoni*, a shipwright of the Venetian Arsenal, because of the need to potentiate and therefore renew the Mediterranean fleet, which was itself a consequence of the growing power of the Ottoman fleet and the increasing presence of pirates; (c) the knowledge of a civil architect, because of the need for architectural changes to the city of Venice in order to pay tribute to the greatness and affluence of the Republic and its empire. In other words, the new needs of Venice represented new practical challenges, and the knowledge required to accomplish them was thus also practical. This cultural turn became increasingly relevant starting back in the early sixteenth century: The *Doge* Loredan, for example, had himself portrayed as bowing while offering to St. Mark the project of the reconstruction of Rialto (Concina 1990, 37). In 1517 Andrea Gritti was elected *Doge* of the Republic after having presented a plan to renew all of the

---

[1] On the development that proceeded from Galileo's first new science up to a complete classical theory about the behavior of materials, see Szabó (1987, 351–402).

DOI 10.1007/978-90-481-8645-7_4, © Springer Science+Business Media B.V. 2010

fortifications on the mainland (Concina 1990, 47–48). The rediscovery of ancient wisdom was related directly to the great architectural, mechanical, technical—in a word: practical—challenges, and to the knowledge they required. During the sixteenth century, it was proposed to build a public library on the square of Venice that was dominated by the Byzantine library of Bessarion. In the words from the first oration of the renowned Venetian humanist Vettor Fausto (1480 ca.–1546), the library would have been a casket of the "secrets of venerable Antiquity" (Fausto 1551, 31r), while Fra Giocondo, the lauded translator of Vitruvius' *De architectura*, presented a proposal for reconstructing Rialto's marketplace in the form of a Greek *agora* (Vitruvius et al. 1513; Concina 1990, 39).

The principle according to which ancient and practical knowledge had to be directly related to each other also led to the decision of the Venetian Senate and the *Collegio della Repubblica* at the beginning of 1526, to commission the Arsenal with the construction of three ships much larger than the usual size. The Senate's intention was to carry out tests on the three ships to investigate several of the problems normally exhibited by the Venetian galleys, such as insufficient maneuvrability in the absence of wind, for example, and a lack of stability when equipped with new heavy artillery. As usual in Venice, the *Collegio della Milizia da Mar* then had to assign the task of constructing the ships to three masters of the Arsenal. This sort of assignment generally was decided not only by representatives of the political and administrative bodies, but also by a variety of masters and other foremen, according to their specialization. The aspiring headmasters had to be ready to present and defend their projects, not only through debate, but also with drawings and models. A discussion lasting over a year resulted in the following assignments for the three test galleys: the first ship to the brothers Matteo and Leonardo Bressan, and the second to Francesco di Todaro, all of whom were famous and experienced masters of the Arsenal at that time. For the third ship the above-mentioned Venetian lecturer in Greek, Vettor Fausto,[2] was chosen. Fausto, who was famous in the intellectual circles of Venice for his philologically accurate translation of Aristotle's *Mechanical Questions* completed in 1517, proposed to the Venetian Senate the construction of a *quinquireme*, whose design was based, he claimed, on the study of the ancient Greeks (Aristotle and Fausto 1517).[3] The Senate granted Fausto free hand to work in one of the Arsenal's typical Venetian shipyards, known as a *squero*, and assigned to him a team of specialized shipbuilders. This was the first documented instance of an intellectual going into a workshop with the purpose of working together directly with professionals of one specific art. In the words of

[2]Archivio di Stato di Venezia, Senato Mar, reg. 21, c. 24. September 29, 1526.

[3]The decision to commission Fausto with the building of a galley and to assign him a shipyard was the output of a convoluted process, which is documented in the diary of Marino Sanuto, a sixteenth-century Venetian nobleman. See in particular: Sanuto and Fulin (1969–1970, 39, col. 322). Aristotle's *Mechanical Questions*, translated by Fausto himself in 1517, address questions concerning ship design as well. In particular, see problems four, five, and six: Aristotle and Hett (1980, 355–361). See also pp. 132ff in this chapter.

Fausto, as he related his experiences about the Venetian Arsenal to his friend Giovan Battista Ramusio (1486–1557), when he went into the *squero* it was like a descent into Hades: "through grottoes shaped by scabrous rocks leaning askant, there, where the tremendous thickness of the subterranean darkness reigns."[4]

Galileo entered the same darkness Fausto did. After completing a sort of training program, he was then able, at long last, to shape part of the knowledge of the shipwrights of the Venetian Arsenal into a more general deductive form. This is what then became the first of his two new sciences: the science of the resistance of materials.[5]

Galileo's science of materials is revealed in the first two Days of his *Discorsi*, and based on three models. The first, the aim of which is to find the resistance to fracture of solid bodies stressed along their longer axis, is called here the rope model. The second is the cantilever model, constituted of a prism or cylinder placed parallel to the horizon and driven into a wall at one of its extremities. The purpose of this model is to find the resistance to fracture along the shorter axis of a solid body fixed to another one. The third, which might be called the "oar model,"[6] resembles the second model, except for the condition that the body is not fixed at one of its extremities to another body, but lies on it at one or more different points, like, for example, a column that rests on a certain number of supports on the floor.

First, Galileo's cantilever model will be analyzed with reference to those aspects which were most relevant for craftsmen such as machine makers. Second, the theoretical model of the same craftsmen concerned with the resistance of materials, as it was commonly explained around the beginning of the sixteenth century, will be described in terms of its Aristotelian origins. Given these two opposite approaches to the discussion, it will be shown how Galileo, also intellectually equipped with Aristotle's *Mechanical Questions*, visited the Venetian Arsenal. It will be shown that the practical knowledge he shared with the Venetian masters, especially with regard to some of the problems upon which Aristotle himself had focused, made Galileo such an expert on nautical issues that he actually became involved in management, working with the executive body of the Arsenal itself. The knowledge that Galileo acquired thanks to this experience served as the basis for his formulation of the cantilever model. Finally, this chapter concludes with an attempt to furnish the historical context from which Galileo's oar model emerged.

---

[4] "[...] attraverso grotte formate da scabre rocce strapiombanti, là dove regna il tremendo spessore delle tenebre sotterranee [...]," from Fausto to G. B. Ramusio, 1530, in *Epistolae clarorum virorum* ... 1586, 128–133. For the importance of this event, see also Concina (1990, 46–70).

[5] As a result of an analysis of some of the sources considered in this chapter, the connection between Galileo's science of the strength of materials and the practical knowledge of the Venetian Arsenal was first described in Renn and Valleriani (2001). Back in 1976, moreover, Thomas Kuhn suggested investigating in such a direction to understand the emergence of Galileo's first new science (Kuhn 1976, 56).

[6] The oar model is discussed on pp. 150ff in this chapter.

Dating Galileo's Work on the Science of Materials

In a letter sent by Galileo to Antonio de Medici on February 11, 1609, in which Galileo discussed his return from Padova to Tuscany, he related his latest scientific results and those he wished to achieve. Among the known results, Galileo also mentioned his first new science:

> And recently I also finished finding all the conclusions, with their demonstrations, related to the forces and resistances of woods of different lengths, thicknesses and shapes: how they are weaker in the middle than at their ends, and how they will support greater weight if this is distributed all along the wood rather than at just one point, and the shape this should have so that it is equally resistant. This science is imperative to build machines and every kind of engine, and yet no one has performed this study to date.[7]

The deductive system of Galileo's science of materials already had been worked out around late 1608/early 1609. The manuscript *On Motion* (Galilei ca. 1602–ca. 1637) contains one folio concerned with the demonstration of a theorem later published in the second Day of the *Discorsi*, the part of that work most relevant to the science of materials (Galilei ca. 1602–ca. 1637, 102v). Watermark evidence exposed by Jochen Büttner in the course of his analysis of the manuscript suggests that the folio stems from Galileo's Paduan period (Büttner 2009), which supports Galileo's statement in his letter to Antonio de Medici.

Galileo certainly worked on some of the theorems of his science of materials in 1633 as well, when he sent some folios of his manuscript to Mario Guiducci (1584–1646) for revision.[8] In conclusion, although the version of Galileo's first new science as it is known today is certainly the result of work that took place between 1592 and 1636, when the final manuscript was ready, it is certain that the most intensive work on the science of the strength of materials had been accomplished while Galileo was resident in the Venetian Republic, that is, between 1592 and 1610.[9]

The Key Question of the Machine Makers

With his cantilever model Galileo tried to identify the resistance to fracture of a given solid body, for instance a prism, of known dimensions and material, and how this resistance is affected when its dimensions are hypothetically increased or

---

[7] From Galileo to A. de Medici, February 11, 1609, in *EN*, X:228–230. For the translation of the entire letter, see pp. 223ff.

[8] Andrea Arrighetti (1592–1672), who received the folios from Guiducci in 1633, also proved the content of one theorem. Galileo gladly accepted his suggestions and introduced them in the final version of the *Discorsi*. For more details, see Andrea Arrighetti to Galileo, September 25, 1633, in *EN*, XV:279–281 and Galileo to Andrea Arrighetti, September 27, 1633, in *EN*, XV:283–284. For the translations of the entire letters, see pp. 270ff and 273.

[9] Bertoloni Meli reached a similar conclusion in Bertoloni Meli (2006, 91), albeit by a different route.

decreased, and when a given weight is suspended from one of the extremities of the body. Thus described it sounds like a highly theoretical field of research, perhaps one originating from a modern physics laboratory. But this is a very misleading impression. Galileo's reflections on the resistance of prisms or cylinders are connected intimately with the problems dealt with by engineers, architects, shipwrights and especially machine makers when using timbers and cylinders of wood to construct machines, ships, weight-bearing structures for buildings, scaffoldings and the like.

Discussions about the commissioning of an engineering device were often held around a functioning model of the apparatus in question. Once the construction of the device in real size had been agreed upon, the next problem to arise usually involved the thickness of the device's components in real size. It was commonly believed that when the dimensions of a component, for example a wooden cylinder, were changed proportionally, its resistance to fracture would remain exactly the same.[10]

However, any carpenter, thanks to the experience he had accumulated, observed that proportionally changing the dimensions of a solid body decreased its resistance to fracture. This gap between theory and praxis was often bridged by early modern engineers, who theorized that the main cause for construction failures was either a lack of experience by the mechanician in calculating the force produced by the compound machine, or other natural hindrances that arose due to the irregularities of real materials in comparison with ideal matter. Lorini, for example, indicated that the mechanicians need to be able to:

> [...] build the proposed machines, and to know not only how to proportionally assemble and rule them, but, with the clarity that one needs, also to know how to find the force with the compass, that is, the multiplication of their levers, so that then, when making the work in real size, one is not defrauded from that force of it, as often happens to those who only trust the ease shown by the small Models, without knowing its necessary grounds.[11]

Complaining that there is no perfect rule as to the materials of which machines consist, Lorini also added:

> But the mind of the Mechanician, who has to guide and order the executors of the work, largely consists in being able to foresee the difficulties which are caused by the diversities of the materials with which one has to operate: and the more he has to be prudent with that because it is impossible to give a certain rule for such accidental hindrances [...].[12]

---

[10]The same resistance is in truth obtained by increasing the dimensions over-proportionally. Such a conception of building was followed and divulged by the Italian architects of the sixteenth century and related to the Vitruvian conception of modular architecture. This issue is extensively analyzed in Valleriani (2009a, especially on pp. 186–190).

[11]"[...] fabricare le proposte machine, e quelle sapere proportionatamente non solo comporre, & ordinare, ma con quella chiarezza, che ancor si ricerca, saper co'l compasso ritrouare la forza, cioè la multiplicatione delle sue lieve, accioche poi nell'effettuar l'opera in forma reale, non si venga a restare ingannati di tal sua forza, come spesso accade a quelli, che confidano solo nella facilità, che mostrano i Modelli piccoli, senza sapere i necesiarij suoi fondamenti" (Lorini 1609, 196).

[12]"E però il giudicio del Mecanico, che deue ordinare, e comandare agli essecutori dell'opera, consiste in grandissima parte nel sapere preuedere le difficultà, che apportano le diuersità delle materie,

The passage from the model of a machine to a full-scale specimen, from the model of a building to the real building, from small to large, was ultimately the most urgent problem for craftsmen such as, for example, machine makers. Success in bridging this gap was what distinguished bad craftsmen from good ones, and amateurs from masters. Galileo's first new science was supposed to help the mechanicians in dealing with such problems.

Galileo's Cantilever Model

With the cantilever model, the second of the models developed by Galileo within the framework of his first new science, he tried to identify the resistance to fracture of a certain given prism of known dimensions and material, when the dimensions of the prisms were hypothetically increased or decreased and when one extremity was fixed, for example, driven into a wall, and a given weight suspended from the opposite extremity.

Galileo framed the cantilever model by considering timber driven into the wall as a lever, the principle of which Galileo expressed in these terms:

> [...] the force related to the resistance is the inverse ratio of the distances which separate the fulcrum from the force and resistance respectively. (*EN*, VIII:152)

Given a prism driven into the wall at one end and a weight suspended at the other (Fig. 4.1), in accordance with the principle of the lever and without taking into consideration the material constituting the prism, the lever would act in such a way that a given weight would cause a fracture at the base of the prism where it is driven into the wall, as governed by the following relationship between the resistance to fracture, the dimensions of the prism, and the suspended weight:

> [...] the magnitude [*momento*] of the force applied at C is related to the magnitude [*momento*] of the resistance, found in the thickness of the prism, i.e. in the attachment of the base BA to its contiguous parts, by the same ratio at which the length CB is related to half the length BA; [...]. (*EN*, VIII:156)

If one considers real prisms, that is, the materiality of the prism, half of the weight of the prism has to be added to the magnitude of the force applied in C, which is represented by weight E. This establishes a relationship between the dimensions of the prism, including its thickness and its weight. The resistance to fracture is represented by half of the thickness, and the cause of the fracture is the weight, either a suspended weight added to the prism's own weight, or the prism's own weight alone.

Galileo then sought to determine how both the resistance to fracture (i.e. the thickness of the prism) and its weight are related, by comparing two prisms of

---

con che si conuiene operare: e tanto più deue in ciò esser cauto quanto che di tali impedimenti accidentali non se ne può dar regola sicura; [...]" (Lorini 1609, 196).

Fig. 4.1 Illustration of Galileo's cantilever model (Galilei 1655, 86)

proportional dimensions but different sizes. Considering weight first, Galileo compared two prisms of the same material, with the same thickness and different lengths, reaching the conclusion that:

> [...] the moments [of gravity] of the forces of prisms and cylinders which have the same thickness but different lengths, bear to each other a ratio double of that between their lengths, that is, they are as the squares of their lengths. (*EN*, VIII:159)

In other terms, the ratios of the weights of the two prisms is given by the ratios of the squares of their lengths.

Next he looked at two prisms of the same length but of different thicknesses. Galileo stated that:

> In prisms and cylinders of equal length, but of unequal thicknesses, the resistance to fracture increases in the same ratio as the cube of the diameter of the thickness, i.e. of the base [...] [,] (*EN*, VIII:160)

which means that the prism's resistance to fracture is given by the third power of the diameter of the base.

Integrating the final two conclusions[13] offers a solution to the old mechanicians' problem of the proportions of machines. If an engineer wanted to retain the structure

---

[13]Galileo first established a relation between weight and resistance to fracture: "Prisms and cylinders which differ in both length and thickness offer resistances to fracture [i.e., can support at their ends loads] which are directly proportional to the cubes of the diameters of their bases and inversely proportional to their lengths" (*EN*, VIII:162–163).

of a mechanical device as it was in the form of a small model and transfer it to the bigger machine, and, more importantly, if that engineer wanted to maintain the same resistance to fracture presented by the small model, then he had to build the mechanical device with increased dimensions so that the thickness of the components was increased by a power of three over the original model. The fact that the weight of the given device, which is the cause of the fracture, increases by a power of two when its dimensions are increased proportionally, constitutes a size limit at which any object will collapse under its own weight.

The Origins of the Renaissance Engineers' Cantilever Model

Before Galileo had published his science of materials, mechanicians were not left to their own devices. As mentioned, engineers, architects, machine makers and shipwrights already had a body of theory. They believed that two machines, different in size but built of components linearly proportional to each other, would also have the same resistance to fracture. Reducing this to the example of the cantilever driven into a wall, they believed that two cantilevers of the same material, different in size but proportional to each other as regards their thickness, length and width, would have the same resistance to fracture as well. If this could not be observed, that is, if the larger cantilever or compound machine was weaker, this was believed to be due to either those unpredictable irregularities of the material or, in the case of compound machines, to the mechanician's lack of experience.

Engineers' opposition The French military architect Antoine de Ville (1596–1674), who also did some of the revision on the first two Days of the *Discorsi* before the work was published, was a vocal opponent of Galileo's theory. Antoine de Ville was a famous military architect, also employed by the Venetian Republic from 1632 to the beginning of 1635. During these years he was in close contact with Fulgenzio Micanzio, who had been the closest collaborator of Fra Paolo Sarpi and was in charge of correcting the proofs of the *Discorsi* and acting as an intermediary for their publication in the Netherlands in 1635.[14] Thanks to Micanzio's mediation, de Ville and Galileo entered into direct epistolary exchange.[15] De Ville, who had published a treatise on military fortification that had been widely circulated and even reprinted several times some years earlier in 1628, read the folios of the first two Days of the *Discorsi*. After several vigorous discussions with

[14]For biographical details about Fra Fulgenzio Micanzio, see Favaro and Galluzzi (1983, II:700–736).

[15]Galileo sent the first folios of his *Discorsi* to Fra Fulgenzio, who organized a reading group. Besides Micanzio and de Ville, the group was constituted of the Paduan mathematician Andrea Argoli (1570–1656), the Venetian engineer Francesco Tensini (1580–1630), Galieo's ex-pupils Paolo Aproino (1586–1638) and Alfonso Antonini (1584–1657), and the Venetian astronomer Marcantonio Celeste. For more biographical details about Antonio de Ville, as well as on the entire correspondence between him and Galileo, see also Vérin (2001).

Micanzio, he decided to write Galileo about his doubts, as he did in a letter of March 3, 1635:

> One says that, since a timber breaks due to its own weight, hence the matter destroys itself and the machine as well due to its gravity, which does not improve the force. One answers that for this there is a rule. But which one, and with which proportion, and with which matter? Since each material is different—the iron supports very heavy hanging weights, the wood carries them placed upright upon it—then which demonstration is able to show all of the imperfections that can be found in the materials, since *there is no science of the singulars*? These are all different, and for these differences or affectations, if we do not want to call them imperfections, one cannot give any convenient rule. And not only is there no rule for all of them, there is no single rule for those materials of the same species. [. . .] And while one does not want to call these things defects of the matter, in any case they make the art faulty, as these accidents cannot be recognized in small machines, but become evident in the large ones since they are increased in force and in weight. [. . .] You also observed that a siphon cannot attract for a height of more than 18 feet, no matter how thick and high it is [. . .]. One must say: this is not a defect of the machine, but of the water.[16]

De Ville was simply not able to accept that other general rules concerning matter could be valid in addition to the law of the lever. He very clearly reported his observations and his experience, according to which bodies similar in proportions but different in dimensions do not have the same resistance to breakage, and undoubtedly saw the cause of this behavior in the imperfections of the materials. Finally, De Ville also quoted a strong epistemological obstacle for the acceptance of Galileo's rules: the idea that a science of the singulars, as de Ville says, is impossible.

Engineers and architects like Lorini, Ceredi and de Ville did not create their theory on the strength of materials. They received it in its present form from the works of the Aristotelian commentators of the sixteenth and seventeenth centuries, who focused their speculations on Aristotle's *Mechanical Questions*, in particular on Question 16. Galileo himself pointed out this aspect in his *Discorsi*.

Aristotle's mechanics in Galileo's *Discorsi* Galileo's *Discorsi* are written in a dialogical form, as a conversation between the speakers Sagredo (the Venetian pragmatic thinker), Salviati (Galileo) and Simplicio (the Aristotelian thinker). Galileo first had Simplicio quote Aristotle's *Mechanical Questions*, needing to introduce the principle of the lever as fundamental knowledge in order to approach the cantilever model (*EN*, VIII:152).[17] The *Mechanical Questions* are then mentioned again between Proposition VI, according to which the moments composed of gravity and the dimensions of two prisms are compared with each other, and Proposition VII, which establishes the physical meaning of the previous statement:

> Among heavy prisms and cylinders of similar figure, there is one and only one which under the stress of its own weight lies just on the limit between breaking and not breaking: so that every larger one is unable to carry the load of its own weight and breaks; while every smaller one is able to withstand some additional force tending to break it. (*EN*, VIII:165)

[16]From A. de Ville to Galileo, March 3, 1635, in *EN*, XVI:221–228. Author's italics. For the translation of the entire letter, see pp. 277ff.

[17]Galileo admitted the temporal priority of Aristotle's formulation of this principle, but he also suggested using the one by Archimedes because he considered it more rigorous.

Given two solid bodies of a certain form and of a certain material, there is one set of dimensions and one set of proportions for the largest possible body. If one maintains the proportions of the smaller body and exceeds this set of dimensions, the body will inevitably fracture under its own weight.[18]

Galileo quoted Aristotle's *Mechanical Questions* in his reasoning because Aristotle's Question 16 approaches a case apparently similar to Galileo's cantilever model:

> Why do pieces of wood, the longer they are, become weaker and bend more while they are being lifted up, and this also when one, which is as short as two cubits, [is] thin, and the other, which is a hundred of cubits [long], is thick? Is it perhaps because the lever and weight and fulcrum are formed, while the length of the timber is being lifted up?[19]

Aristotle considered the process which leads to the bending and the fracture of a cantilever to be a mechanical one, which therefore can be explained by appealing to the law of the lever formulated at the beginning of his text. Therefore, the longer the timber is, the weaker it is. The fulcrum is where the hand is placed, and this is toward one extremity of the timber. As this text clearly shows, although Aristotle was comparing solids of the same matter and shape, only their length is taken into consideration in a quantified form and not all of their dimensions, as the key problem of the machine makers required. In the *Discorsi* Galileo suggested, therefore, that his cantilever model be compared with the one emerging from the Aristotelian tradition as it was passed down and transformed in two later commentary works: by Giovanni di Guevara in 1627 (Aristotle and Guevara 1627) and by Giuseppe Biancani (1566 ca.–1624) in 1615 (Aristotle and Biancani 1615),[20] authors with whom Galileo was personally acquainted, and who transformed Aristotle's cantilever model such that it could address the practical problem of developing a model into a full-scale machine.

---

[18]Galileo's specified that once the maximum thickness of a prism, given its length, is found, "each smaller [prism] [...] will be able to resist some additions of new violence, in addition to that of the own weight." (*EN*, VIII:166). Such a statement is evidently particularly concerned with the problems of machine makers. In fact, once the maximum length for the given thickness is known, it becomes possible to assemble components by decreasing their length such that the total weight cannot cause the device to collapse.

[19]Translation based on the Greek critical edition Aristotle and Bottecchia Dehò (1982). The translation is one of the results of the workshop *Q.XVI* held at the Max Planck Institute for the History of Science in Berlin on August 2007. For more details, see Valleriani (2009a, 197–198).

[20]Biancani is cited in a note on the copy of the manuscript of the *Discorsi* destined to become a failed publication in Prague. The manuscript was sent to Giovanni Pieroni, who was supposed to print the book far away from the Roman censors. The publication of the *Discorsi*, in fact, was first attempted through the mediation of Giovanni Pieroni, the emperor's military engineer. A friend of Galileo, he never succeeded in accomplishing this task because of the various professional obstacles placed in his path, which obliged him to travel frequently. Galileo had part of his manuscript copied—the whole first Day and part of the second—and then sent it to Pieroni, adding some further notes by hand. The note concerned with Biancani's work is published in *EN*, VIII:165, n 1. The copy of the manuscript of the *Discorsi* sent to Pieroni, which was then returned to Galileo after Pieroni's failure, is now at the Biblioteca Nazionale Centrale of Florence, Banco Rari, A. 5, p. 2, n. 13. For an introduction to this manuscript, see *EN*, vol. VIII, *Avvertimento* of A. Favaro, pp. 20 ff.

Giuseppe Biancani The Jesuit Giuseppe Biancani (1566–1624) first met Galileo in Padova at some point during Galileo's stay there.[21] He was involved in two debates concerned with Galileo's work. The first was about Galileo's solution to calculate the height of the lunar "protuberances,"[22] and the second about the priority of the discovery of sunspots in 1613.[23] Despite some ambiguities, especially on the occasion of the debate of 1613, Biancani always professed himself to be Galileo's friend. According to his publications, Biancani was a representative of the scientific view as promoted by the Peripatetics after the Counter-Reformation. However, as Ugo Baldini showed (Baldini 1992, 217–250) by analyzing the work of the censors of the Company of Jesus, Biancani always tried, unsuccessfully, to provide positive evidence for the "new Galilean science," especially as concerned the astronomic system and Galileo's work on floating bodies.[24] Biancani's two main works are the *Loca mathematica* of 1615, part of which is a commentary on Aristotle's *Mechanical Questions* and, in 1620, a work on the sphere, *Sphaera mundi* (Biancani 1620).

Giovanni di Guevara Not many details about Giovanni di Guevara are known. Descendant of a noble Spanish family that had emigrated to Sicily, he was born in Naples. Later, in his capacity as General of the Congregation of the Theatine Clerics Regular Minor, he became a close collaborator of Cardinal Barberini, with whom he also spent some time in France. Pope Urban VIII also sent him as *legatum a latere* to Spain to visit King Phillip IV. At the beginning of 1627 he was appointed Bishop of Teano, where he remained until 1636, when he returned to his native Naples. In 1626 Guevara and Galileo met in Florence and in Bellosguardo,[25] Galileo's first residence after returning from Padova to Tuscany. On this occasion they engaged in discussions concerning the problem known as "Aristotle's wheel," Question 24 in his *Mechanical Questions*.[26] In 1623 Guevara was appointed to evaluate whether it was "appropriate" to publish Galileo's *Il Saggiatore* and subsequently make his recommendation to Cardinal Barberini, who ultimately gave his *placet*.[27] In 1622 Giovanni di Guevara published two books, *Horologio spirituale di Prencipi* and *De*

---

[21] G. Biancani to C. Grienberger, June 14, 1611, in *EN*, XI:126–127. This letter was sent by Grienberger to Galileo who made a copy of it by hand.

[22] Galileo to C. Grienberger, September 1, 1611, in *EN*, XI:178–203 and the letter cited in the previous note.

[23] G. Biancani to G. A. Magini, May 17, 1613, in *EN*, XI:509.

[24] Biancani intended to publish in his *Loca mathematica* a chapter on the science of floating bodies—*Brevis tractatio de iis quae moventur in aqua unde caput ultimum de caelo explicabitur*—where he substantially supported Galileo's *Discorso intorno alle cose che stanno in su l'acqua*. However, the censor Camerota prohibited the publication of this chapter (Baldini 1992; Ceglia 1997). On the controversial relationship between Biancani and the Jesuit censors within the general framework of the prohibition to teach in a way contrary to Aristotelian physics from the beginning of the seventeenth century onward, see Blackwell (1991, 148–153).

[25] G. di Guevara to Galileo, July 17, 1627 in *EN*, XIII:368–369.

[26] G. di Guevara to Galileo, November 15, 1627, in *EN*, XIII:377–378. For the translation of the entire letter, see pp. 268ff.

[27] M. Guiducci to Galileo, April 18, 1625, in *EN*, XIII:265–266.

interiori sensu, sending Galileo a copy of them in 1626 as a token of his friendship.[28] In late 1627 Guevara released for publication his last work, a commentary on Aristotle's *Mechanical Questions*,[29] and urgently requested that Galileo summarize their discussion about Question 24 Galileo answered quickly, in January 1628,[30] providing some general considerations and promising to send more details. Although he never did so, Guevara went ahead with printing and in April 1629 sent Galileo two copies of his work, which fails to cite Galileo at all in its lengthy commentary on Question 24 (Aristotle and Guevara 1627, 205–224). Galileo cited and criticized Guevara's commentary in his *Discorsi*, however, though not as it concerns Question 24 but rather number 16, namely Aristotle's study on the behavior of the cantilever.

Biancani's cantilever model Giuseppe Biancani is quoted in the manuscript of the *Discorsi* in a note at the margin of Proposition VII. The note explicitly quotes Biancani's *Loca mathematica* and the page: "car. 177,"[31] where Question 16 is discussed. Biancani considered two timbers, one shorter than the other (Fig. 4.2). After having quoted Aristotle, according to whom the longer should bend more than the shorter, Biancani wrote that this happens:

> because in reference to the larger [timber], the further the weight of the same timber, which is in A, is away from the fulcrum B, the more it pushes downward, in comparison to the smaller timber.[32]

In accordance with the law of the lever and considering the weight of the timber as placed on the extremity opposite to the one where the fulcrum is, Biancani concluded correctly that, the longer the timber is, the more it should bend. This is correct and still in agreement with both Aristotle and Galileo. However, the question is whether the resistance to bend or to fracture remains the same by proportionally increasing the dimensions of the timber. In other words, the height and the depth

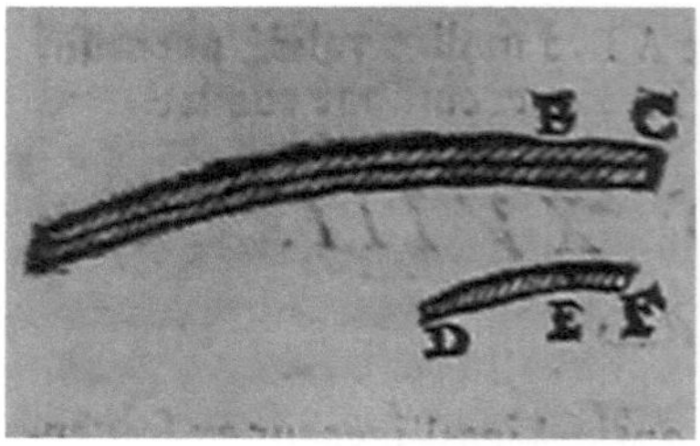

Fig. 4.2 Illustration of Aristotle's cantilever model in Giuseppe Biancani's commentary work of 1615 (Aristotle and Biancani 1615, 177)

[28] G. di Guevara to Galileo, November 21, 1626, in *EN*, XIII:341–342.

[29] G. di Guevara to Galileo, November 15, 1627, in *EN*, XIII:377–378. For the translation of the entire letter, see pp. 268ff. Guevara's commentary work is Aristotle and Guevara (1627).

[30] G. di Guevara to Galileo, January 24, 1628, in *EN*, XIII:389–390.

[31] Biancani's discussion about Question 16 starts at car. 176.

[32] "[...] quia in maiori onus ipsius ligni, quod circa A, deorsum premit lo[n]gius distat ab hypomoclio B, quàm in minori ligno" (Aristotle and Biancani 1615, 177).

of the timber had to be taken into account mathematically as well. Biancani, whose commentary is characterized by a certain brevity and schematism, could not avoid addressing this topic so relevant for all of the machine makers of his day:

> [...] I believe that, if between the length of the larger timber and its thickness there were the same proportion as between the length of the smaller timber and its thickness, so that it would be divided by the fulcrum with the same relation, they would then bend in the same way, since the weights would have the same relation to the distances to the fulcrum [...].[33]

In the same words as Biancani, Simplicio introduces the reasoning which leads to Proposition VII in Galileo's *Discorsi* (*EN*, VIII:164).

Di Guevara's cantilever model Guevara took into consideration two objects, a long lance or spear and a short branch of weaker matter (Fig. 4.3). Initially he neglected to compare the two objects as concerns the materials of which they are constituted, stating that the spear will bend more than the branch:

> [...] because the weight located in B is at a distance from the fulcrum C greater than that between E and the same [fulcrum] F; and since it weighs more, it bends downward, gradually starting from the straight line, on which it was when it stood or lay at the bottom.[34]

Guevara simply applied the law of the lever, thus obtaining the first result that the longer the timber is, the more it bends. But he introduced the relevant difference of considering two objects of different materials. In fact, Guevara had a very peculiar theoretical explanation in mind for the "bending." He considered the solid bodies to be constituted of particles and then tried to accommodate this view with the principle that "nature acts going through all degrees." Therefore, a lance can bend only gradually and this means that the particles constituting its lower part,

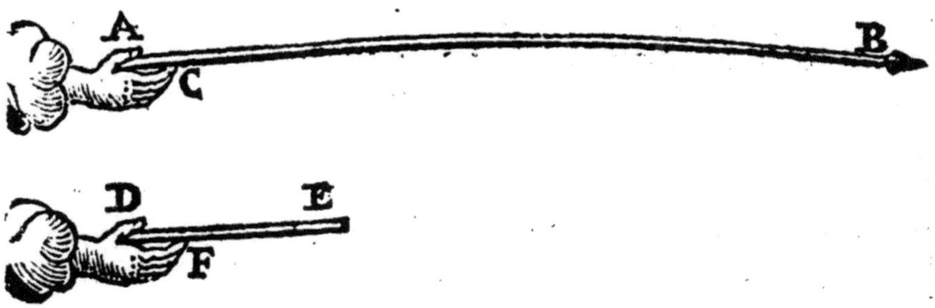

Fig. 4.3 Illustration of Aristotle's cantilever model in Giovanni di Guevara's commentary work of 1627 (Aristotle and Guevara 1627, 164)

---

[33] "[...] existimo, quod si maioris ligni longitudo ad eiusdem crassitiem haberet ea[n]dem proportionem, quàm minoris longitudo ad eiusdem crassitiem, sicq[ue] vtrumq[ue] esset ab hypomoclio in eadem ratione diuisum, fore, vt vtrunq[ue] eodem modo inflecteretur, quia haberent pondera eandem rationem ad distantias ab hypomoclio [...]" Aristotle and Biancani (1615, 177).

[34] "[...] pondus tamen constitutus in B magis distat à fulcimento C, quàm quod co[n]stituitur in E ab ipso F; magisq[ue] propterea grauitat, & inclinat deorsum, paulatim recedendo à rectitudine, quam stans, vel in solo iacens habebat" (Aristotle and Guevara 1627, 164).

the concave one, must be able to condense (*constipatio/condensatio*), whereas the particles constituting the upper part, the convex one, to rarefy (*laxatio/rarefactio*). Finally, according to the law of the lever, the longer timber bends more because of its weight; in particular, it bends more than a branch because, although the material constituting the branch is weaker than that of a spear, it is also lighter. But this means that a short but thick object, because its thickness makes it heavy, should also bend, which is obviously wrong. In order to cover this case, Guevara introduced the theory of the particles, according to which:

> If the shortness of the timber is compensated by a great thickness, that same thickness on the other hand, on account of the greater number of particles, some of which must condense and some others rarefy, will hinder when that bending takes place.[35]

Guevara's theory is complete. He first introduced a precarious example in which two objects constituted of different materials are taken into consideration. By explaining their behavior only according to the law of the lever, he ran the risk of reaching the incorrect conclusion that short but very thick objects should bend. To avoid this conclusion he introduced an almost "atomistic" view of the solid bodies, obtaining the important result that the thickness of the object could be determined as resistance against bending or fracture. In this sense also Guevara transformed the Aristotelian cantilever model in order to take into account in a quantified way the other dimensions of solid bodies, so to approach the key problem of the machine makers (Valleriani 2009a, 193–196).

The question of the machine makers, in fact, which involved two objects that were constituted of the same material, different in dimensions and similar in proportions, was pressing to Guevara as well. Conscious that his model could not determine the resistance of the timbers elevated parallel to the horizon, he introduced an external assumption: by maintaining the proportions, the resistance to bend or to fracture should remain the same. Contrary to Biancani, however, Guevara also watched craftsmen at their work, observing that:

> [...] one does not sufficiently see the correspondence that longer and shorter bend more or less equally easily.[36]

Given his assumption and the right observations, Guevara concluded his comment by searching for a possible explanation for this "observational gap." He added:

> Probably it can be said that one observes first that there are dispositions of the matter, so that in itself it is heavier or lighter, denser or more rarefied, stronger or weaker. Thus on these [dispositions] frequently it depends that some bodies acquire more facility toward the

[35]"[...] si brevitas ligni compensetur magna crasssitiei, obstabit ex alio capite ipsamet eadem crasssities propter maiorem multitudinem partium, quarum aliæ constipari, aliae autem laxari debent cum sit ipsa inflexio" (Aristotle and Guevara 1627, 165).

[36]"[...] æquè facilè inclinetur magnum, ac paruum, seu longum, ac breue, non satis videtur constare" (Aristotle and Guevara 1627, 165).

inclination from the greater length than difficulty from the greater thickness. [And] others, indeed, behave in the opposite way.[37]

Guevara had a mixed theory which could have been applied similarly to his assumption that, by maintaining the proportions, two bodies different in dimensions show the same resistance to bend, and to its opposite. He chose the first assumption, and concluded, in order to explain contradictory observations, that there are so many characteristics of matter that its behavior cannot actually be determined. Accordingly:

[...] the proportion, that makes easier or more difficult the bending with one sort of wood, does not have the same effect with another sort [of material] or with the same sort of wood, and lead iron or steel. The reason why nobody can determine anything without physical proof is thus due to the disposition of the matter and the different proportions, which, in various ways, lead to greater or smaller magnitudes of the bodies.[38]

Galileo recognized in the *Discorsi* that Guevara had accomplished an important step toward the comprehension of the cantilever model, and probably was referring to the fact that Guevara had been able to determine the two factors relevant for describing its behavior, namely longitudinal dimensions and thickness. However, Guevara proceeded by making an assumption that contradicted his observations, offering the indeterminateness of the nature of matter as an explanation for this gap.

In the *Discorsi*, Salviati, Galileo's spokeman, admitted to Simplicio that he had shared this conviction with Guevara and Biancani for a given time until, after certain and very different observations, he began to believe that it was a mistake. Therefore the young Galileo, Biancani and Guevara, and most of the architects and engineers all agreed in considering two objects of the same matter and with the same shape, and with different but proportional dimensions, to be equally resistant.[39] As will be shown, Galileo achieved his new theory thanks to the practical knowledge the shipwrights of the Venetian Arsenal shared with him, but it is on the basis of Aristotle's *Mechanical Questions* that Galileo was able to spend such a fruitful period at the Arsenal. Galileo went to the Arsenal primarily to research what were known as the Aristotelian Nautical Questions.

---

[37]"Probabiliter tamen dici potest, spectandum primò esse qualitatem, ac dispositionem materiæ, vt si grauior, aut leuior; densior, aut rarior; fortior, aut imbecillior in se sit. Nam frequenter ex ijs pendet, vt nonnulla corpora plus facilitatis ad se inclinandum acquirant ex maiori longitudine, quàm difficultatis ex maiori crassitie: Alia verò contra" (Aristotle and Guevara 1627, 165–166).

[38]"[...] proportio, quæ auget facilitatem, aut difficultatem inflexionis in vna specie ligni, non auget in alia sicut non æquè in ligno, ac ferro plumbo, aut calibe. Quare nihil determinari potest quo ad hoc nisi perspecta, vt diximus dispositione materiæ, variaq[ue] proportione, quæ diuersimodè iuxta maiorem, aut minorem corporum magnitudinem operatur" (Aristotle and Guevara 1627, 166).

[39]Many professionals considered this conception to be an original Aristotelian one, although the original argument of Question 16 considers only the dimension of the length of the solid body. This is the consequence of the use made by early modern commentators of ancient scientific texts as a basic theoretical structure for the generation of new knowledge during the Renaissance (Valleriani 2009a).

Galileo at the Arsenal: The Aristotelian Nautical Questions

When Galileo began visiting the Arsenal, he probably did it with the precise aim of using his visits to further investigate a certain group of Aristotelian Questions concerning shipbuilding and navigation. These are the Questions 4,[40] 5, 6, and 7.[41]

The textual evidence of Galileo's research at the Arsenal on the basis of Aristotle's Nautical Questions consists of mere fragments, later collected by Galileo's pupil Vincenzo Viviani and published by Favaro in the *Edizione Nazionale* (*EN*, VIII:609–610). Unfortunately, it is unclear exactly when Galileo wrote these fragments, and no watermark analysis is yet available.[42] These fragments are grouped under the title *Nell'arte navigatoria* (*On the Art of Navigation*). They primarily concern the functioning of several components of ships.

Question 5 Aristotle's Question 5 on the functioning of the rudder is the following:

> 5. Why does the rudder, which is small and at the end of the vessel, have such great power that it is able to move the huge mass of the ship, though it is moved by a smaller tiller and by the strength of but one man, and then without violent exertion? Is it because the rudder is a bar, and the helmsman works a lever? (Aristotle and Hett 1980, 355)

---

[40]Question 4 will be discussed last.

[41]Galileo's manuscript *Delle macchine*, which was introduced in the previous chapter, begins with an introduction that traces the *Mechanical Questions* in such a way to dispel any doubts about Galileo's familiarity with the Aristotelian text since 1593. In 1598, moreover, Galileo held a public course on that text (*EN*, XIX:120). The first reference to this text in one of his printed publications dates back to 1612, in the *Discorso intorno alle cose che stanno in su l'acqua* (*EN*, IV:57–140). Furthermore, throughout Galileo's correspondence it is evident that Galileo's confrontation with Aristotelian mechanics certainly continued until at least 1638, when he informed Elia Diodati, the Medici Ambassador in Paris, that he would like to write a book of *Problemi spezzati* (*Broken Problems*) in the wake of Aristotle's *Mechanical Questions* and *De incessu animalium*. For more details, see Galileo to Elia Diodati, January 23, 1638, in *EN*, XVII:262. Galileo's unpublished treatise constituted of "interrupted problems" was in fact begun, but remained incomplete. It consists of twelve problems voiced in the form of indirect questions. Their answers were written by Galileo's son, Vincenzo Galilei, partly under his father's guidance. Galileo and Vincenzo Galilei's treatise of *Problemi spezzati* is published in *EN*, VIII:598–607. Galileo's intention to write such a treatise is also revealed in the following letters: Galileo to M. Bernegger, July 15, 1636, in *EN*, XVI:450–452, especially p. 452, where Galileo called them *Problemi naturali e matematici*, Galileo to E. Diodati, November 7, 1637, in *EN*, XVII:213, where they are called *Problemi spezzati, fisici e matematici*, Galileo to G. B. Baliani, January 7, 1639, in *EN*, XVIII:10–13, especially p. 13, where they are called *Problemi e questioni spezzate*.

[42]Galileo's fragments related to nautical issues represent a small group of fragments among all of those collected by Viviani. In general, most of the fragments take the form of questions about either observations made by Galileo himself or topics suggested by other authors. The topics of the fragments are quite diverse, and most of them seem to be memoranda for further research or problems whose solutions Galileo included or wanted to include in his writings.

Galileo's fragment on the same topic reminded him of research to be accomplished to answer the following question:

> Which is the use of the rudder and how, with it, one turns the vessel with so much ease. (*EN*, VIII:609)

Galileo's question evidently follows along with that of Aristotle, showing that he effectively made use of Aristotle's text. Unfortunately it is not possible to compare the answers the two authors gave for this question because Galileo never published a single word on the topic, although he did write a whole *Discorso sul timone* (*Dialog on the rudder*) which is now lost. Today we have no choice but to believe the words of Galileo's friend Giovanni Ciampoli (1589–1643), who read the text in Rome in 1625, declaring it to be a "very noble dialog."[43]

A letter written by Niccolò Aggiunti (1600–1635) in 1634 offers further testimony that Galileo was considered as an expert on the functioning of the rudder.[44] His former pupils Niccolò and Ludovico Aggiunti proposed to him the following problem:

> How can one make a boat go from one side of a river with a very rapid current to the other without moving anything but the rudder of said boat?[45]

For the problem (Fig. 4.4, left), which was presented by Ludovico, Niccolò suggested that the only solution would be to pass a rope through a ring placed at the

Fig. 4.4 *Left*: Illustration of the problem formulated by Niccolò Aggiunti and sent to Galileo in 1634 (*EN*, XVI:50). *Right*: Illustration of the same problem, but formulated by Bernardino Baldi and published posthumously in 1621 (Baldi 1621, 48)

---

[43] For evidence that Galileo wrote a dialog on the rudder and that Ciampoli possessed it, see G. Ciampoli to Galileo, February 15, 1625, in *EN*, XIII:254 and G. Ciampoli to Galileo, December 28, 1625(4), in *EN*, XIII:295.

[44] N. Aggiunti to Galileo, February 22, 1634, in *EN*, XVI:49–50. For the translation of the entire letter, see pp. 274ff.

[45] From N. Aggiunti to Galileo, February 22, 1634, in *EN*, XVI:49–50. For the translation of the entire letter, see pp. 274ff.

boat's bow, and fix that rope on the two banks of the river. Unfortunately Galileo never responded to this problem because of the many obligations he had to fulfill due to his troubles with Roman censors during this period, but even this scant evidence clearly shows that Galileo took this topic under consideration seriously enough to dedicate a *Discorso* to it, and thus also to make himself known as an expert concerning the functioning of the rudder. The problem suggested to Galileo by the Aggiunti brothers is also interesting because it appears in the same formulation in the commentary on Aristotle's *Mechanical Questions* that was written by Bernardino Baldi (1553–1617) and published posthumously in 1621 (Fig. 4.4, right).[46]

Question 6 Aristotle's Question 6 is the following:

> 6. Why is it that the higher the yard arm, the faster the ship travels with the same sail and the same wind? Is it because the mast acts as a lever with its base in which it is fixed as a fulcrum? (Aristotle and Hett 1980, 361)

Aristotle considered the system constituted of vessel, mast and sail placed at its top to be a lever. As the bottom of the mast was conceived as the fulcrum in this system, a higher sail meant that the force of the wind was applied further away from the fulcrum, such that the same wind produced more movement of the vessel on the sea. Following Aristotle, Galileo first asked himself:

> Whether it is true what Aristotle says, that is, that the higher the sail, the stronger it pushes the vessel; and whether this happens because of the reason adduced by him, and taken from the lever [,] (*EN*, VIII:609)

and then formulated the problem in a more specific way:

> Which is the use of the very small sail placed over the box [along the mast] of the ship. (*EN*, VIII:613)

Finally Galileo found the solutions to both his questions:

> Sail, though small, placed very high, helps to sustain the ship when it goes inclined [,] (*EN*, VIII:611)

and:

> How childishly wrong is Aristotle when he assigns the reason why the sail placed higher pushes the vessel more. (*EN*, VIII:611)

Galileo first investigated (by observing practical experts in shipbuilding, like the mastmakers and shipwrights of the Arsenal, and by posing questions on navigation to figures such as Venetian admirals and coxswains) the use of the upper sail, discovering that its role is decisive when the ship has to navigate "against" the wind, when a particular use of rudder and sails causes the ship to continue moving at an inclined angle. Finally it becomes clear that what Aristotle said, in the words of

[46]There is no evidence as to whether Galileo was familiar with this and other works of B. Baldi. The fact that N. Aggiunti did not quote the origins of the problem sent to Galileo seems to suggest that this question was circulating without any specific paternity. For an extensive analysis of Baldi's commentary on Aristotle's *Mechanical Questions*, see Becchi (2004).

Giovanni Battista Benedetti, another early commentator of Aristotle, "*verum non est*," (Benedetti 1585, 155) because:

> [...] the higher the sail that is struck by the force of the wind, the more the ship's prow is submerged in the water[47]

and therefore, the more the wind blows, the slower the vessel would move.

Although the analysis concerning Question 6 is also supported by a few documents only, it does show that Galileo must have been in contact with professionals of the art in order to understand the use of the upper sail. Its use for a peculiar navigatory method, however, was known by Aristotle as well and therefore by the readers of his work.

Question 7 Aristotle's seventh question, in fact, asks:

> 7. Why is it that, when the wind is unfavorable and they wish to run before it, they reef the sail in the direction of the helmsman, and slacken the part of the sheet toward the bows? Is it because the rudder cannot act against the wind when it is stormy, but can when the wind is slight and so they shorten sail? (Aristotle and Hett 1980, 361)

The art of navigation had certainly changed over the centuries, and this especially because ships were built differently. For example, in Aristotle's day the famous Greek *trireme* had an external rudder on one side,[48] whereas the Venetian galleys of the sixteenth century had the rudder placed at the back of the ship and in the middle. Galileo therefore, interested in the special method of sailing the Venetians called *a orza* (windward), felt the need to reformulate the questions in a more articulate way. Four other fragments, which Galileo presumably tried to forward to the masters of the Venetian Arsenal, each address detailed questions. The questions, on the topic of *Del navigare a orza* (*Navigating hauling to the windward*), are the following:

> How can one navigate with the same wind in different directions. (*EN*, VIII:609)
>
> If it is possible to move against the wind, or at least to keep oneself in one place without being pushed back, and how. (*EN*, VIII:609)
>
> How by navigating to the windward one can hold the ship straight toward the place where one wishes to arrive (*EN*, VIII:609).
>
> With which artifice one navigates almost diametrically against the wind, moving by staying on the sides[49] (*EN*, VIII:611).

Galileo's research about the motion of the ship focused on its propulsion: he started with the rudder, and then considered the wind as a propulsive force, taking

---

[47] "[...] quanto altius est velum, vi venti impulsum, tanto magis proram ipsius navis in aquam demergit" (Benedetti 1585, 155). Unfortunately, in this case, too, no other comments by Galileo can support this analysis. However, because of the obviousness of Aristotle's mistake, and since all of the early modern commentators on this question accord with this critique, it is supposed that what Galileo called the "childish mistake" corresponds to what was universally accepted at his time and expressed by Benedetti.

[48] For more details on ancient shipbuilding, see Sherwood (1997). For an analysis of early modern commentaries on Aristotle's Question 7, see Rank (1984, 41–46).

[49] Navigation *a orza*, because of the inclination of the ship, obliged a great part of the crew to be stationed on the opposite side in order to counterbalance the hull on the sea.

this model as a point of departure to investigate methods of sailing. Finally, to complete the context, he considered the oars. In this case, too, Galileo first followed Aristotle.

Question 4 Aristotle's fourth question deals with the functioning of the oar as propulsive device of ships:

> 4. Why do the rowers in the middle of the ship contribute most to its movement? Is it because the oar acts like a bar? For the thole-pin is the fulcrum (for it is fixed), and the sea is the weight, which the oar presses; the sailor is the force which moves the bar. In proportion as the moving force is further away from the fulcrum, so it always moves the weight more; for the circle described from the centre is greater, and the thole-pin, which is the fulcrum, is the centre. The largest part of the oar is within in the centre of the ship. For the ship is broadest at this point, so that it is possible for the greater part of the oar to be within the sides of the ship on either side. Therefore the movement of the ship is caused, because the end of the oar which is within the ship travels forward when the oar is supported against the sea, and the ship being fastened to the thole-pin travels forward in the same direction as the end of the oar. The ship must be thrust forward most at the point at which the oar displaces most sea, where the distance between the handle and the thole-pin is greatest. This is the reason why those in the middle of the ship contribute most to the movement of the ship; for that part of the oar which stretches inside from the thole-pin is greatest in the middle of the ship (Aristotle and Hett 1980, 355).

Aristotle considered the oar as a first-degree lever. The sea is the weight, whereas the fulcrum is the thole pin, on which the oar rotates. Because of the form of the ship, which is broader in its middle, Aristotle concluded that the internal part of the oar placed in the middle of the ship was longer than that of the oars toward stern and bow. Thus, according to the law of the lever, the oars in the middle of the ship should cause a greater effect as regards the propulsive force. However, although this explanation seems quite acceptable at first glance, it conceals some important difficulties.

A great debate arose around this topic during the second half of the sixteenth century, which clearly showed Aristotle's mistake. The solution was first provided by the famous Portuguese cartographer Pedro Nuñes, who published it initially in Basel in 1566 and then in Coimbra in 1573. When Galileo arrived in Padova, however, he was not familiar with Nuñes books. Thus he began, armed with his copy of the *Mechanical Questions*, by writing notes for future investigations.[50] Galileo wrote:

> And if it is true that those who row in the middle of the galley, row more than the others at stern or at bow, also for the reason of the lever (*EN*, VIII:609).
>
> Why are the benches of the galleys placed at oblique angles (*EN*, VIII:613).
>
> On the operations of the oars, and how not all of the force of the oarsmen is employed in pulling the oar, while the ship moves (*EN*, VIII:613).

[50]Nuñes work was known by Galileo in 1615 at the latest, when Giuseppe Biancani published his commentary on Aristotle's *Mechanical Questions*, reporting on Nuñes' solution to Question 4 in its entirety. For more details about the reception of this work by Nuñes, see the introduction to the reprint of Nuñes' work written and edited by Henrique de Sousa Leitão (Leitão 2000).

> The force which moves is employed completely only when it is applied to a mobile at rest; but when it [the mobile] has already received the motion, then only the excess of the moving virtue is that which works. Because of this it happens that while a coach is at rest, the horses need greater effort to move it than they do to preserve its motion (*EN*, VIII:613).

The first fragment cites Aristotle directly and amounts to an exact translation of the Stagirite scholar's question. The second shows that Galileo was already observing the rowing units[51] of the galleys, which effectively were constituted of oblique benches. The third coincides with the main point of Nuñes' argument to show Aristotle's mistake. As Nuñes and Galileo clearly stated, an oar is not a simple lever, as Aristotle considered it, but a lever which moves together with the ship. The fourth, finally, shows what became a typical distinction made by many Aristotelian commentators during the Renaissance, that is, the distinction between employing the force when the ship is at rest and when the ship is already in motion.[52]

Ship in motion and at rest The distinction between the last two cases was also typically related to a discussion concerning a natural phenomenon that apparently occurred often on wooden vessels: the formation of barriers consisting of a peculiar type of shellfish, in such a way that the motion of the ship was hindered. This shellfish was always designated by the same word in both Latin and Italian, namely *remora*.[53] Galileo was eventually considered as an expert not only on ship motion and stability, but also on shellfish and their effects on the motion of ships. In 1621, Galileo's friend Giulio Cesare Lagalla (1571–1624), professor of philosophy at the first chair of the *Collegio romano* in Rome, wrote Galileo the following:

> I am writing some pamphlets on philosophy and among them *De simpatia et antipathia*. And I need to think about the remora that hinders the ship in its movements. I try to reduce the cause of this effect, not to an occult cause, but to the obstacle that it could present to the ship, since the ship is in equilibrium within a liquid element where the smallest hindrance can cause the greatest effect, as we can see in the steelyard how each small difference of weight along the line lifts up a great quantity and greatly varies the motion in the center. And this can easily happen with the remora, in part because of the slowness of its fluid, by means of which it adheres so strongly to the keel or to the rudder of the ships, since it is a kind of conch or sea-snail, as Plinius says, half a foot in size, and because it has the fins of the conch projected outside and scattered in such a way that it seems to have feet, as Aristotle says, one can assume that it can cause hindrance to the motion of the ships in the water, the more because Plinius accredits the same effect to every kind of conch. Therefore, before I write this thought of mine, I wanted to kindly ask you for your opinion. So please

---

[51] A rowing unit of a Venetian galley was constituted of bench, oar, thole pin (and therefore protection). A large galley had from thirty-two to forty-six rowing units.

[52] The distinction between moving what is already in motion and moving what is stationary is also the main subject of Aristotle's Question 31 (Aristotle and Hett 1980, 405–406).

[53] The *remora* is a shellfish considered to have special influence on both the motion of ships and on pregnant women. Among the many ancient sources that mention this shellfish, the most relevant are Aristotle, *De Hist. Anim.*, Lib. 2, Ca. 14 and Plinius, *De Hist. Nat.*, Lib. 9, Cap. 25 and Lib. 32, Cap. 1. In modern Italian the word is still in use: a person who has *remora* is one who is indecisive or hesitant, like a ship which is being rowed but does not move.

> do me the favor of considering it and see whether one can determine it with mathematical reasons. If Your Lordship approves it, I will write it based on your authority.[54]

Unfortunately no answer from Galileo has survived, although there is good circumstantial evidence indicating that such an answer did exist.[55] Lagalla's letter testifies that Galileo was also considered an expert on the way a ship moved, how the oars worked, and on the stability of vessels.

Did the Venetian Arsenal Employ Galileo?

Galileo went to the Arsenal to address questions to shipbuilders. At least some of these questions were derived directly from the *Mechanical Questions* of Aristotle, as if they were a sort of schematic plot to guide his observation of the work of the *Proti*, the shipwrights of the Arsenal. This sort of Aristotelian apprenticeship at the Arsenal eventually made him a scientific authority among the members of this institution's executive body. During the first years of Galileo's stay in Padova, moreover, he befriended Giacomo Contarini, the Commissioner of the Arsenal.[56] Through the mediation of Pinelli, finally, Galileo was asked by Contarini to assist in resolving a specific problem concerning the rowing units of the galleys.

Galileo's involvement in the Arsenal Contarini's question was probably stated in the following terms: does it cause a difference in reference to the propulsive force performed, whether the support of the oars, which is assembled together with the side protections of the ship, is located inside or outside the live part of the vessel? Above deck a Venetian galley constituted of a live part and a dead part. The live

---

54 From G. C. Lagalla to Galileo, July 30, 1621, in *EN*, XIII:72–73. Author's italics. For the translation of the entire letter, see p. 263.

55 Although no evidence directly shows that Galileo's answer really existed, there are several indications suggesting that it did: first, because the correspondence between Galileo and Lagalla is quite copious; second, because the books by Lagalla that Galileo possessed and which are now preserved among the Galilean inheritance at the Biblioteca Nazionale Centrale of Florence are richly annoted in their margins, testifying that Galileo occupied himself with them; third, because Lagalla was, through his deep scholasticism, very closely acquainted with many members of the *Accademia dei Lincei*, as was Galileo; and fourth, because Galileo himself intended to support Lagalla for the chair of philosophy of the University of Pisa after Papazzoni's death (1614). Although Lagalla held such an important position as the first chair for philosophy at the *Collegio romano* for thirty years, and although it seems that as a physician he had an almost revolutionary approach to surgery, only two biographical papers on him can be found (Gallo 1986 and 1987). Many commissioned researches at the Barberini Collection in the Vatican Library and on the collection of his main pupil, Leone Allacci (1586–1669), at the Biblioteca Vallicelliana in Rome, failed to uncover such an *opuscolum* in response to Lagalla's request for an opinion, which presumably contains Galileo's view of the functioning of the oars and a discussion about the analogy of the movements of the ship to a steelyard.

56 Galileo and Giacomo Contarini first met thanks to the cultural circle around the patron G. Vincenzo Pinelli, resident in Padova, who helped Galileo obtain his chair in Padova. Contarini may have been aware of Galileo's geometrical talent since 1589, when it was first attempted to obtain that chair for Galileo. For more details, see B. Zorzi to B. Valori, December 2, 1589, in *EN*, X:42.

part was located on the hull of the ship, while the dead one was the above-deck enlargement obtained by the construction of wings, or superstructures, on the sides of the ship. The protections were those handrails surrounding the above-deck area, within which the thole pins were assembled. Contarini, facing the technical possibility of enlarging the above-deck area of the galleys, and presumably inspired by Aristotle's fourth question, was probably asking himself and Galileo whether increasing the distance between the middle of the ship and the thole pins (the supports), would have changed those relations governing the propulsion power of the galley (the force-resistance ratios).

The emergence of Galileo's cantilever model Thanks to Galileo's written answer, the problem can be regarded in technical detail. On March 22, 1593, Galileo wrote to Contarini:

> Concerning the need to apply more or less force in propelling the vessel forward, it does not make any difference if the oar lies on the live or dead part of the deck, since all other circumstances are the same. And the reason is that, since the oar is practically a lever, as long as force, support and resistance divide it with the same proportion, it will operate with the same vigor, and this is a universal and invariable proposition. And I do not believe that making the wings in the galley will achieve anything but the ease of having more space for the soldiers and convicts, who otherwise could not be seated in rows of four or five per oar, especially toward stern and bow, if there were no wings. But if they could sit and row both in one way and in the other way, I do not necessarily believe that placing the protection inside or outside the live part of the galley would make any difference if, however, the oar is divided with the same proportion. And I do not see anything that could hinder or facilitate the rowing other than placing the protection further away from or closer to the handle: the closer it is, the more one can apply force. And the reason is the following, a reason that has perhaps not been investigated by anyone else: The oar is not a simple lever like any other one, indeed, there is a great difference for the following reason. Ordinarily the lever should have a mobile force and a mobile resistance and a support at rest, but in a galley support, force and resistance move. It follows from this that *support and resistance are the same* because when the blade of the oar is placed in the water, the water becomes the support, and the protection becomes resistance. But when the oar moves the water, in this case it becomes the resistance, and the protection is the support. And since, when the support is fixed, the whole force is applied to move the resistance, if the oar is immersed so that the water becomes almost immovable, then most of the force is employed to propel the vessel. On the contrary, if the oar is immersed so that the water is moved easily by the blade, then one is not able to apply the force to move the boat. And since the greater the length of the part of the lever is toward the force, the more easily one can move the resistance, when the part of the handle is very long, the water will be moved more easily, and hence its support will be weaker and one will propel the vessel less. On the contrary, when the same part between the protection and the force is shorter, then it will be more difficult to move the water with the blade and consequently, since it is needed as support, it is more solid and one is able to propel the vessel with more force. And one concludes that, the closer the protections are to the handle, the stronger the force can be applied in propelling the vessel, as the water is not able to be moved so easily with a blade very distant from the protection by a force close to the same protection. Hence, in such a case, the water functions more as support than resistance. All of this is very evident from experience.[57]

[57] From Galileo to G. Contarini, March 22, 1593, in *EN*, X:55–57. Author's italics. For the translation of the entire letter, see pp. 214ff. This letter was evaluated for the first time in Renn and Valleriani (2001).

For both Galileo and for Aristotle the oar is a lever, but whereas it is a first-degree lever for Aristotle, Galileo considers the oar *almost* as a lever. In particular, if the blade of the oar sinks deeply into the water, then the water can be considered to be the fulcrum of the lever, because, when the weight of the water displaced by the blade is very heavy, the ship, by means of its connections to the thole pins, is moved more efficiently than the water displaced. On the contrary, if the blade of the oar does not sink deep into the water or if the handle is long, then the weight of the water displaced is not heavy enough to function as a fulcrum and therefore it becomes resistance, while the thole pin is the fulcrum, as Aristotle said. The great difference, however, is that, according to Galileo, when the thole pins are the fulcra, mostly the water should move and not the ship, whereas, according to Aristotle, the same view with the thole pins as fulcra should explain the movement of the ship and not of the water.

In closing Galileo remarked, first, that the placement of the thole pins in relation to the longitudinal center of the ship does not change the propulsion power of the galley; and second, that the deeper the blade lies in the water, i.e., the closer the oarsmen are to the protection, the more the propulsive force of the oarsmen is transformed into the motion of the ship. Galileo did not yet see the possibility that the oars could also be longer and arranged on the ship in a way that it remains possible to have blades that sink deep into the water, but his theoretical approach was different than Aristotle's, shifting toward a model according to which the fulcrum of the lever is at the extremity opposite to where the force is applied. This is the theoretical framework within which he later developed his cantilever model, after having learned some qualitative data about the robustness of the oars from the professionals at the Arsenal.

Galileo's Apprenticeship as a *Proto*

From the perspective of the master of shipwrights, Galileo was neither right nor wrong. He was simply too abstract and ignorant of further real and relevant aspects. The letter with which Contarini replied to Galileo's answer introduced him to those aspects, launching what is called here his apprenticeship as a shipwright (*Proto*).

Practical knowledge's criticism Encouraged by Galileo to open his mind to "such mechanical problems," Contarini opened his answer with the statement that, "the oars which are being used are not proportioned to the body of the vessel,"[58] and, in his opinion, if the right proportion between them—oars and the body of the ship—were found, the problems related to the agility and velocity of the vessel would be solved. In a way apparently unrelated to this statement, Contarini continued by considering human force and its application. The better way for the oarsman to apply force is to push and pull the oar by keeping the handle in front of the

[58] G. Contarini to Galileo, March 28, 1593, in *EN*, X:57–60. For the translation of the entire letter, see pp. 216ff.

breast, and by moving while holding a position parallel to the horizon. To satisfy this condition and, obviously, to be able to operate the oar in the most efficient way, which means dipping it as deeply as possible into the water, the oar must therefore be very long. But the longer the oar, the heavier it is, that is, the longer the handle must be because of the increased number of oarsmen needed on the vessel to operate the heavy oar. But the length of the handle of the oar, Contarini continued, is given by a certain ratio to the width midships above deck; and the entire length of the oar is deduced by means of a ratio to the handle.[59] Thus, keeping in mind that a certain space is needed on a ship for benches, and for a gangway between them to position or move goods, artillery and soldiers, as well as space for soldiers toward the protection behind the last oarsmen, Contarini first pointed out to Galileo that the superstructures were imperative and then, above all, reminded him that long-handled oars are particularly important "because the handle not only moves the pole of the oar, which is outside the protection, but also acts as a counterweight for the mentioned oar." And calculating the force performed by the single oarsmen,[60] Contarini reached two conclusions:

> Therefore what one says cannot happen, that the longer the handle, the easier it is to move the water. And therefore its support will be weaker and the vessel will be propelled less [...].
>
> [...] it is certain that with a short handle, one will never have force both to steer the oar and to row it.[61]

Contarini concluded by summarizing and reordering his fairly scattered list of thoughts: (1) the oar must be long; (2) the oarsmen rowing toward the middle of the ship perform two kinds of movements, upward in pushing the blade as much as possible under the water level, i.e. closer to the ship, and forward; (3) the oarsmen closer to the protection, since they perform only one movement, that is, forward, apply most of the propulsive force; (4) the superstructures are relevant not only because they allow long oars to be fitted, but especially because the long handles allow more oarsmen per bench.

Contarini's reply seems to be a list of interrupted points, not always connected with each other and, what is more, relevant to Galileo's letter only in part. In its opening Contarini wished to find the right proportion between the oars and the body of the galley in order to improve the agility and the speed of the vessel. However, he concluded by stating only that a longer handle would improve the propulsive force

---

[59]The "practical knowledge" of the shipwrights of the Venetian Arsenal was codified in part in the form of sets of ratios, one for each ship model. Given some main measures and the model of the ship, the ratios provided a method to obtain the measures of all other components of a ship. As concerns the handle of the oars, its length had to be the half of the width of the ship midships above deck.

[60]The way Contarini suggests calculating the force applied by each oarsman at a single oar is based on a comparison between the virtual circles drawn by the blade and those drawn by the points of the handle where each oarsman works.

[61]From G. Contarini to Galileo, March 28, 1593, in *EN*, X:57–60. For the translation of the entire letter, see pp. 216ff.

of the rowing unit. In this sense Contarini addressed Galileo's conclusion according to which the longer the oar is, the more easily the water is moved and therefore the less propulsive force is exerted. Contarini also reminded Galileo that a large Venetian galley is such a big machine that not even all of the oarsmen along the same oar could be considered to work in the same way. He said that if the oar is very long, and if one considers only the work of the oarsmen positioned at its extremity in the middle of the ship, they would not be able to operate the oar lever or move the water so easily, because they are busy with not only one movement but two, forward and upward. Thus Galileo was right in principle, but only if the movement were abstracted to the extreme such that all of the force performed by the different oarsmen is considered to be a unique force applied only at the extremity of the oar lever. But an oar with a short handle is neither steerable nor rowable in practice.

The official inquiry Galileo's letter and Contarini's response seem to be a nearly isolated case, especially if only Galileo's published works are considered. A situation where an early modern manager quite versed in practical activities like Giacomo Contarini, then Commissioner of the Venetian Arsenal, "disturbs" the then professor for mathematics at Padova through Pinelli's mediation, merely to request of him a written opinion on a personal issue concerning the placement of the ships' protection seems quite unlikely, however. In fact, this exchange of opinions about the placement of the oars above deck was not an isolated or causal event, but emerged from the context of an official inquiry led by one of the ruling bodies of the Venetian Arsenal.

On February 9, 1592, the *Collegio della Milizia da Mar*—Committee for the Navy—promoted an official inquiry, which ended over one year later in June 1593. The inquiry was iniated by the *Savij* ("the sages") in order to find solutions for a spectrum of problems ranging from ship design and shipbuilding issues, to related expenditures, and architectural changes to the Arsenal itself, which were related to the former issues. Anyone considered relevant to this inquiry had to answer under oath by submitting a written document, personally signed. Giacomo Contarini, who, as Commissioner of the Arsenal, was in charge of the organization of the inquiry, kept a copy of many, probably all written documents produced as a consequence of this inquiry.[62] Both Galileo's letter and Contarini's reply are still kept in this register of documents.

The points of the inquiry Contarini related the order of the Committee of the Navy in full and the points to which the selected persons had to respond. The points are:

> [1.] in which way one can remedy the lack that the large galleys have in reference to the rowing unit, so that, on occasion, they can be rowed without being pulled.
>
> [1.1.] whether one has to enlarge the superstructures of those galleys.
>
> [1.2.] which quality of oars and of which length will be necessary to use.

---

[62] The documents produced during the inquiry and collected by G. Contarini are preserved in a bundle entitled *Fabrica di galee* (Contarini 1592–1593).

> [1.2.1.] whether one has to provide those galleys with two oars per bench, or with one.
>
> [1.3.] The expenditure which could be caused by this change.
>
> [2.] The way to bring outside those galleys from the Arsenal, in which the superstructures were to be enlarged.
>
> [3.] Beside this, speaking about everything which could seem to anyone to be of public relevance.[63]

Large galleys The inquiry concerned large Venetian galleys. This kind of ship had been built for the first time at the end of the fifteenth century, but only occasionally; its construction became systematic during the second half of the sixteenth century. The typical Venetian ship built in the Arsenal before and after the advent of the large galley was the "thin galley." This was a ship normally employed for military purposes, whereas the large galley was normally built for trade. However, the increasing power of enemy fleets, especially of the Ottoman one, and the fast development of fire artillery, obliged the Venetians to equip their ships with more and more powerful artillery. More powerful, larger and heavier artillery also required a more capable, resistant and larger ship, namely one large enough to host more oarsmen than were used in thin galleys, so that it was possible to push the vessel forward with greater speed. In the face of this challenge, the Venetians began using the large galleys for military purposes as well.[64] During the famous battle of Lepanto in 1571, for example, the large galleys turned out to be the ace up the Venetians' sleeve. Deployed transversely against the huge Ottoman fleet, they discharged their fire power while the fast thin galleys attacked the single ships from the sides.

Propulsion of large galleys There was a problem with large galleys. They were not able to move by means of their own propulsion and therefore, in the absence of wind, their own rowing normally needed to be complemented by towing by one or more thin galleys. Although on the occasion of the battle of Lepanto the slowness caused by this propulsion method had no negative effects on the success of the battle, from the protocols of the inquiry it is clear that:

> [...] one can say that, on another occasion, they [the large galleys] could rather hinder than help [...] since they cannot follow a thin navy without being towed, they could be a cause of slowness and [for this reason] endless good occasions could be missed. and anyone who has only thin galleys could never accept the battle, because it is sure that, since it is convenient to use this vessel [the large one] for that enemy, he would never reach the enemy; and if

---

[63]"che modo si deve tenere per rimediare al mancamento che hanno le galee grosse nella vuoga si che si possano in occasione vuogar senza remurchio," "se si devono allargar le postizze ad esse Galie," "che qualità de remi, et di che longhezza sara necessario adoperare," "se si devono accommodar esse galee à doj remi per banco, o, à uno," "La spesa che potesse andar in detto accommodamento," "Il modo di cavar poi esse galee dall'Arsenal in caso che si dovesse allargar le postizze," "Discorrendo oltra di cio intorno a tutto quello di piu, che gli paresse poter esser di pubblico servitio" (Contarini 1592–1593, 1v). Author's enumeration.

[64]The masters of the Arsenal and its executive body, constituted of Lords and Commissioners, distinguished between large galleys for trade and those for military purposes. In detail, the galleys for trade were "rounder," that is, a little bit wider than those destined for the military fleet. However, it was no rarity for trade galleys to be armed and sent as part of the military fleet or *vice versa*.

> one has the navy constituted of thin and large galleys, it will be useful for him to tow them always, since one is sure that he cannot win only with thin galleys unless they are of the same number as the enemy ones. and in order to address this lack, it will be prudent to use this sort of ships [large galleys] upon the condition that their imperfections are removed.[65]

Unless the dimensions of the military fleet constituted of thin galleys were increased, and this was impossible for the Arsenal of Venice at the time because of the lack of material, the use of large galleys would have been unavoidable. But if the imperfections which made it necessary to tow them had not been solved, such galleys could have been more a reason for military failures than a help for victorious conclusions.

Understanding why a particular model of ship could not be pushed forward by means of its own propulsion, especially within a tradition of centuries of successful shipbuilding like that of the Venetian Arsenal, was no trivial issue. Whether it depended on general or particular design issues, or construction techniques, or choice of materials, or rowing unit construction or position, it was a very puzzling problem indeed. Seventeen years after the battle of Lepanto, however, its cause was better determined and the inquiry to find a solution initiated. As implied in point 1. of the inquiry, the necessity of towing the large galleys was supposed to be a consequence of a design problem related to the dimensions of the superstructures of the vessel, and perhaps to those of the rowing unit constructions built upon them and, in turn, with the oars—their quality, length and number per bench.

Contarini opened the inquiry by repeating the points upon which he consulted with Galileo:

> [...] one has therefore to consider the instrument which makes her [the ship] go, which are the oars, as well as the force of the man who has to use those oars. For as concerns the oar which is used at present, it is not proportioned to the ship [...], which, if she [the ship] had larger superstructures could be given more [longer] handle, and consequently a longer pole, which would find the water far away from the ship, and so the oar would move slower and the force that the man would need to apply would be natural, because it would not deviate from the path followed by pulling to the breast and one could also place more men there [...]. Thanks to this remedy, there is not a single expert who cannot understand that one would provide a solution to this hindrance of the slowness by making this ship go, if not so well as the good thin galleys, at least as well as the mediocre ones, without which [the thin galleys] one could never advance, and never offer their service without being towed, and to

[65]"[...] si puo dire che possano esser un altra volta piu tosto di impedimento che di aiuto [...] non potendo star dredo una armata sotile senza esser remurchiati, possono esser causa per questo di tardanza di far perdere infinite occasioni buone. et chi havera galee sotil solamente potria a sua voglia non accettar mai la battaglia essendo sicuro, che l'inimico che conveniva valersi di questo vassello dovendolo remurchiar non lo arivara mai; et havendosi armata composta di galee grosse et di sotili si convenira star sempre sul remurchio, essendo sicuri di non poter vincere con l'armata di galee sotili solamente che non saranno per numero come quelle dell'inimico. et per proveder a questo mancamento sara sempre prudenza valersi di questa sorte de navilij grossi quando se gli levino quelle imperfezioni" (Contarini 1592–1593, 7r).

position them at the appropriate locations thanks to their own [propulsion], and to perform really great operations. [66]

Galileo's letter is evidently especially concerned with point 1.1. In fact, one of the solutions that the inquiry suggested addressing was the possibility of enlarging the superstructures of the ships, on whose sides the thole pins of the oars were positioned. Thus, Contarini was probably wondering whether, by enlarging only these, and without changing the rowing units, the oars or the number of oarsmen, they could expect some advantages. Indeed, this would have been the cheapest solution; point 1.3. of the inquiry signifies the relevance that the amount of expenditures entailed in the suggested solutions were required to have. Unfortunately for the treasuries of the Arsenal, however, Galileo was right in at least one point. Enlarging only the superstructures while leaving the rowing unit unchanged would not have affected the amount of propulsive force applied.

Definition of the problem The masters of the shipwrights (*Proti dei Marangoni*) were charged with supervising the construction of the ships. Because of their knowledge, experience and authority, their testimonies were considered to be of supreme importance. The first master of the shipwrights, interrogated by the delegate and nobleman Mocenigo during a meeting of the Committee dedicated to the inquiry, defined the problem clearly in the following terms:

> [...] since the door of the Arsenal is narrow we are obliged to settle the measure of the superstructures not according to the mouth[67] of the galley, as one should do, if one wants to work in a good way: but because of the narrowness of the door, the superstructures in use now have a relevant problem such that one cannot use them [the galleys] as one would like to do on certain relevant occasions because of the shorter handle they have now [...].[68]

66"[...] s'hanno da considerar cosi l'instrumento che lo fa caminare, che sono i Remi, come la forza dell'huomo che ha da adoprar essi Remi. Quanto al remo che si adopra al presente non è proporzionato al navilio [...] che se havesse le postizze più larghe si daria piu ziron, et per consequenza piu asta, la qual trovaria l'acqua lontana dal navilio, et il Remo andarebbe piu piano, et la forza che bisognasse che l'huomo vi ponesse sarebbe naturale, perche non passarebbe al tirar il petto, et si potrebbono anco metter piu huomini [...]. con questo rimedio di allargar le postizze non è alcuno intendente che non conosce che si provvedera a questo impedimento della tardanza facendo caminar questo navilio, se non tanto quanto le galee sotil buone, almeno quanto le galee mediocri, senza le quali non si potra mai andar avanti, et potranno servir senza remurchio, et mettersi alli sui luoghi da se stesse, et far grandissime operazioni" (Contarini 1592–1593, 7r). In this document, which starts at folio 6v and ends at folio 7v, and from which this quotation is extracted, the author is not cited. But two reasons suggest that this text was the one proffered by Contarini at the *Collegio della Milizia da Mar* of Venice: first, because it is the only writing collected by Contarini that does not bear any name; second, because both of the deeply significant similarities between Contarini's reply to Galileo and this document, and of the presence of points like, for example, the one concerned with the movement of the oarsmen, which are not to be found in any other text submitted on the occasion of the inquiry.

67The mouth of a galley was the width of the ship measured midships above deck.

68"[...] essendo la porta dell'arsenale stretta semo di necessita di a far la mesura delle postizze non dalla bocca di essa galia come si doveria far a far ben: ma dalla strettezza della porta del Rastello, onde sono uscite dette postizze con difetto d'importanza non potendogli con così poco ziron come

Shipbuilding method Every kind of ship and boat was built according to a traditional set of ratios among the measures of their main components. Once the longitudinal length of the ship was decided upon, for instance, all other main measures, including the width, the superstructures, the handles of the oars and so on, could be calculated by simply applying the appropriate set of ratios. The situation concerning the large galleys presented problems even more serious than the ones discussed by Contarini and Galileo: Not even the traditional ratio between the mouth of the galley and the handle of the oar was respected.[69] Because of the narrow door of the Arsenal, the large galleys were equipped with superstructures that were narrower than what their appropriate and traditionally given set of ratios required. Morever, the dimensions of the mouth of the galley also determined the length of the handles of the oars. The handles were therefore shorter than prescribed by the traditional rules. The handle, in turn, was the term of comparison for the entire length of the oar: the ratio between the handle and the entire oar had to be 1:3. For this reason point 2. of the inquiry requests an opinion about a way to bring the galleys out of the Arsenal if they were to be built with those superstructures prescribed by the traditional building ratios. The master of the shipwrights agreed with Contarini. The problem of the galleys was represented by their oars, because they, that is, their handles, were too short.

The entire inquiry contains twenty-two written answers and the subsequent protocols and deliberations of the Committee. No opinion expressed in those delivered documents is in disagreement with the master of the shipwrights. All of them wanted the large galleys to be equipped with wider superstructures to make longer handles possible.[70]

---

portano al presente adoperarsi come bisognarebbe nelle occasioni d'importanza [...]" (Contarini 1592–1593, 2v).

[69]Contarini's last letter indeed already contains the suspicion that the propulsion problem of large galleys depended on an erroneous ratio between the mouth of the ship and the length of the handles of the oar. On that occasion, however, Contarini did not further analyze this point.

[70]The inquiry also investigated whether it would have been better to substitute the rowing unit equipped with one oar with another equipped with two of them. Although the opinions directly concerned with this point of the inquiry will not be taken into consideration here, the following brief explanation of this issue could be helpful for the general understanding of the way the Venetian shipbuilders worked. According to the traditional sets of ratios governing shipbuilding activity in Venice during the sixteenth and seventeenth centuries, a rowing unit equipped with one oar per bench of a certain and given length could be substituted by a different rowing unit with two or more oars, whose lengths were a certain ratio of the oars designed for a rowing unit with one oar. However, since in multiple-oar rowing units other factors had to be altered, such as the height of the benches, the Venetian masters of the Arsenal did not have any rule to foresee whether the two corresponding rowing units, the first equipped with one oar and the second, for example, with three, would effectively perform the same propulsive force. This was the main reason why the few masters who faced this opportunity remained conservative, suggesting that one large galley be prepared with a rowing unit equipped with two oars so that this construction could be tested. All other documents submitted propose retaining the rowing unit with a single oar per bench. First, because it meant lower costs in terms of the carpentry, material and screws needed and, second, because of the need to leave more space on the ship free for artillery and soldiers.

The problem of the "longer oar" Enlarging the superstructures of the galleys would have meant providing the galleys with longer oars. This operation, therefore, definitely involved considerations about the resistance of the oar. Galileo, who had already constructed the theoretical framework concerned with the cantilever model while discussing with Contarini how the oars functioned, found among the masters of the Arsenal the knowledge he needed to determine the ratio between the dimensions and weight of a solid body that define its resistance to fracture.

The Committee decided to experiment by changing the superstructures on one large galley already present at the Arsenal, which required new superstructures anyway because the old ones had rotted. Since it was a large galley built for trade and, therefore, had a mouth of 24 Venetian feet, the handle of the requested oar had to be 12 feet and the whole oar 42 feet, amounting to almost 14.6 meters! Since no large galley had ever been built according to the traditional set of ratios concerning the handles of the oars, and since these ships were the largest ships ever built in the Arsenal, the problem was that the shipwrights had no experience with oars of the requested length, for the simple reason that such oars had never been built. The inquiry therefore first had to investigate whether there was material at hand that could somehow prove useful for this purpose.

The written documents that have survived testify clearly to the institution's embarrassment in the face of this seemingly less significant technical challenge. But it was far from easy to provide a fitting solution. The first reason was the chronic difficulty the Venetians had in procuring material for shipbuilding, especially due to the loss of the Balkan territories, their main source of wood, after the Ottomans became established in the Balkan regions close to the Adriatic coasts.[71] In particular, the Arsenal did not have any warehoused oarage long enough to make the desired oars. Further, with one exception, all of the persons interrogated agreed that none of the forests at the Republic's disposal could supply the requested oarage. Therefore the only practicable solution was to use oars that had already been prepared, by attaching a piece of wood to the side of the handle in order to achieve the desired length.

None of the documents produced by the inquiry mention a word about the method used to join the poles together. The masters who were asked for their opinion as to whether such joints would have been resistant enough did not seem to have any trouble with it. However, this was not the case with regard to their opinions about another problem that was a consequence of this joining together pieces of oarage, namely, that the thickness of the oars would not have been changed. In particular, since the oars stored at the Arsenal had a length of 36 Venetian feet (just over 12.5

[71] The Ottomans' conquest of the Balkan region started very early in the fifteenth century, but did not begin to affect the Venetian economy seriously until after the Venetian military defeat at Corinth in 1465. Although the conquest continued to the doors of Vienna, the peace treaty between the Ottomans and the Venetians, signed in 1479 and then ratified in 1503, ensured the Venetians some of the materials needed from the Balkan woods through commercial trading with the Ottomans. This was no longer the case toward the end of the sixteenth century, however, so that materials were provided from regions which belonged to the Republic and were located on the Italian peninsula, in particular from those regions known today as Friuli and Venezia-Giulia.

meters),[72] the joint method meant that the Arsenal masters would have to lengthen the oars by more than two meters while maintaing the same thickness. This was a serious point of contention. Contarini's register reports eight opinions *contra* and seven *pro* the joint solution because of this aspect.

On March 16, 1593, a few days before Galileo wrote, Nicolò Balbi, a delegate of the *Collegio della Milizia da Mar*, submitted his written opinion after an exploratory visit to the Arsenal, stating that:

> [...] if the superstructures are enlarged toward the outside by only one foot, one should consequently lengthen the oarage so that the [oar] 36 feet long would become 40 feet and that of 38, then 42, which would become so weak that with a little bit of exertion, or a slightly greater number of people, it would break, or the handle would come to the breast of those who row it, [and] therefore nothing good could be made, because the oarsman would have more effort in making himself free from the impetus of that handle from the breast than being applied in rowing, as is very well known to all those who are expert in this profession. Not to mention that, every storm on the sea, and every little bit of effort they had, would break the oars into [several] parts.[73]

Here Balbi took into consideration the possibility of joining a pole to the handle of the stored oars, but considered the resulting oar too weak. However, Balbi did not limit himself to a mere criticism of this solution. He also considered the possibility of sawing brand-new oars of the desired measures, could the needed wood have been found. He specified that:

> If one wants to let them saw with that length in the wooden of the Archduchy, one should let them remain so thick, and heavy, that not four men, and not even six would be able to use them, neither for long [time], nor for short [time].[74]

In conclusion, Balbi seemed to have found a true practical, or even structural, limit for shipbuilding with wood. He stated that if one wants to follow the traditional set of ratios, then large galleys should be built with decreased measures, as efficient oars 42 feet long would be so thick and thus so heavy that the number of oarsmen that would fit in a ship whose mouth was in the correct proportion to the handle of that oar would still be not able to row it.

Solution of the problem and the content of a new science Balbi's visit to the Arsenal must have been a very interesting one. It is impossible to extract more details about his investigation, like the specialization of the masters with whom

---

[72] 1 Venetian step = 1.738 m = 5 feet: 1 Venetian foot = 34.76 cm.

[73] "[...] tirando in fuora le postizze un sol piede, bisognarebbe sussequentemente slargar il palamento a tale che quello di piedi 36 verrebbe ad esser di piedi 40, et quello di 38 di 42, li quali [...], s'indeboliria si fattamente che ogni poco di sforzo, o di maggior numero de genti si scavazzaria, overo ch'el zirone veniria al petto di cui lo vogassero, per il che non potrebbe mai far cosa buona, et sendo che maggior sarebbe la fatica del galiotto nel liberarsi dall'impeto di esso ziron dal petto che non sarebbe quella del vogar, si come è benissimo noto a tutti quelli che intendono quella professione. Lascio da parte, che ogni borasca da mare, et ogni poco di straccolo che havessero andarebbero tutti in pezzi" (Contarini 1592–1593, 9v).

[74] "Volendone mo far tagliar di quella longhezza nelli boschi de Arciducali, bisognaria farli tenir si fattamente grossi, et pesanti, che non solo 4 huomini, me ne anco sei potranno, ne per molto, ne per poco adoperarli" (Contarini 1592–1593, 9v).

he spoke and what they showed him. However, a good idea of how Balbi's visit took place can be pieced together on the basis of another testimonial collected in Contarini's register. This evidence is particularly relevant not only because of its content, but also because of the person who has given it: the master of the oar makers (*Proto de Remeri*), in charge of the production of oars at the Arsenal, and therefore the most representative voice from the front of the practical knowledge concerning oars.

Upon interrogation in March 1593, the *Proto de Remeri* Christoforo de Zorzi gave a precise report about the measures and numbers of the oars currently warehoused, clearly pointing out not only that there were no oars 42 feet long, but, in particular, that if the joint solution had been chosen, the oarage at disposal would have sufficed for very few galleys. But then, in the written document submitted, in consideration of the joint method to obtain oars 42 feet long, he first added that:

> [. . .] if one makes the oarage greater it will happen that it will be longer, and as the Oar will be greater it will have no force, because it will be tender, and will make it impossible for the galley man to row, because the handle will hit him on the breast; [. . .].[75]

Neither did the master in charge of oar production like the solution represented by the joint method. Instead of taking into consideration the eventuality of finding the right wood for longer oars, however, he made some suggestions about the lengths which can be obtained by using the joint method with the oarage already at hand in order to avoid the problem of the weaker oar:

> [. . .] but one can well change the galleys which are there at present[:] but that which has an oar of length of 36 feet, and a handle of 12 feet can be changed so that it has an oar of 39 feet and a handle of 13, and that which has an oar of 38 feet and a handle of 12 [and] 1/2 feet can be changed so that it has an oar of 40 feet and thus a handle of 13 [and] 1/3.[76]

Mr. de Zorzi was equally unconvinced of the practicability of making 42-foot oars. However, he determined the exact length to which the warehoused oars could be extended using the joint method, and did so in a very interesting way for two different kinds of oars, 36 feet long and the other 38, and further presented two correspondent measures which, compounded with the original lengths of the oars, do not amount to any linear proportionality. Whereas the 36-foot oar could be lengthened to up to 39 feet, that is, a three-foot pole could be added to it, in reference to the other oar, 38 feet long and certainly thicker than the first, at most a 2-foot long pole could be joined. Although these numbers do not agree with the rules later given by Galileo in his *Discorsi*, they certainly show that the master of

---

[75]"[. . .] facendole maggiori il palamento convenira esser di maggior longhezza et come il Remo sara maggiore non havera forza, che sara tenero, et fara che il galiotto non possa rogare, che il ziron li dara nel petto [. . .]" (Contarini 1592–1593, 4v).

[76]"[. . .] ma ben si potra accomodar le galie che sono in esser al presente quella pero che ha il remo de longhezza de piedi 36, et il ziron sia pie 12 se potra accomodarle che habbi il remo de pie 39 che havera il ziron de pie 13, et quella che ha il remo de pie 38 che ha il ziron de pie 12 1/2 si potra accomodarla che habbi il Remo de pie 40, che havera il ziron pie 13 1/3" (Contarini 1592–1593, 4v).

the oar makers knew perfectly that by increasing the length of the oar linearly, its thickness had to increase over-proportionally—and he could even quantify how.

Soon after his arrival in Padova, and thanks to his involvement in the official inquiry promoted by the *Collegio della Milizia da Mar* of Venice, Galileo not only generated the core idea of what would become his first new science, but even encountered a practical and quantified example of how this idea works. First, Galileo revolutionized the way to apply the lever model by considering the water, that is, the extremity opposite to the one where the moving force is applied, as the fulcrum. In fact, if the problem is no longer the propulsive force performed by the oar, but the resistance of the oar to fracture, then Galileo's words contained in his letter to Contarini—"fulcrum and resistance are the same"—took on a very precise meaning related to his cantilever model: Fulcrum and resistance are placed at the same point, where the cantilever is driven into the wall. Galileo in fact used the same concept and the same word—*resistenza*—to denote both the weight to be moved and the resistance to fracture. According to this view then, Galileo rightly suggested to Contarini that the handle of the oars be shortened. But in his reply Contarini taught Galileo that the same result must be obtained with long handles. Yet this solution was doomed by the problem of lacking material and the consequent practical considerations by the masters of the Venetian Arsenal, according to whom simply lengthening the oars would have presented a problem related to their resistance. The representatives of the practical knowledge of the Arsenal, that is the masters, the members of the administrative and organizational executive body of the Arsenal and the members of the Navy Committee of Venice, finally offered Galileo centuries of experience, constituted of not only qualitative statements, but even quantitative indications, which Galileo integrated into his new cantilever model.

Galileo's Masterpiece: The Oar Model

The cantilever model was useful to Galileo in reference to the way the oars work in order to establish the circumstances under which one could achieve the most efficient propulsion. This was when the handle is very short, the force is considered to be applied only at its extremity, and the blade is immersed deeply into the water. Eventually, due to the impracticability of such a model in the context of the way the rowing unit functioned, Galileo formulated a third model in the context of his science of materials. Under certain aspects, this final model seems to be an attempt to furnish a method of solving the problem of the "longer oar" that emerged during the inquiry.

Proceeding toward such a third model, which might be called "oar model," Galileo showed that a cylinder driven into a wall, the size of which is at the limit over which the cylinder would break, can be twice as long, keeping the same thickness, when it is removed from the wall and leaned on either one support at its middle point or two supports at its extremities. And then Galileo proposed to:

> [...] find whether the same force or weight which produces fracture when applied at the middle of a cylinder, supported at both ends, will also break the cylinder when applied at some other point nearer one end than the other [...] (*EN*, VIII:173).

Considering the necessity of long handles, Galileo considered a model which more closely resembles that of the oar, although it is not identical. According to this model, the cylinder is supported by one point in between the extremities and the force is applied at its ends. Indeed, this is almost the same model that allows the analysis of the resistance of an oar supported by a thole pin. However, for this model Galileo did not take into consideration the materiality of the cylinder, as he did in the cantilever model. Galileo did not consider the weight of the cylinder or oar itself. For this reason, although this model corresponds geometrically to that of the oar, it cannot be considered as a definitive answer to the doubts of the Venetian masters. In the case of the cantilever model, too, Galileo began his considerations by first disregarding the weight of the cylinder driven into the wall, but then integrating this data into the model so as to avoid addressing only the behavior of an ideal cantilever. In the case of the oar model, however, Galileo was no longer able to integrate this data and so proceeded by investigating how an ideal lever works in the case of the oar model. In this sense, to conclude, Galileo's answer to the *Proti* concerns only an ideal oar.

Translated into the language of the master of the oar makers, Galileo investigated how the resistance to fracture changes when the thole pin is not at the middle point of the oar, but positioned toward one of the two extremities, with the thole pin placed at a distance of one third of the length, measured from the handle end. Galileo searched for a general rule first, considering the case in which the support can be moved infinitely toward an extremity. He postulated that, if the support were shifted infinitely toward an extremity, the force needed to break the whole cylinder, which must be applied at that same extremity, must also increase infinitely; therefore (Fig. 4.5):

> [...] as the fulcrum F approaches the end D, we must of necessity infinitely increase the sum of the forces applied at E and D [the two extremities] in order to balance, or overcome, the resistance at F (*EN*, VIII:175).

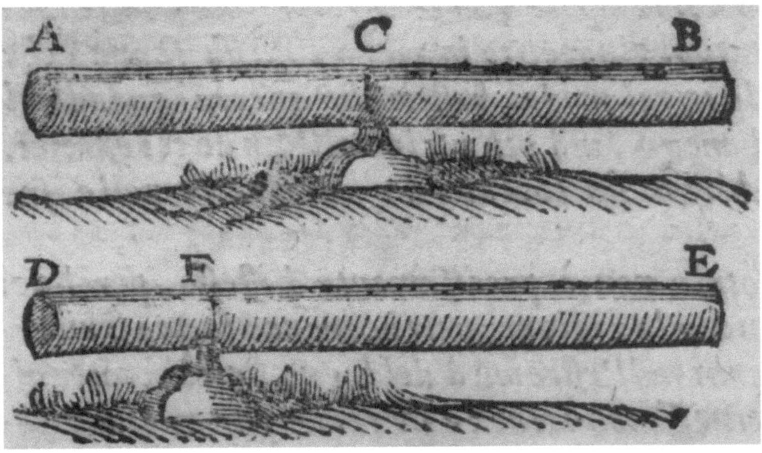

Fig. 4.5 Illustration of Galileo's oar model (Galilei 1655, 102)

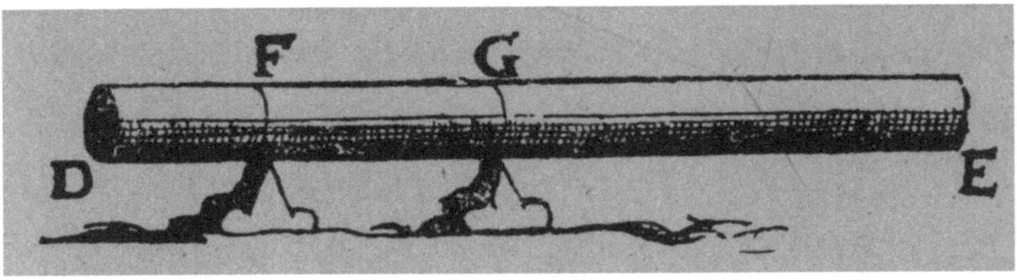

Fig. 4.6 Illustration to explain the behavior of the oar model (*EN*, VIII:176)

Since, however, the problem of the "longer oar" concerned the resistance of oars whose thole pin is placed at one precise point, in the *Discorsi* Galileo's Venetian friend Sagredo seems to ask Salviati for this specific solution:

> But it would be better if Salviati were to show us by just what proportion the forces must be increased in order to produce a fracture as the fulcrum is moved from one point to another along one and the same wooden rod (*EN*, VIII:175).

In the manuscript prepared by Galileo to try to have the *Discorsi* printed in Prague and sent to Giovanni Pieroni, Galileo added an explanation by his own hand, which ultimately was not included in the final version, although it would have improved the understanding of the text, possibly because its inclusion would have required preparation of an additional illustration (Fig. 4.6). This explanation offers the solution sought by Sagredo by using the oar model, proceeding by comparing two cylinders of the same dimensions but whose supports are displaced at two different points:

> If upon a cylinder one marks two points at which a fracture is to be produced, then the resistances at these two points will bear to each other the inverse ratio of the rectangles formed by the distances from the respective points to the ends of the cylinder (*EN*, VIII:176).

Galileo's answer is only partially satisfactory. First, because it does not consider the cylinder's own weight. Second, and more importantly, because the *Proto* would have probably preferred to know a way to compare two oars of different lengths—one with which he had lifelong experience, and the other with which he was not familiar—than two equal oars with fulcra displaced at different points. Moreover, Galileo also disregarded Contarini's view of the oar, according to which one cannot consider the force as being applied only at the extremity of the pole on the handle end, but rather at the number of points represented by the hands of the oarsmen. This is relevant not only for the explanation of how the oar functions, but also for the determination of its resistance.

Did Galileo Become a *Proto*?

Galileo's masterpiece was not a success. It certainly would not have sufficed for him to become a master. However, it provided him with the opportunity to found a new science of the strength of materials. The contact with the practical knowledge that proved so fruitful for the foundation of the first new science took place in the Venetian Arsenal; the professionals involved were shipbuilders; and the questions discussed concerned primarily shipbuilding and the functioning of the ship as a propelled mechanical device. However, Galileo's first new science was first perceived as an attempt to found a science of the singulars, that is, a science able to take into consideration all irregularities and accidents of each kind of material, or even of each solid body. The strongest opposition, on the basis of this perception, came particularly from engineers and craftsmen like the machine makers. This apparent paradox arose because Galileo superseded the Aristotelian vision, which was shared by most of the engineers, on the basis of the practical knowledge of the Arsenal.

Galileo used the *Mechanical Questions* as a sort of tabular research program to be accomplished by visiting the Venetian Arsenal and investigating the work, methods and accumulated experience of the shipbuilders. It has been already shown (Valleriani 2009a) how the first early modern theory of the strength of materials was the result of a process of integrating the practical knowledge of the architects into the Aristotelian arguments of Questions 14 and 16 of the *Mechanical Questions*, namely of those arguments related to the resistance of materials and the cantilever model. Galileo's first new science is, in turn, the result of integrating aspects of the practical knowledge of the Arsenal into the early modern Aristotelian theory of the strength of materials. Galileo's first new science, therefore, is not merely founded in the practical knowledge Galileo shared at the Venetian Arsenal, as if this theory had suddenly occurred to him while visiting the shipyards. Among the shipwrights Galileo found knowledge that challenged the Aristotelian doctrine, that is, the only theoretical approach that Galileo had at his disposal to start with. In conclusion, the generation of new theoretical knoweldge is the result of the investigations of an Aristotelian engineer.

Chapter 5
Pneumatics, the Thermoscope and the New Atomistic Conception of Heat

In Galileo's day, once important scientific questions had been formulated, they often found wide circulation by means of letters or notes sent to friends and acquaintances. Today's name for one such scientific problem is "Bardi's problem," for the question had been presented to Galileo by Count Bardi di Vernio (1534–1612).[1] The problem, and even more so its solution, represent a paradigmatic logical model for the period before instruments had been invented to measure temperature. The problem suggests investigating why a person feels cold when he goes into a body of water like a river during the summer, and even colder when he comes out, but, going back into the water, finally feels comfortable. There were several variations of this problem after his first formulation. For example, the condition was often added that, before the person goes bathing, he spends time in the shade where he feels neither cold nor hot and, when he comes out of the water, he returns to this shady location.[2]

Galileo's solution to the more elaborated version of Bardi's problem, on the basis of those empirical data provided by the human senses, is the following:

> The problem is to be solved in the following way. In a room we have a tub full of water, and this has been there, for example, 15 days long: one person comes, takes off his clothes and goes into the tub: it is clear that he feels much colder in that water than he felt before he entered it; from which one can conclude that, if air and water are placed in the same location, that is with the same heat or coldness, the water will be always perceived to be colder than the air. We therefore say that, if the air has 2 degrees of cold, the water has 10 of them: hence another water which has only 6 of them, appears cold in comparison with

[1]Bardi's problem was formulated with the help of Father Grienberger at the *Collegio romano* in 1614 and presented ("recited," as they called it) at the institute's weekly meeting by Bardi himself. Bardi sent Galileo the text with the problem as an attachment to a letter: Giovanni Bardi to Galileo, June 20, 1614, in *EN*, XII:76–77. See also Francesco Stelluti to Galileo, June 28, 1614, in *EN*, XII:78 and Giovanni Bardi to Galileo, July 2, 1614, in *EN*, XII:79–80. For Galileo's relationship with Giovanni Bardi, especially concerning the frameworks of the Jesuits in Rome and their mutual connections to Christoph Grienberger, see Blackwell (1991, 135–137).

[2]There is textual evidence of both formulations of Bardi's problem in Galileo's legacy. The first formulation is reported in a textual fragment, and the second is one of the problems Galileo intended to discuss in his unpublished treatise of *Problemi spezzati*. For the fragment, see *EN*, VIII:610. For the formulation of the problem in the treatise, see *EN*, VIII:599. The treatise of the *Problemi spezzati* was introduced in the previous chapter. For more details, see p. 132.

M. Valleriani, *Galileo Engineer*, Boston Studies in the Philosophy of Science 269,
DOI 10.1007/978-90-481-8645-7_5, © Springer Science+Business Media B.V. 2010

the air that has 2 of them, but much hotter in relation to the water that has 10 of them. Now, given this, the person who goes bathing in the Arno, when he is naked in the shade, enjoys the moderate coolness of the air, which has only two degrees of cold; but when he goes into the water of the Arno, he feels its cold which is of 6 degrees (I say 6 degrees and not 10 because the ardent Sun, which hits it over a distance of many miles, took away 4 of them [degrees of cold] from it [the water]); and therefore, in comparison with the air, which has only 2 of them, the water seems very cold to him. This person then gets out of the Arno and returns to the shade, wet and covered with a very thin veil of water so slight that, as soon as the person is under the tree in the shade, the water will have already obtained the 4 degrees of cold back, taken away from the Sun. Hence from the 6 that it had before, it gains 10, so that the person who bathed no longer feels 6 degrees of cold but 10: and therefore while he stays under the tree he feels extreme cold. But if he then goes back diving and enters the water, which has 6 degrees of cold, losing four degrees of cold, it seems to him as if he had entered a mild bath (*EN*, VIII:599).

Galileo was wrong in considering the temperature of water exposed to the rays of the sun to be lower than that of the air over the water. However, this was the general conviction before the thermoscope, or the thermometer, was applied to such investigations.[3] Once the thermoscope appeared, Galileo, based on the knowledge of pneumatics that he shared with Italian Renaissance hydraulic engineers of his day, also turned to the question as to how such a pneumatic instrument actually worked. This research then led Galileo to an atomistic conception of heat, which, once it was further developed in the form described in his *Il Saggiatore* of 1623, allowed him to address Bardi's problem again, this time coming up with a completely different solution.

Although Galileo never published anything directly concerned with the thermoscope or the investigations using or testing it, he nevertheless expended great effort in working with such an instrument. Most of the evidence of such research is represented by letters and fragments, most of which have never previously been evaluated by historians, so that even Eduard Dijksterhuis felt obliged to state that Galileo analyzed and postulated a heat doctrine in his *Il Saggiatore* as a consequence of a casual event, which he did not describe further (Dijksterhuis 1983, 473–475).

The thermoscope was the first instrument built to measure temperature. In Galileo's day, however, no modern concept of temperature had yet been formulated. The semantic instruments used to speak about temperature were "degrees of cold" and "degrees of heat," which were sharply distinguished definitions. Once the thermoscope was equipped with a scale, it became a thermometer. Galileo is one of several, who, more or less simultaneously at the beginning of the seventeenth century, and in different geographic locations, "invented" the thermoscope: the first instrument that could be used to obtain information about the degrees of heat and

[3]Galileo's solution to Bardi's problem, as given in the treatise of the *Problemi spezzati*, was not written before he invented the thermoscope, for the outline of such a treatise was compiled by his son during the final years of his life. Although the history of this fragment is unknown, its argument is relevant as a paradigmatic example of a method for solving problems related to temperature before the invention of the thermoscope.

cold without appealing to the human senses. The thermoscope circulated for about ten years before being transformed into the thermometer. Thus, the thermoscope represents a small but crucial link in the chain back to the earlier era, when people judged temperature only on the basis of their own senses, and the current era, when we cannot even conceive of life without such an instrument and the information it provides.

But the thermoscope was not really invented: more accurately, it was the result of a conceptual reshaping process which took place at the beginning of the seventeenth century. The thermoscope is a pneumatic device that functions on the basis of the phenomenon according to which the air contained in a vessel dilates when the temperature of the device's surroundings increases, and contracts when the temperature decreases. Considered only as a pneumatic device rather than as an instrument to measure temperature, this instrument is a very old technical realization. It is not even possible to determine when such an instrument appeared for the first time. Certainly, such devices became extremely common during antiquity up to the Hellenistic era, for works on pneumatics, such as those by Philo of Byzantium and Hero of Alexandria, clearly show how this natural phenomenon was used to power plenty of pneumatic devices, many of them conceived as kinds of trick fountains. Theoretical speculations about those principles on the basis of which pneumatic devices work have their origins in antiquity as well. The earliest surviving textual evidence is a poem composed around 460 BC by Empedocles of Agrigentum, entitled *On Nature* (Philo of Byzantium and Prager 1974, 5–6). The appearance of the thermoscope, therefore, is the result of a process of reconfiguration of an old device. The process of reshaping the ancient pneumatic device into an instrument for measuring temperature is also closely related to the reception and transformation of ancient pneumatics that took place in Italy during the Renaissance, which focused mainly on the work of Hero of Alexandria.

Moreover, pneumatic devices that worked on the basis of the same phenomenon as the thermoscope were very common in Galileo's day. Such instruments were used in the medical field, for example, or simply kept as fashionable objects. Bleeding cups, milk pumps for nursing mothers, *calendaria*[4] and many sorts of fountains and water gardens were probably the most common devices among the instruments of the thermoscope type. Considered from this perspective, therefore, it is hardly appropriate to speak of the invention of the thermoscope. The thermoscope is an ancient device, which was conceptually reshaped in order to meet needs and *desiderata* that emerged between the end of the sixteenth and the beginning of the seventeenth centuries, and remain established today.

To fully understand how the appearance of the thermoscope led Galileo to his atomistic conception of heat, and how this could happen essentially because of

[4] *Calendaria* were pneumatic devices, often hung on the outside of doors, which, thanks to the daily motion of the liquid upward and downward, were considered to be a kind of time-keeping device.

Galileo's sharing pneumatic science with his contemporary hydraulic engineers, first the details of the thermoscope's composition will be shown. Second, the emergence of the instrument will be described briefly, along with the context within which it was first applied. The thermoscope and its first applications for scientific purposes soon challenged some aspects and core principles of the Peripatetic doctrine concerned with natural motion, such as the Aristotelian processes of condensation and rarefaction. The instrument presented a further challenge not only because of the empirical data it provided, but also and especially because the Aristotelian doctrine apparently was not able to provide a satisfactory explanation for the way the instrument worked. This point already had been recognized before the thermoscope appeared, and had led to the formulation of new pneumatic principles as a consequence of the process of reception and transformation of ancient pneumatics, mainly through the work of hydraulic engineers. Ultimately, the new pneumatics of Renaissance engineers was the background from which Galileo's new pneumatic principles emerged through his studies about the functioning of the thermoscope. Once this whole process has been described, it will be shown how Galileo's first atomistic conception of heat was formulated around 1619, while he was attempting to lay the theoretical foundations for pneumatic devices powered by heat sources, like the thermoscope. Finally, this section will consider the development that led Galileo to his second atomistic conception of heat, as published in *Il Saggiatore* in 1623, and present Galileo's updated solution to Bardi's problem.

The Thermoscope

The first thermoscopes (Fig. 5.1) are instruments constituted of a small vase full of water at the bottom, from which a thin pipe vertically emerges, the upper part of which normally ends with a bowl. The bowl and the upper part of the pipe are empty or, more accurately, filled with air. When the air is heated, for example, by exposing the bowl to the rays of the sun, the air expands, pushing the water downwards. When it is cooled it contracts, pulling the water upwards as it tends to create a vacuum. The expansion and contraction of air are consequences of changes not only in temperature, but also in pressure. In fact, the first thermoscopes were actually a sort of thermo-baroscope. Recognition of this characteristic, still unknown at Galileo's time, led to the second generation of instruments, known as the liquid in glass thermometers.[5]

[5] The variability of air pressure became known toward the mid-seventeenth century, and the invention of the liquid in glass thermometer apparently can be attributed to Ferdinando II, Grand Duke of Tuscany. For a detailed historical view of the emergence of the liquid in glass thermometer, see Knowels Middleton (1966, 27–39). For the history of the barometer, and especially, the discovery of the sensitivity of air to atmospheric pressure, see Knowels Middleton (1964, Chapters 3 and 4).

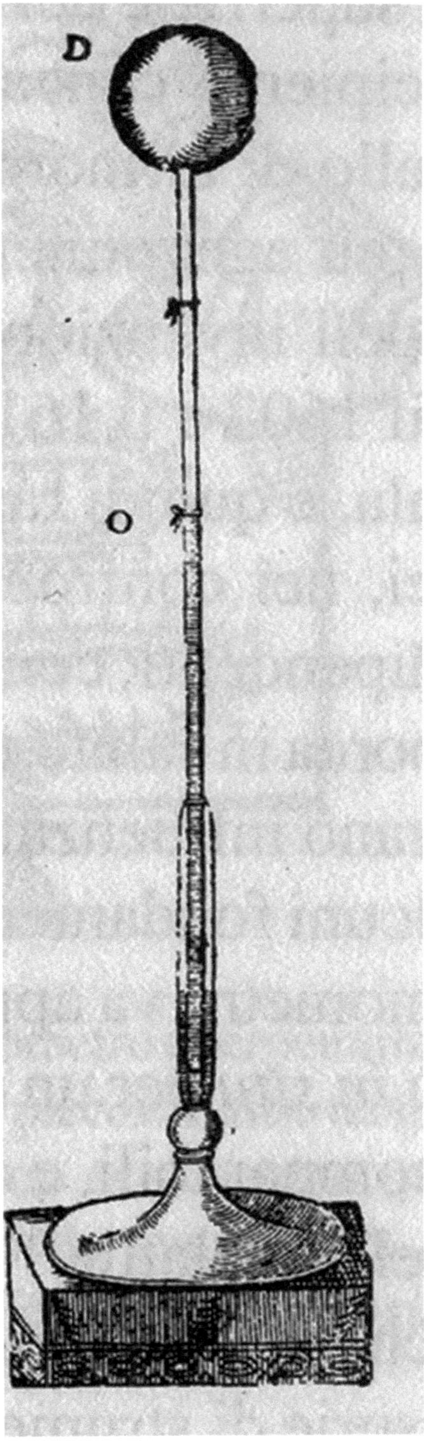

Fig. 5.1 Example of an early thermoscope (Sanctorius 1646, col. 29-30)

The thermoscope as a *perpetuum mobile* If the components of the thermoscope are implemented appropriately, the thermoscope can be built in very large sizes. When a large thermoscope is placed outdoors with the part containing air covered, the instrument simply shows a kind of "perpetual" movement of the water. For this reason instruments like the thermoscope were initially known best as *perpetuum mobile*.[6] The sixteenth and seventeenth centuries are distinguished by the search for a perpetual motion machine, a search encouraged by the aristocracy's promises of generous rewards for successful inventors. Especially in northern Europe, the appearance of these instruments is often related to the attempt to have them recognized as perpetual motion machines. The only difference between a perpetual motion machine and a thermoscope, besides the larger dimensions required for a more marvelous effect, was in fact that the "inventors" of the perpetual motion machines kept silent about the principle on the basis of which their devices worked, generally claiming that the motion of the water corresponded to the flow and ebb of the seas.

The Emergence of the Thermoscope

Many historians have dedicated parts of their works to the emergence of, first, the thermoscope, and then the thermometer. Most, if not all of these studies, focus mainly on questions of priority, although it is quite impossible to show who invented the thermoscope first.[7] In fact, the thermoscope probably started circulating in market squares, from the stalls of which it was transformed into a scientific instrument by people like Galileo, who applied it to their research. In the literature of the previous century, however, there is general agreement about the first four men who are supposed to be "inventors" of the thermoscope: Galileo, Sanctorius Sanctorius (1561–1636), Robert Fludd (1574–1637) and Cornelius Drebbel. Usually forgotten in the history of the thermoscope and of the thermometer is Giovan Francesco Sagredo. In part together with Galileo and in part alone, he is the author of the most important developments of this instrument during its early years.

Galileo's early use of the thermoscope Discussing the appearance of a device of the "flow and ebb" sort with his friend Cesare Marsili (1592–1633) in Bologna in 1626, Galileo himself reports that he has been familiar with this sort of device after having made "a similar amusing device" twenty years earlier when he was in Padova. Thus, Galileo had worked with such instruments as far back as 1606.[8]

---

[6] As mentioned, such a device was also called *calendarium*.

[7] See, for example, Favaro (1966, I:193–212), Caverni (1972, I:265–298), and Hellmann (1920).

[8] Galileo to Cesare Marsili, April 25, 1626, in *EN*, XIII:319–320. Galileo told Marsili that it was a device which works on the basis of the principle according to which air expands when heated and contracts when cooled. Since Marsili also reported that the author of the device called it "Flow and Ebb *perpetuum mobile*"and suggested using salt water, Galileo added that this was a trick to conceal the truth. This letter has never been cited as evidence for Galileo's construction and use of the thermoscope.

Fig. 5.2 Replica of Galileo's thermoscope (Istituto e Museo di Storia della Scienza, Firenze. Inv. 2444)

In a letter to Fernando Casarini of 1638, Benedetto Castelli (1577–1644), one of Galileo's most important pupils, wrote:

> [...] I remembered an experiment showed to me already more than thirty-five years ago by our Lord Galileo, which was that, taken a small decanter of glass with the size of a small chicken egg, and a neck about two spans long and as thin as a grain stalk, and once the mentioned small decanter was well heated by means of the palms of the hands, he turned it upside down with its mouth into a vase placed below, where there was a bit of water. By making the small decanter free from the heat of the hands, the water immediately started ascending along the neck and went over the level of the water of the vase for more than one span; that effect then was used by the same Lord Galileo to build an instrument to examine the degrees of heat and cold (Fig. 5.2).[9]

Galileo had demonstrated the pneumatic phenomenon to Castelli earlier than 1603. However, being able to set up such an experiment does not amount to Galileo's conceiving of such a device as a thermoscope. This instrument cannot be considered "invented" until somebody decides to perform the demonstration described

[9]From Benedetto Castelli to Ferdinando Cesarini, September 20, 1638, in *EN*, XVII:377–380.

by Castelli with the purpose of measuring temperature. Castelli himself stated that Galileo made such an instrument "later."[10]

One can therefore conclude that Galileo applied the fundamental pneumatic phenomenon to build the thermoscope during the period between 1603 and 1606.

Drebbel's *perpetuum mobile* The Dutch machine maker Cornelius Drebbel,[11] to whom the invention of the thermoscope is also often ascribed, never built a single thermoscope. In his day Drebbel was famous for his astronomic clocks and his *perpetuum mobile* of the flow and ebb variety. Contrary to what is often believed, Galileo came into indirect contact with Drebbel and was aware of his work by 1610 at the latest.[12] Drebbel made his first perpetual motion machine in 1604 for King James I of England, who highly appreciated the device. In 1610 Drebbel moved to Prague, to the Court of Rudolph II, Emperor of the Holy Roman Empire, for whom he assembled the same device which had made him famous in England. Drebbel's "great machine" was constituted of a semicircle of glass, which contained water or another more visible liquid. A tight-fitting lid then was used to cover the other components, namely those containing air. At the beginning of Drebbel's stay in Prague, on October 1610, the Tuscan ambassador at the court of the Emperor, Giuliano de Medici, wrote Galileo that "a Flemishman is there and pretends to have found the perpetual motion machine."[13] Giuliano de Medici described the machine, adding that Kepler would not accept that it was a *perpetuum mobile* until he understood the principle on the basis of which the machine works, which Drebbel took care to keep under wraps.[14]

The name Drebbel came to Galileo's ears again in 1612, when his former pupil Daniello Antonini (1588–1616) wrote him from Brussels, where he was serving as a military officer. Antonini learned that James I had a *perpetuum mobile*, namely the

---

[10] A third indication for Galileo's invention of the thermoscope is represented by a passage in Viviani's *Racconto*. Viviani relates that Galileo invented the thermoscope, that is "those instruments of glass, with water and air, in order to distinguish the mutations of heat and cold and the variety of temperatures of [different] places," during the first years of his stay in Padova, after 1592 (*EN*, XIX:607).

[11] For an exhaustive overview of the life and works of Cornelius Drebbel, see Tiere (1932).

[12] In 1611 Drebbel asked Giuliano de' Medici to provide two pieces of Galileo's glass in order to have them polished and made into telescope lenses for the emperor. According to the Tuscan ambassador, His Caesarean Majesty was spending a great deal of time with Drebbel investigating technical contrivances. For more details, see Giuliano de' Medici to Belisario Vinta, November 14, 1611, in *EN*, XI:234, and Giuliano de' Medici to Belisario Vinta, November 11, 1611, in *EN*, XI:235.

[13] Giuliano de' Medici to Galileo, October 18, 1610, in *EN*, X:448–449.

[14] Galileo asked the Tuscan Ambassador for more details about the device later in 1610 (Giuliano de' Medici to Galileo, November 29, 1610, in *EN*, X:478–479). At the end of the same year Martin Hastal, probably an ex-pupil of Galileo, who was also in Prague and was usually involved in the scientific life of the court, allowed Drebbel and Galileo to meet indirectly and entertain each other with conversation on the *perpetuum mobile*. For more details, see Martin Hastal to Galileo, December 19, 1610, in *EN*, X:491–492. Without great success, Favaro tried to collect more details on Martin Hastal, who certainly met Galileo and was very familiar with Venice and Padova. For more details, see Favaro and Galluzzi (1983, I:600–606).

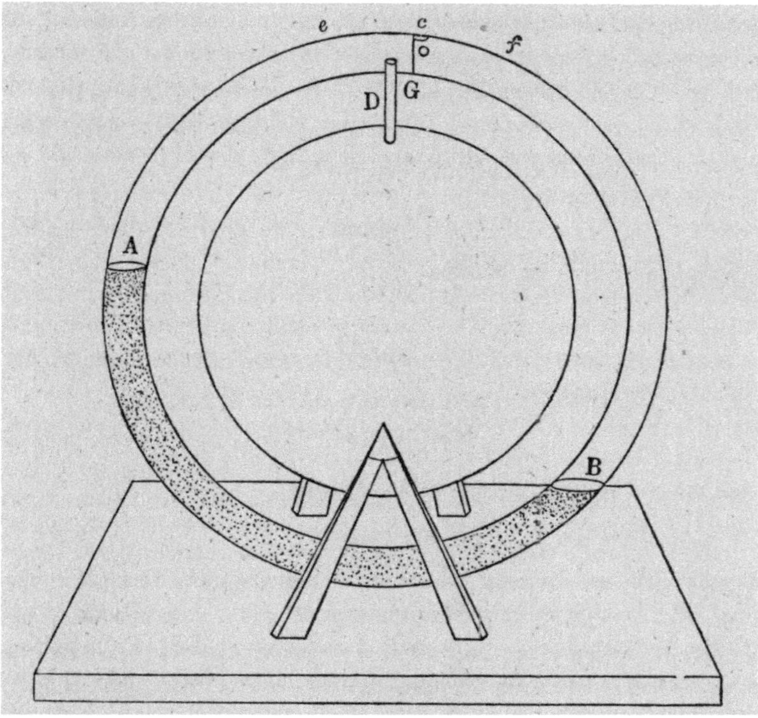

Fig. 5.3 Daniello Antonini's drawing of Cornelius Drebbel's *perpetuum mobile* (*EN*, XI:275)

one from Drebbel (Fig. 5.3). Informed about the shape and dimensions of the device, he immediately understood its function and constructed a similar one. As he wrote Galileo, he had grasped how to make one of those machines "thanks to the experiments with the small decanter," that Galileo performed in Padova when Antonini was his private student.[15] Antonini made a perpetual motion machine which was straight and vertically positioned rather than circular in shape. The prince under whom Antonini was serving immediately wanted to see the machine, and Antonini decided to give it to him as a gift rather than asking for money or privileges. Like Antonini, Drebbel never used his instrument to measure temperature and therefore he did not "invent" the thermoscope.[16]

[15]Daniello Antonini to Galileo, February 4, 1612, in *EN*, XI:269–270. Unfortunately, it is not possible to know with any precision when Antonini took private lessons from Galileo. Otherwise this would amount to further evidence relevant for dating Galileo's first use of the thermoscope. For the translation of the entire letter, see pp. 227ff.

[16]Knowels Middleton tried to change the meaning of this conclusion, to which he also arrived, adding that Drebbel certainly could have made a thermoscope if he had wanted to do so. Yet this remark does not seem terribly useful. For more details, see Knowels Middleton (1966, 21).

Sanctorius' early use of the thermoscope Five months after Antonini's letter, in 1612, Giovan Francesco Sagredo communicated to Galileo that their mutual friend Agostino da Mula had been at the fair of the patron saint of Padova, where he saw:

> [...] an instrument by Lord Sanctorius with which one measures the cold and the heat with the divider. Finally he communicated to me that it is a large bowl of glass with a long neck [...].[17]

Sanctorius Sanctorius was a doctor, practicing mostly in the Venetian Republic, but also in some other Italian and European countries thanks to his renowned foundation of what the history of medicine calls the Iatromechanical School (or Iatrochemical or Iatrophysical). A member of Galileo's Venetian circle of friends, he was particularly well acquainted with the family Morosini, in whose house the group of scholars often met.

In the first edition of his *Commentaria in artem medicinalem Galeni*, published in 1612, Sanctorius presented the instrument for the first time:

> I wish to tell you about a marvellous way in which I am accustomed to measure, with a certain glass instrument, the cold and the hot temperature[18] of the air of all regions and places, and of all parts of the body; and so exactly, that we can measure with the divider the degrees and ultimate limits of heat and cold at any time of day. It is in our house in Padova and we show it very freely to everybody.[19]

Probably in response to this public invitation, Agostino da Mula went to Sanctorius' house in Padova to see the instrument, as Sagredo told Galileo. Hence, in 1612 Sanctorius had at his disposal an instrument made of glass, constituted of a great bowl with a long neck. The water was placed in the lower part. As the instrument certainly lacked any scale, the levels of liquid along the neck were recorded using a divider and therefore a ruler to measure its opening.

Fludd's thermoscope The Welsh doctor Robert Fludd had practiced mostly in Oxford, apart from six years traveling through continental Europe starting in 1598. In 1609 he became a member of the College of Physicians, where he practiced until his death.[20] Fludd aspired to construct a cosmic theory, in the context of which the effects of light and darkness and of heat and cold were to play a relevant role. He published, in 1617, an account of the pneumatic phenomenon of air expansion caused by heating (Fludd 1617, 30). Sherwood Taylor was able to show that both Fludd's apparatus and its description were taken from Philo of Byzantium's *De ingeniis spiritalibus*, a manuscript he may have possessed (Sherwood Taylor 1942). Fludd did not publish the first illustration of his thermoscope until 1626 (Fludd

---

[17]From G. Francesco Sagredo to Galileo, June 30, 1612, in *EN*, XI:350. For the translation of the entire letter, see pp. 229ff.

[18]Sanctorius used the concept of "temperamenta" which does not correspond exactly to the English "temperature." This may mean that not only the degrees of cold and heat, but also the general climatic status of the environment are taken into consideration.

[19]Sanctorius 1612, Part III, Cap. LXXXV, Particula X, col. 62. Translation from Knowels Middleton (1966, 9).

[20]For an extensive work on Robert Fludd's work and life, see Sherwood Taylor (1942).

1626), and he polemized against the many people who considered themselves to be the inventors of the instrument.

Telioux's thermoscope The *Bibliothèque de l'Arsenal* in Paris holds a manuscript written in Rome in 1611, by a certain Telioux (Telioux 1611). In the manuscript a thermometer equipped with a scale is illustrated. The analyses by J. A. Chaldecott (Chaldecott 1952) and Knowels Middleton (Knowels Middleton 1966, 10–13) concur with the conclusion that Telioux did not test the instrument himself, but rather was reporting something he had heard or seen about it. If this conclusion is true, in 1611 the thermoscope presumably was about to become a very common instrument, and thus already known outside the Venetian circle to which Galileo and Sanctorius belonged.

Excluding Drebbel from the ranks of those who apparently first used the device to measure degrees of cold and heat, it seems certain that Sanctorius had a thermoscope in 1612, Fludd published his first description of a thermoscope in 1626 and, at the beginning of the seventeenth century, Galileo was already analyzing those pneumatic phenomena he would later use as the basis for building his thermoscope.[21] Telioux's manuscript, however, provides hints that the thermoscope was probably not "invented" by people like Sanctorius and Galileo. It may be that they merely picked up a technical curiosity that was already in circulation, perhaps conceptually reshaped by vendors on the market square, and applied it for scientific purposes. In any case, by 1624 the thermometer was very well known and probably a standard product sold in many workshops and markets (Leurechon 1624, Problem LXIX).

From the Thermoscope to the Thermometer

In 1638 Fludd published an illustration of a scaled thermometer (Fludd 1638, 2). In the same work, he also gave an explanation on how to determine a scale for the instrument (Fludd 1638, 4). Sanctorius presented his thermometer in 1630 (Sanctorius 1630, Chapter LIII, col. 762). In order to obtain a scale for the instrument, he first determined terms of comparison for its extremities (hottest and coldest) so that he could divide the scale as he wished. He found those terms in the "coldest snow" and in the "hottest fire of a candle."

The development extending from the first thermoscope to a thermometer equipped with a scale can be followed in greater detail, however. In fact, it was during this phase that Galileo and, above all, his friend Giovan Francesco Sagredo, played a significant role and, perhaps for the first time, arrived at the idea of providing the instrument with a scale.[22]

---

[21] On the basis of the epistolary exchange between Galileo and Giovan Francesco Sagredo, one can conclude that Galileo ascribed to himself the paternity of the invention of the thermoscope and that Sagredo acknowledged it. For more details, see G. Francesco Sagredo to Galileo, May 9, 1613, in *EN*, XI:505–506. For the translation of the entire letter, see pp. 231ff.

[22] Telioux's manuscript, which is dated 1611, shows an illustration of a thermometer. However, the text does not refer to that picture which, therefore, may have been added later.

Sagredo's first thermometer The epistolary discussion between Galileo and Sagredo concerned with the instrument lasted for almost three years, from 1612 until April 1615. In the same letter in which Sagredo informed Galileo about the Sanctorius thermoscope in 1612, he added that:

> [...] I immediately applied myself to producing some of them very exquisitely and beautifully. I make the ordinary ones with an expenditure of four Lira each, that is, a small water vase, a small decanter and a glass siphon. My method of production is such that I can assemble up to ten of them within one hour. The nicest one I made was produced by means of a small flame. It has the size and the shape of the one in the drawing included here,[23] with all its parts. I am waiting to hear that you have made *mirabilia magna*.[24]

Sagredo first became aware of the existence of Sanctorius' thermoscope on the occasion of the celebration of the patron saint of Padova on June 13, and his letter is dated June 30. During a period slightly longer than two weeks, therefore, he was already so expert that he was able to assemble ten thermoscopes in one hour. After this short time, too, he already attempted to build instruments of several shapes, sizes and quality, in reference to both the material and its manufacture. Moverover, Sagredo wanted to obtain a thermoscope from which information about the temperature could be read more easily, that is, he presumably wanted to abandon the method of measurement by means of a divider, as required by the first of Sanctorius's instruments, to achieve another one provided with a scale. One year later Sagredo had a new thermometer able to show temperature differences as great as "100 degrees" between one room of his house and another.[25] Although Sagredo left completely unclear the method he used to provide the instrument with a scale, almost two months later, in July 1613, he triumphantly wrote Galileo that he had succeeded in nearly perfecting the instrument![26]

Sagredo's standardization of the scale In 1615 Sagredo felt confident enough to shift his attention from the method of constructing the instrument to the conception of the thermometer itself. In other words, he performed experiments which would have resulted in an instrument provided with a scale, whose characteristics were somehow communicable. He informed Galileo that:

> [...] two days ago, when it snowed, my instrument displayed 130 degrees more heat in this room than what [it showed] two years ago during a time of very rigorous and extraordinary cold. The same instrument, immersed and buried in the snow, displayed 30 degrees less, that is, only 100.[27]

[23]This drawing is now lost.

[24]From G. Francesco Sagredo to Galileo, June 30, 1612, in *EN*, XI:350–351. Author's italics. For the translation of the entire letter, see pp. 229ff.

[25]G. Francesco Sagredo to Galileo, May 9, 1613, in *EN*, XI:505–506. For the translation of the entire letter, see pp. 231ff.

[26]G. Francesco Sagredo to Galileo, July 27, 1613, in *EN*, XI:544–545. For the translation of the entire letter, see pp. 233ff.

[27]From G. Francesco Sagredo to Galileo, February 7, 1615, in *EN*, XII:140. For the translation of the entire letter, see pp. 239ff. Sagredo's observation corresponds to what Sanctorius published

Sagredo performed long series of experiments, recording all outputs, in order to provide a scale with communicable terms of comparison:

> But then, immersing [the instrument] in snow mixed with salt, it displayed a further 100 [degrees] less. I believe that it really displayed even less, but one could not see it because of the hindrances [caused] by the snow and the salt. Since, at the hottest point of the summer, it had displayed 360 degrees, one can see that salt added to the snow increases the cold by as much as one third of the difference between the greatest heat of the summer and the greatest cold of the winter.[28]

Sagredo chose two terms of comparison for the hottest and coldest, the two extreme points which, in reference to his own scale, were reached by the liquid over two years of observations. Finally, in order to compensate for their incommunicability, he provided the observational tables of data collected over the same period.[29]

Galileo's and Sagredo's technical tests Once a scale had been achieved, Sagredo was all the more motivated to continue with this field of research. He investigated ways to improve old thermoscopes and thermometers and searched for new and more efficient shapes.[30] He interrogated Galileo about his thermoscope and research with it and discovered that he had attained an even more advanced state of knowledge than his friend had.[31] Both Galileo and Sagredo noted that the ascending motion of the water along the neck was irregular, that is, it did not always show the same characteristics given the same situation. These irregularities were certainly caused in part by the fact that the thermoscope was actually a thermo-baroscope, that is, because it was sensitive to atmospheric pressure. When they believed, therefore, that they were dealing with two identical situations for which they supposed the instrument must show the same degree of cold or heat, in fact they were neglecting the effect of atmospheric pressure, which may have varied significantly.[32]

Irregularities in the ascending motion could have had other causes, however. They were also caused by different viscosities of the water, for example, or of the other liquids often used in the thermoscope.[33] Because of this problem,

---

fifteen years later. Sagredo, like everybody else during his time, thought in terms of degrees of cold and of degrees of heat.

[28] From G. Francesco Sagredo to Galileo, February 7, 1615, in *EN*, XII:140. For the translation of the entire letter, see pp. 239ff.

[29] Knowels Middleton pointed out the accuracy of Sagredo's observation: "If we might take the extreme summer and winter temperatures in Venice in those years to be about 34° and −5°, this would bring the mixture of ice and salt to a −18°, a very likely value" (Knowels Middleton 1966, 10).

[30] G. Francesco Sagredo to Galileo, March 15, 1615, in *EN*, XII:156–158. For the translation of the entire letter, see pp. 241ff.

[31] G. Francesco Sagredo to Galileo, April 11, 1615, in *EN*, XII:167–170. For the translation of the entire letter, see pp. 244ff.

[32] In order to create identical situations to which the thermoscope could be applied, one could, for example, apply the flame of a candle.

[33] Wine was often used instead of water in order to improve the visibility of the scale and so to increase the ease of measurement.

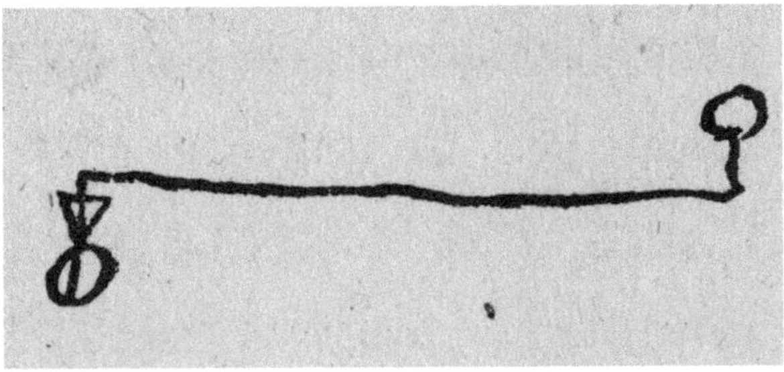

Fig. 5.4 Illustration of Sagredo's bent thermoscope (*EN*, XII:168)

Galileo experimented with thermoscopes equipped with internally broader necks, as Sagredo confirmed he had as well. After these experiments Galileo went back to using thin-necked thermoscopes, while Sagredo, who dedicated more time to the observation of the behavior of the instruments, after comparing observations, concluded that instruments equipped with broader necks worked more accurately.

Sagredo further informed Galileo that he had performed several experiments with different sorts of instruments: one also equipped with a small upper decanter of air and a long neck, where the water ascended and descended along the neck without escaping from it; another one provided with a neck bent twice so that the motion of the liquid was horizontal rather than vertical, probably believing that one cause of the irregularities of the ascending motion of the liquid was that the water was forced to move in a motion that was not natural, the natural one being downwards toward the center of the Earth (Fig. 5.4). After this phase of experimentation Sagredo abandoned these differently shaped instruments, recognizing that the best ones were those equipped with a broad vertical neck.[34]

Finally, Sagredo applied himself to investigating other aspects of the movement exhibited by the liquid in the instruments:

> The best and most perfect instruments I made were with a neck as broad as a finger, referring to the internal part of the neck over which I had blown at the furnace of Murano a vase whose volume corresponds to three or four glasses, using the mentioned instrument in the way Your Lordship writes. In this way I have had three of them made in different sizes, and which have worked now for almost three years in such harmony with each other that it is marvelous. These I have observed for over almost one year, one, two, three, four, five, six, up to eight times a day, with such correspondence that from those observations I have achieved a table of correspondences and equations among them. First I have seen that they work with the absolute same proportion, during both extreme heat and extreme cold, so that each time

[34] A further suggestion by Galileo was to decrease the height of the scale. This was rejected by Sagredo. One of the irregularities demonstrated during ascending motion was that different heights were reached by the liquid in supposedly identical circumstances. Galileo's advice, therefore, certainly provided a method to make the error less evident. Sagredo justified his decision to oppose Galileo by explaining that his suggestion had no theoretical foundation. For more details, see G. Francesco Sagredo to Galileo, April 11, 1615, in *EN*, XII:167–170. For the translation of the entire letter, see pp. 244ff.

> I see one of them I guess, by using the table, the degree of the other two, sometimes with a variation of give or take two or three degrees.[35] [...]. So, changing these instruments a little because of the slightest occurrence, they change more or less depending upon whether they are more or less exposed to those occurrences, either because they are close to the apertures of the room, or to the persons, or to the lights, etc. Moreover, since some of them have thicker and some thinner glass, it is conceivable that not all of them change over the same period, but, should some alterations in the temperature of the surroundings occur, the thinner one is the first to sense and show it. Concerning the instruments with a very thin neck, as those of Your Very Excellent Lordship, you should accept that the viscosity of the water and of the wine also causes variation. Therefore I decided to use instruments of such a size, that, when one takes away the lower vase, the neck empties out.[36]

Since Sagredo and Galileo believed that the irregularities showed by the thermoscope were due to particular characteristics of the instrument itself, Sagredo sought a way to obtain high-quality measurements by constructing a set of instruments of different sizes proportioned to each other so that he could apply all the thermometers simultaneously for each measurement.

Empirical Data Provided by the Thermoscope

The thermoscope and the first thermometer were investigated by Sanctorius, Galileo and Sagredo: three friends, all of whom were involved in the Venetian cultural circle during the first fifteen years of the seventeenth century. During this period the instrument experienced wide diffusion because of its application to scientific purposes. From its inception the thermoscope was also applied both in medicine and in what could be called a preliminary stage of modern meteorology.[37] The first consequence of the application of the thermoscope to scientific problems was the discovery that many empirical data provided by the human senses are incorrect. As will be shown, this represented a first challenge to Aristotelian doctrines.

Early scientific use of the thermometer At Galileo's time meteorology was not yet established. At this early stage, meteorological phenomena were often discussed together with completely different subjects like, for example, those concerned with the temperature of the interior of animals. Moreover, empirical observations had yet to be organized, and not even the necessity of standardizing the instrument had been

---

[35]Considering that the hottest day of the summer corresponded to 360°, 2 or 3° of difference were small fractions of 1°C.

[36]From G. Francesco Sagredo to Galileo, April 11, 1615, in *EN*, XII:169. For the translation of the entire letter, see pp. 244ff.

[37]The oldest evidence of the use of the thermometer in the field of meteorology in modern terms is Leurechon (1624). See, in particular, Problem LXIX. Modern meteorology emerged when networks were created that were able to collect data from standardized instruments. The first network in this sense was created by the *Accademia del Cimento* during the second half of the seventeenth century. On the emergence of modern meteorology and the role played by the first scientific networks, see Daston (2008). For a general introduction to the work of the *Accademia del Cimento*, see Knowels Middleton (1971) and Boschiero (2007).

fully recognized. Giovanni Battista Benedetti achieved the recognition of the former complex of problems concerned with temperature, in particular, in his *Diversarum speculationum mathematicarum et physicarum liber*, published in 1585 (Benedetti 1585).

Benedetti's work, which was written before the thermoscope came into common use, is devoted primarily to critiques of the opinions of the Peripatetics. In a chapter entitled *De raro et denso nonnulla, minus diligenter à Peripateticis perpensa* (Benedetti 1585, 191–194), Benedetti discussed a series of phenomena for whose solutions the thermoscope would have provided a great impulse. Summarized, the relevant problems are the following: (1) why animals' breath can be seen in winter; (2) why, if one draws water from below ground during the winter, it emits steam; (3) why subterranean water is hotter than that above the surface during the winter; (4) whether animals' stomachs contain more heat during the winter than during the summer; (5) why, if during the summer cold water is introduced into a vase of glass or silver, it "sweats"; (6) what the cause of the winds is; (7) why fog remains at rest in the place where it originated.

Aristotle's processes of condensation and rarefaction Obviously some of the questions listed by Benedetti are phenomena that would have been investigated in the framework of modern meteorology, but most certainly were not. What these apparently different phenomena had in common at the beginning of the seventeenth century were the principles on the basis of which they were explained by the Aristotelian commentators. These principles were those of condensation and rarefaction, as Aristotle described these processes in his *Meteorology* (Aristotle and Goold 1987, 369A10–369b3).[38]

Condensation and rarefaction are processes on the basis of which natural motions occur. They take place when matter, which exists in a particular form, such as water, for example, changes its form, for instance, into vapor (air). The extension of volume, which is a quality of the form and not of the matter, increases while passing from the form of water to the form of air. In this sense, when a process of rarefaction takes place, this involves only the form and not the matter. It is not appropriate to speak of the rarefaction of matter. The rarefied body then becomes lighter than the body from which it originated, and this is the reason why it eventually moves upwards. The process of condensation works in the opposite direction. Obviously, any process of rarefaction is accompanied by an increase in temperature; and processes of condensation by a decrease. According to Aristotle, condensation and rarefaction always imply a change in form, that is, a change in volume, a change in temperature, a change in weight and therefore a change in natural location. The

[38]In his *Meteorology*, more precisely, in the fourth book, Aristotle approached the issue of element transformation also from the perspective of corpuscolar visions (Newmann 2001, 145–153). Although the fourth book of Aristotles' *Meteorology* became fundamental within the framework of Aristotelian alchemy, this part of the doctrine does not seem to have played any role in Galileo's research.

most convincing application of this doctrine was to furnish an explanation of the water cycle.

Galileo's scientific investigation with the thermoscope Galileo, too, investigated those typical phenomena discussed by the Aristotelian commentators, some of which were listed by Benedetti, on the basis of the Aristotelian principles of condensation and rarefaction. For example, he applied his mind—but not yet his thermoscope—to the widely discussed phenomenon of steam from well water during the winter. Before the thermoscope was applied, it was commonly believed that well water is hotter during the winter than during the summer, both because this corresponded to the sensation of immersing a hand into the water, and because during the winter such water gives off steam. In his critical mind and on the basis of solid observation, Galileo had already noticed:

> That the smoking of the waters of the wells during the winter does not come from the heat, it is manifest: because the linens which dry under the Sun, during the winter give out smoke and during the summer not; one sees breath during the winter and not during the summer, etc. (*EN*, VIII:636)

In February 1615, Sagredo, after having applied the thermoscope, wrote Galileo:

> With these instruments I clearly see that the water in our wells is much colder during the winter than in the summer. For my part, I believe that the same thing happens in live fountains and subterranean locations, although our senses consider these in a different way.[39]

According to the Aristotelian doctrine, ice was the result of the process of condensation applied to water. Thus ice was supposed to be colder and heavier than water.[40] However, since ice does not sink into the waters, for example, of a river, as it was supposed to, the measurement of the temperature of the ice, the water and the surrounding air when this phenomenon happened was considered to be one of the most urgent open questions when the thermoscope finally appeared.

An undated text fragment by Galileo shows how Galileo himself applied his mind to this extremely topical issue:

> The very cold air of the north wind is colder than the ice and the snow: to confirm this, one can bring close to the instrument, during those weathers, some snow or ice, and the wine will evidently descend. Moreover, to confirm this further, a vase full of water and introduced into water will not freeze over, but it will do so if placed in the air. Moreover the waters of the rivers should freeze at the bottom where they are more distant from the air, and not at the surface, where they are very close to the air, but the contrary happens, hence etc. (*EN*, VIII:635)

---

[39]From G. Francesco Sagredo to Galileo, February 7, 1615, in *EN*, XII:139. For the translation of the entire letter, see pp. 239ff.

[40]The Aristotelian doctrine was evidently contradicted by the observation that, when ice forms in the rivers, this happens not at the bottom of the river but on its surface, and it does not then sink, as it would have to if it were heavier. However, this contradiction proved not to be an obstacle to considering Aristotle's doctrine as correct, for it was a functioning explanatory theoretical structure for many other natural phenomena.

Sagredo, too, devoted some of his research with the thermoscope to this issue. In May 1613 he informed Galileo that:

> With these [instruments] I speculated about several marvelous things, like, for example, that the air in winter is colder than the ice and snow, that the water now [May 9, 1613] seems to be colder than the air, that a small amount of water is colder than a large amount, and other such similar perceptions. Our Peripatetics cannot give any resolution for these, to such an extent that some of them (among whom is also our Gageo[41]) are so far off track that they do not yet understand the cause of the first operation, since they believe that one should see an opposite effect, because, since heat (as they say) has an attracting virtue, it should be that, when the vase is warmed, it pulls the water toward itself.[42]

The data provided by the thermoscope were not in agreement with those expected from the application of the Aristotelian theories.[43] But this was not yet enough to declare such theories and principles, like those underlining the processes of condensation and rarefaction, to be wrong. The appearance of the thermoscope challenged the Aristotelian doctrines in a much deeper way, however. In particular, when the thermoscope appeared, an important process of reception and transformation of ancient pneumatics had already taken place. It is the combination between the output of the process of transformation of the general theoretical principles of ancient pneumatics, mainly undertaken by engineers during the second half of the sixteenth century, and speculations about the functioning of the thermoscope itself, which not only presented a profound challenge to some aspects of the Aristotelian doctrine, but also led Galileo to his atomistic conception of heat. It is at this point that Galileo took advantage of the knowledge of pneumatics he had learned from engineers, especially during his youth.

The Reception of Ancient Pneumatics

The thermoscope is a pneumatic device powered by a heat source. When the heat source is applied, the air dilates or, in Aristotelian terms, it rarefies. Thus, not only was the thermoscope applied to study those natural phenomena which were traditionally explained on the basis of the processes of condensation and rarefaction, but its own functioning was explained on the basis of these same processes.

According to Aristotle, the process of the maturation of fruits could be described in terms of condensation and rarefaction. When the thermoscope appeared, Galileo tried to formulate a theory on the process of maturation of fruits by comparing it

---

[41]Gageo is a sarcastic Venetian distortion of the name Gaio. Sagredo is referring to the Aristotelian philosopher Bernardino Gaio, whom he knew personally.

[42]From G. Francesco Sagredo to Galileo, May 9, 1613, in *EN*, XI:506. For the translation of the entire letter, see pp. 231ff.

[43]The discovery that air can be colder than ice was still reason for great research efforts by the members of the *Accademia del Cimento* toward the mid-seventeenth century, who ideated many experiments to analyze the phenomenon of freezing.

with the functioning of the thermoscope itself. Galileo's idea is preserved in the form of a fragment:

> In the same way, that is, from the operations of heat and of cold, all of the fruits and crops mature. Because, if we consider the structure and the construction of these, we first see that the grape is constituted of berries, or we want to say blisters, and this is seen apparently in the grape where each berry is a blister; it is similar in pomegranates, figs, watermelons and others: since these blisters are full of humors, when the heat of the Sun comes, it presses them out and empties them, pushing out some of that humor so that they are withered in the evening; but when night falls and the air cools down, those blisters fill themselves with new humor, and more than they had sent out the day before, and therefore those blisters become much more capacious; and by means of this alteration they mature *making the same effect that the instrument makes*:[44] which is confirmed by the fact that, in the morning, they are very hard (*EN*, VIII:635–636).

Fruits mature just as the thermoscope works, with the only difference that berries can become more capacious whereas the glass of the thermoscope cannot. The heat of the day rarefies the air in the blisters of the grapes, pushing out their humor. When air condenses (contracts), the plant can produce its humor again. The only condition is that the plants have to produce a little bit more humor each day, which is then correspondingly pushed out by the increasing temperature from spring through late summer.

Although Galileo's fragment is undated, it is nevertheless possible to recognize that his idea of the way the thermoscope works expressed in the period between 1612 and 1615 showed an important distinction from that expressed in this fragment. Between 1612 and 1615 Galileo actually accepted the assumption about the functioning of the thermoscope that had been formulated by hydraulic engineers encountering ancient Hellenistic works on pneumatics for the first time during the Renaissance.

The reception of Hero's *Pneumatics* In Galileo's day in Italy, the reception of Hellenistic pneumatics was for the most part limited to the pneumatics of Hero of Alexandria. Hero's *Pneumatics* was probably originally written during the first century AD. This work contains a collection of technical applications and a theoretical introduction to pneumatic principles, revealed in the *proemium*.[45] Considering the longest version of Hero's *Pneumatics*, as Schmidt reconstructed it philologically (Hero and Schmidt 1899), there are only four technical applications that work on

[44] Author's italics.

[45] Schmidt's German translation of Hero's *Pneumatics* is considered here as a reference work: Hero of Alexandria and Schmidt (1899). In the following: Hero and Schmidt (1899). As Schmidt was able to show, there were numerous Greek manuscripts of Hero's work circulating in Italy during the early modern period. Not all of the circulating manuscripts contained the integral work, but most of them did include the theoretical introduction by Hero. Schmidt was also convinced that not all of the seventy-nine devices he listed were originally Heronian. He also considered the valve to be a hydraulic machine. Although the tremendous relevance of the Heronian valve for technological development cannot be denied, it is not, however, an application of a pneumatic principle. For more details, see Hero and Schmidt (1899, *Supplementum*, pp. 3–53).

the basis of the same principle as the thermoscope, that is, the principle according to which the volume of air changes when the temperature changes.

The versions of Hero's work circulating in the geographic areas Galileo frequented during his lifetime did not correspond with Schmidt's reconstruction. As regards the transmission of Hero's *Pneumatics* in Latin, one translation was relevant for Galileo: the translation by Federico Commandino, published posthumously in 1575 (Hero and Commandino 1575). Although this was not the final version Commandino intended to publish, because he died before it was completed, it is nevertheless held to be the *editio princeps* which eventually served as a great impulse for the process of diffusing Hero's *Pneumatics*. Its sixteenth-century propagation in Italian was indeed based largely on Commandino's translation. In 1582 Bernardo Davanzati (1529–1606) translated only the *proemium*,[46] while the first Italian translation of the entire work was accomplished by Oreste Vannocci Biringucci, also in 1582 (Hero and Vannocci Biringucci 1582).[47] In 1589 the Italian translation by the hydraulic engineer Giovan Battista Aleotti appeared. He was employed by the Duke of Ferrara (Hero and Aleotti 1589).[48] Aleotti's translation, which is very rich in technical applications, some of which were very new and introduced explicitly for the first time, is particularly relevant because of the commentaries he wrote on Hero's theoretical explanations. Finally, Alessandrino Giorgi translated Commandino's work into Italian with the explicit intent of making the *editio princeps* more understandable, since Commandino's published translation was not the final version and therefore still contained many convoluted passages. Giorgi's Italian translation was published in 1592 (Hero and Giorgi 1592).[49]

Pneumatic technology Concerning pneumatic applications, most of Hero's descriptions are of ludic devices that reflect the extremely advanced status of pneumatics during the entire Hellenistic era. Most of the devices he described were decanters, designed to accomplish a wide variety of tasks: Hero's *Pneumatics* includes a description of a sort of automatic wine dispenser, for example, and of awe-inspiring devices like doors that open without being pushed. The technology Hero employed, however, was the same technology applied to machines like the water lifting machines that supplied whole cities.[50] In Galileo's day, far more than

[46]Davanzati's translation of Hero's *proemium* was transcribed and published much later, in 1862. The manuscript seems to be lost (Hero et al. 1862).

[47]Published in http://www.echo.mpiwg-berlin.mpg.de/content/pratolino/sources/ Accessed October 2009. Transcription by M. Valleriani and T. Werner. For a detailed analysis of Vannocci Biringuccis's theoretical commentary work on Hero's *Pneumatics*, see Valleriani (2007). Oreste Vannocci Biringucci was a nephew of the famous engineer Vannoccio Biringuccio, author of *De la pirotechnia*, published in 1540.

[48]For an extensive study on Giovanni Battista Aleotti, see Fiocca (1998).

[49]Giorgi resolved to have the typographer of Commandino's work print his work as well, and even used the same engravings for the illustrations.

[50]For an introduction to Hero's *Pneumatics* and Hellenistic technology and science, see Russo (2001). Hero also left a work concerned with the construction of the automata. The technology employed for the functioning of these ludic devices, a technology able to exploit hydraulic

the ludic aspects of pneumatics could be experienced in everyday life. One instrument, which had been in use for many centuries and still was during his day, was the cupping glass for bleeding. This instrument was constituted of a vase divided horizontally into two parts. Once the lower part was warmed up, the glass was applied to the skin on the side of the upper part. Cooling the glass down slowly, the contraction of the air in the lower part was transmitted to the upper one by means of an air valve, which could be opened at will. Thanks to this contraction, then, the blood was sucked out, apparently with quite beneficial effects. The cupping glass for bleeding and the thermoscope thus work on the basis of the same principle. Another popular device in Galileo's day, whose functioning followed the same principle, was a milk pump for nursing mothers.[51]

Though the interest in pneumatics, which experienced a very high degree of articulation during the Hellenistic era, never disappeared during the Middle Ages, it certainly grew enormously from the thirteenth century on, and especially during the early modern period (Valleriani 2007), as apparent in the intensification of translations and commentary works at this time. In the publications, however, the ludic aspects seem to prevail. In fact, it was very easy to apply pneumatic principles to conceive water games, and the best locations to place such games were clearly gardens. In 1615, the famous architect and engineer Salomon de Caus (1576–1630), who oversaw the construction of the Palatina Garden in Heidelberg, published his *Les raisons des forces mouvantes* (Caus 1615), in which he described several machines and new pneumatic devices, among them some powered by heat, including a fountain that works using the rays of the sun (Fig. 5.5).

Pneumatic theoretical principles The reception of Hero's *Pneumatics*, moreover, also brought to general attention Hero's theoretical framework, on the basis of which the functioning of these devices is explained, and which is expressed in Hero's *proemium*. According to Hero, air is material and constituted of particles; among the particles are interstitial vacua, which can become larger or smaller due to the action of external factors (Hero and Schmidt 1899, 4–5). Hero focused his explanation in particular on air's capacity to contract. He gave the example of a sphere into which one blows forcefully and then closes its opening with a finger. If one then immerses the sphere into water and removes the finger from the opening, it is possible to detect a certain amount of air exiting the sphere violently. According to Hero, if there were no interstitial vacua in the body of air, it would not have been possible to blow more air into the sphere. The violence of the exiting air is due to the tendency of air to return to its natural state, that is, to the natural dimensions of the vacua. The interstitial vacua cannot only contract, but also enlarge. If, for example, the air is sucked out from the same sphere, according to Hero it is easy to detect that

energy, was the same one used for mills. In particular, see Russo (2001, 152–159). For a general introduction to ancient technology, see also Schürmann (1991).

[51] The instrument for nursing mothers is cited and described by Giovanni Battista Aleotti. For more details, see Hero and Aleotti (1589, 8), Valleriani (2007), and on pp. 177ff in this chapter.

Fig. 5.5 Solar-powered fountain (Caus 1615, Table 22)

a greater vacuum was "pulling" against the exit, because the vacua tend to return to their natural dimensions (Hero and Schmidt 1899, 8–11).

When a source of heat is added, the Heronian system becomes a little more complicated. When air is heated, it becomes a sort of corrupted body, because of the action of the element Fire, so that the air particles become thinner and eventually exit their container through the pores of the material. For this reason, the interstitial vacua are supposed to become larger because they "compensate" for the reduction in volume caused by the loss of particles. When the heating process stops, the enlarged interstitial vacua tend back to their natural state and thus "pull" the surrounding matter (Hero and Schmidt 1899, 10–11).

Although Hero postulated the existence of vacua between the particles that constitute a body, his general framework remained Aristotelian. For Hero the natural state of a body was also connected directly with its natural place. The world in which these phenomena take place is still the Aristotelian sublunar world, where bodies are constituted of the four Elements and their changes are related to a change in their natural position. Heating air or sucking it out from a sphere are violent actions that interrupt the natural motions of these bodies. Finally, Hero completely refuted the idea that an external vacuum, one which is not related to particles constituting a body, can exist.

When the Heronian interstitial vacua change their dimensions, a change in temperature,[52] a change in the extension of the volume, and a motion take place, as in the case of the Aristotelian principles. The only difference is that Hero did not introduce the distinction between form and matter, which means that when all the changes take place, there is still no change in form. This is what actually happens in pneumatic devices. According to Aristotle, the contraction of air is an aspect of the more general process of condensation, which brings a body to change its form. While this solution was still absolutely convincing with reference to meteorological phenomena such as rain, Renaissance engineers soon discovered that it was not sufficient to describe and explain how air can contract but still be air, that is, not enter into any general physical process of transformation, such as Aristotelian condensation.

Transformation of Hero's theoretical principles Since Hero's conception very clearly challenged the Aristotelian one, concerned with similar phenomena explained on the basis of the principles of condensation and rarefaction, most of those who translated and commented on Hero's work during the Renaissance could not avoid facing his theoretical framework. Once large-scale diffusion of Hero's text had been achieved, a theoretical debate arose because the intellectual equipment of engineers and professionals like Oreste Vannocci Biringucci, Giovanni Battista Aleotti, Galileo and Giovan Francesco Sagredo consisted primarily of Aristotelian doctrine, which appeared able to furnish an explanation of the same phenomena. Hero's principles contradict Aristotelian doctrine not only with regard to the principles of condensation and rarefaction, but also, and especially, because they suppose the existence of interstitial vacua. Engineers like Oreste Vannocci Biringucci and Giovanni Battista Aleotti, in particular, deliberately contributed to the theoretical debate that arose during the propagation of Hero's ideas.[53]

Aleotti's conception of heat The last word on this matter, among the Italian engineers of the sixteenth century, fell to Giovanni Battista Aleotti. His new interpretation of the functioning of pneumatic devices powered by a heat source, in particular, led him to abandon both the Aristotelian and the Heronian theoretical approaches. Aleotti explained his theory by means of an example, that is, with a direct description of the pneumatic milk pump for nursing mothers:

> These [the women] take a glass cruet with a neck on the upper part, which is wide enough to be able to contain the nipple of the breast, and they warm up its body [of the cruet] very well by means of fire until the heat, penetrating the thinness of the glass through the pores, pushes the air out from it and fills the body of the cruet with a very thin vapor, and when

---

[52]According to Hero's principles, however, a change in temperature takes place only when the pneumatic device works by applying a heat source.

[53]For a detailed description of the theoretical positions assumed by engineers during the process of receiving Hero's *Pneumatics*, and of the way such a process ended up, first, in a transformed science of pneumatics, and also with the abandonment of Aristotelian principles, see Valleriani (2007, 2009b).

> the mentioned body [the cruet] is warm enough, they immediately place the opening of the neck of the cruet at the breast, placing the nipple into it, and since that thin igneous vapor cannot remain in that [the cruet], it exits through the vacua of the glass, through which it [the vapor] penetrated it [the glass], and so begins rising upwards to its place, although from the air around, it is transmuted into aerial substance, and since through these *meatus*,[54] which are very thin, the air cannot enter, and since the vacuum cannot exist, that body, which cannot stay empty, immediately pulls milk from that breast, and by emptying it [the breast], it [the cruet] fills itself, and when it is completely full, it ceases pulling [...].[55]

While Hero's heat corrupts and makes particles of air thinner, so that they escape from the vase, in the case of Aleotti heat is a thin vapor which penetrates the vase and pushes away the air simply because heat is a body, which needs space and occupies a volume. This is what could be called a mechanical conception, because heat is seen as an object with a certain extension and ability to move other bodies. Since, after all, heat is a body, when it escapes the glass, milk is pulled in order to avoid the generation of vacuum. Aleotti therefore formulated a new conception of heat in order to determine the pneumatic principle, on the basis of which pneumatic devices powered by a heat source work.

Galileo as a Pneumatic Engineer

Galileo was well aware of how pneumatics developed among engineers during the Renaissance. First of all, he had the opportunity to share in such a development thanks to what has been called his apprenticeship in pneumatics. The first chapter showed how close Galileo's connections were to the workshop of the engineer Bernardo Buontalenti. In 1569 Buontalenti was commissioned by the Grand Duke Francesco I to construct the entire garden of Pratolino.[56]

The garden of Pratolino According to Sgrilli (d. 1755), an engineer in charge of maintaining the garden of Pratolino who wrote a complete description of it in 1742 (Sgrilli 1742), at every corner, in every artificial cave, by every statue there was a water game, where visitors could be suprised by jets of water. Water games

[54] Plural of the term "meatus," used in archaic English with the meaning of "small openings," "holes," "pores."

[55] "Queste pigliano una ampolla di vetro con il collo tanto nella parte superiore largo, che sia cappace del capitello della mammella, et riscaldano con il fuoco di essa il corpo ben bene fin che il caldo penetrando per li vacui la sottigliezza del vetro ne scaccia l'Aria riempiendo il corpo dell'ampolla di sottilissimo vapore, et quando è ben bene riscaldato detto corpo subito si pongono la bocca del collo dell'ampolla alla mamella dentro imponendovi il capitello, et perche quel sottil vapore igneo non puo star ivi renchiuso se n'escie fuori per quei vacui del vetro per gli quali entrò, et per levarsi in alto al suo luogo s'invia Se ben dal circomposto aria è trasmutato in sostanza aerea, et perche per questi meati, che sottilissimo sono non vi puo entrar l'aria non potendo esser vacuo subito quel corpo che, non può star voto tira da essa mammella il latte, et votando la viene à riempir se stesso, et ripieno a fatto, non piu tira [...]" (Hero and Aleotti 1589, 8). Author's italics.

[56] There is a massive body of literature concerning the garden of Pratolino. Due to their full coverage of the subject and historical sources, the texts that deserve mention above all others are Ulivieri and Merendoni (2009), Zangheri (1987), Dezzi Bardeschi (1985). For specific studies concerning pneumatics and the garden of Pratolino, see Valleriani (2007, 2009b, 2010, Forthcoming a).

Fig. 5.6 Fountain-parkway in the garden of Pratolino (Engraving of Stefano della Bella in Sgrilli 1742)

were implemented in every architectural element, including stairs, where water was sprayed out against pedestrians from each step (Sgrilli 1742, 12). The garden was then enriched with innumerable fountains and baths, and even a hot water supply (Sgrilli 1742, 14). One boulevard, 292 meters long, was lined with fountains that could project water not only vertically, but also from one side of the boulevard to the other so that the water formed a parabola, a few meters long, under which people could walk comfortably (Fig. 5.6) (Sgrilli 1742, 22). Hydraulic organs (Valleriani 2010) were installed both inside the villa of the garden and outside (Sgrilli 1742, 7 and 20). Gigantic systems of automata, also powered by means of hydraulic energy, rounded out the Heronian program.

Buontalenti's interest in pneumatics was not limited to mere practical realizations. Both Davanzati's and Vannocci Biringucci's translations of 1582 were actually completed at the request of the famous Tuscan engineer, who was about to start working on the garden of Pratolino after having completed its villa in late 1580 (Valleriani 2007). It also seems, finally, that Buontalenti built a perpetual motion machine which was able to lift up great quantities of water without the aid of any mechanical device. Although there are not enough details to conclude that such a device was of the flow and ebb variety like Drebbel's, this seems quite probable (Fara 1988, 204–207).

Galileo as a designer of pneumatic devices While Tuscan engineers were building first the garden of Pratolino, and then the one of Boboli, the young Galileo approached the science of pneumatics in the workshop as it was received by hydraulic engineers. This is then the material background, which, as special problems like the meteorological ones emerged, served as the framework for Galileo's work on and with the thermoscope. However, his attention to the science of pneumatics was not limited to his youth and the period when he constructed the thermoscope. Galileo remained a renowned and recognized expert on pneumatic devices throughout his entire life.

First, Galileo certainly possessed two of the mentioned translations of Hero's *Pneumatics*. He had Commandino's translation, as is clear from the comparison between Commandino's book and a letter Galileo wrote in Padova in 1594 that was sent to Alvise Mocenigo in Venice, in which he described an oil lamp.[57] The other translation Galileo possessed was the one by Giorgi of 1592, as emerges from the partial inventory of Galileo's library published by Favaro (Favaro 1866, 54).

As early as 1611 Galileo served as a consultant on pneumatics to Antonio de Medici, who ordered him to describe the design and the functioning of a fountain to Francesco Maria del Monte in Rome.[58] The hot phase of Galileo as a pneumatic designer started some time later, however. In late July 1613, Galileo sent Sagredo some flasks of his red wine as a gift, which, according to Sagredo, was so good that he could not keep himself away from it. Thanks to Galileo's wine, Sagredo became a pneumatic designer first:

> After the arrival of the very precious wine of Your Lordship, and with this heat, my intellectual purpose lies in measuring that heat while I have cold drinks. [...]. I also found: a decanter that, when the wine passes through it, cools down immediately, and if needed, warms up; some glasses in order to drink it with ice, and another where, once the wine is introduced into it, one can see how many degrees of cold it has taken, and it can also be used to drink; an inkwell that preserves the ink in this hot weather so that it does not dry up, become thick, or make the pen too wet, which is cheap and lasts a long time. After drinking two glasses of the wine of Your Lordship, these inventions came to me so now I hope that, as soon as I have drunk only one of your flasks, I will have invented divine things.[59]

Unfortunately, we shall never know what divine things Sagredo invented that night. The reception of pneumatics started bearing fruit at the beginning of the seventeenth century, not only thanks to engineers commissioned with the construction of marvelous gardens, but also to everyone else who approached this science and its technology with an open mind. In this context, the conceptual reshaping of the ancient pneumatic device in terms of an instrument to measure temperature allowed devices to be developed that were capable of showing and even controlling the temperature of liquids.

Galileo learned from his friend Sagredo that many applications of the same principle could be realized besides the thermoscope; he never forgot this lesson. In

[57]Galileo to Alvise Mocenigo, January 11, 1594, in *EN*, X:64–65. For the translation of the entire letter, see pp. 219ff.

[58]For more details, see F. Maria del Monte to Antonio de' Medici, April 8, 1611, in *EN*, XI:83–84.

[59]From G. Francesco Sagredo to Galileo, July 27, 1613, in *EN*, XI:545. For the translation of the entire letter, see p. 233.

1627 he was still overseeing the construction of devices like the ones suggested by Sagredo. On June 27, 1627, the military officer Baglioni Malatesta (1491–1540) wrote Galileo from Pesaro that:

> Since I have known that the Very Excellent Lord Don Carlo Barberini has a glass, invented by the high mind of Your Lordship, which shows the degrees of heat and cold that one drinks, I came to the wish of having a drawing of it.[60]

It is impossible to know whether Galileo appropriated Sagredo's invention and told Barberini it was his own. However, Baglioni's letter makes evident that at the time Galileo was also known as a designer of pneumatic instruments. Galileo's response to Baglioni shows that he offered to have the glass built in Florence and dispatched to him.[61] Presumably, Galileo already had at his disposal a network of such experienced craftsmen in building pneumatic devices that it cost him less time and effort to have the glass made and send it than to prepare a written description. Taking advantage of Galileo's generosity, Baglioni asked for two glasses instead of one; at the end of the year, both arrived safe and sound in Pesaro.[62]

The thermoscope, his activity as a pneumatic consultant and the production of drinking glasses showing the temperature of the liquid within are evidence that substantiates Galileo's experience as a designer of pneumatic devices.[63] Galileo presumably was able to accomplish these tasks on the basis of the reception of Hero's work, which he experienced indirectly in the workshop of Buontalenti, who was busy building the garden of Pratolino during this period. In conclusion, these are all of the material conditions which allowed Galileo to investigate the theoretical background of not only the thermoscope, but also of that new pneumatic science which resulted from the process of reception and transformation of Hellenistic pneumatics.

The Functioning of the Thermoscope

When Daniello Antonini, Galileo's former pupil, wrote in 1612 to report that he had built a perpetual motion machine like that of Drebbel in Brussels, he also concluded that, as he learned from Galileo, the thermoscope is a mechanical instrument:

> [...] because I know well that there is no difference between this motion [of the *perpetuum mobile*] and that of a water mill apart from the cause of motion, which is seen by everybody [in the water mill], whereas in this case it is not.[64]

---

[60] From Malatesta Baglioni to Galileo, June 16, 1627, in *EN*, XIII:363.

[61] Malatesta Baglioni to Galileo, July 17, 1627, in *EN*, XIII:367–368.

[62] Malatesta Baglioni to Galileo, December 12, 1627, in *EN*, XIII:380.

[63] Galileo was still active as a pneumatic engineer in 1635, when he sent to his friend Micanzio in Venice a design for a fire hydrant. For more details, see Fulgenzio Micanzio to Galileo, Septemer 15, 1635, in *EN*, XVI:310–311.

[64] From Daniello Antonini to Galileo, February 4, 1612, in *EN*, XI:270. For the translation of the entire letter, see pp. 227ff.

Galileo's 1615 explanation of the thermoscope Galileo's precise opinion about the way the thermoscope works, as Antonini learned from him, probably corresponded to the conception he formulated in 1615. This is the first evidence that testifies directly to Galileo's opinion about the way the thermoscope worked, and was formulated by Galileo upon Sagredo's request. Galileo's response was lost, but not so Sagredo's comments on it:

> I understood your opinion about the way those instruments function [. . .] and I would even dare to say [it is] also true, *if it were not for the reason that in itself it is not evident to the senses*. [. . .]. But it satisfies the mind much more than the arguments of the Peripatetics: If, because of the external heat, the air that is inside the warmed glass bowl evidently dilates so that it pushes out the water, it is easy to believe that the heat penetrates the glass. Once it has penetrated there in greater or smaller quantity, it requires more or less space. Since it [the space] cannot simultaneously contain the air and the soft and igneous spirit, the air is obliged to exit the space. In addition, when the external environment cools down, it is believable that the igneous spirit, which is overabundant in the bowl, exits until it equilibrates with the environment. Thus, since the space that contained it becomes empty, the air is obliged to follow, and water or wine after it.[65]

Galileo started by considering the temperature of the air outside the bowl of the thermoscope. When this becomes hotter, the heat penetrates the glass, and, since the latter was already full of air, more space should be produced: The air is pushed down and therefore the water descends. Galileo supposed therefore that heat, in the form of igneous spirit, is a body with a certain and definite extension, so that it would be able to cause mechanically the motion of other and different bodies.

Galileo's opinion is extraordinarily similar to Aleotti's, from which it presumably was adopted. From the end of the sixteenth century on, pneumatic devices were explained either on the basis of the Aristotelian principles of rarefaction and condensation, or on the basis of those principles formulated by the engineers—in particular, those formulated by Aleotti. Galileo, therefore, clearly influenced by the knowledge of the engineers, decided to abandon the Aristotelian approach, at least temporarily, in favor of the mechanical explanation. Because of Sagredo's critique, Galileo later abandoned the mechanical explanation as well and tried to formulate a new pneumatic theory, into which he also integrated a new version of the Aristotelian processes of condensation and rarefaction.

Sagredo's critique and Galileo's experiments Sagredo did not consider Galileo's conception to be evident to the senses, and his explanation was indeed highly abstract, apparently impossible to show in the literal meaning of the word. Galileo took such criticism very seriously and started performing new pneumatic experiments, never before evaluated by historians. The first thing he did after read-

[65]From G. Francesco Sagredo to Galileo, April 11, 1615, in *EN*, XII:167–168. Author's italics. For the translation of the entire letter, see pp. 244ff. In the same letter, Sagredo described an experiment he performed to demonstrate the existance of the void.

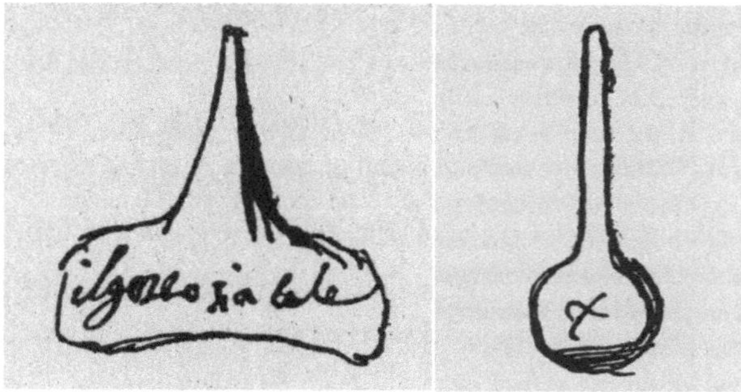

Fig. 5.7 Galileo's notes on for pneumatic experiments to make the igneous spirit visible (*EN*, XII:170)

ing the stirring letter he received from Sagredo was to annotate, on the same folio, a couple of experiments clearly intended to make his opinion evident to the senses:

> 1) Hold an empty decanter over the fire and from the mouth (which has to be very narrow) observe by means of an air valve[66] whether the igneous spirit exits continuously (Fig. 5.7 left).
> 2) Introduce into decanter x a very small quantity of wine, ink, quicksilver, etc. Then, place it over the fire, observe whether the mentioned [things] are consumed, etc. or what it makes (Fig. 5.7 right).[67]

These two experiments were intended to make the igneous spirit evident to the senses and presumably to investigate its characteristics. Obviously Galileo was not able to observe the igneous spirit exiting the flask, although under certain circumstances the air valve might have exhibited some movement. In the second experiment Galileo probably applied an air valve as well. There is no historical evidence on the results of these experiments, but the subsequent progression of Galileo's research on this topic seems to show that they turned out to be inconclusive.

Galileo's second explanation of the thermoscope If the igneous spirit could not be shown to the senses, if its existence could not be verified, it was difficult to affirm that the igneous spirit is a body with a certain extension, which occupies space, and which is able to move other bodies. Probably because of this dramatic problem, Galileo changed his approach completely and formulated a new pneumatic theory. This new formulation is preserved in a textual fragment written perhaps in 1619 or

[66] The use of an air valve, specific to such an experiment, testifies further to Galileo's familiarity with pneumatic technology.

[67] Notes in Galileo's hand in the margin of the letter from G. Francesco Sagredo to Galileo, April 11, 1615, in *EN*, XII:170. Author's enumeration. For the translation of the entire letter, see pp. 244ff.

shortly thereafter.[68] Galileo's formulation of his new pneumatic theory about the way devices powered by a heat source work is the following:

> At the schools of the philosophers, it is approved as a true principle, that a characteristic of cold is the tightening, and of heat the rarefying. Now, this accepted, be given that the air contained by the instrument has the same temperature[69] as the rest of the air in the room where it is placed; and since these two bodies have the same specific gravity, one does not push away the other [...]. But if the air which surrounds the bowl cools down, because of the displacement of some cooler bodies, the heat particles [*calidi*] contained in the air within the bowl, since they are in a medium [which is] heavier than they are, will ascend, and that air will become cooler than before; and so, because of the principle just mentioned, it will tighten and occupy a smaller space: hence (*ne detuur vacuum*) the wine will ascend to occupy the space left free by the air; and then, having warmed up that air, since it rarefies and needs more space, it will push down the wine, which, since it is heavy, gladly will surrender that place to it; hence it follows that cold is nothing but a loss of heat (*EN*, VIII:634–635).

Galileo first redefined the Aristotelian principles of rarefaction and condensation. Since rarefaction means less weight, and condensation more density and more weight, introducing a cold body into the arrangement intially makes the air in the room colder, that is, heavier. The process of communicating the temperature between the air in the bowl and the air of the room is not specified, but movements are explained in terms of specific gravity. The colder surrounding air is a heavier medium which, therefore, pushes away and upward not that air which is in the bowl, but those components of it which make it lighter (and hotter), in order to obtain equality in terms of specific gravity.[70] These components, finally, are *calidi*, a concept that is intended to specify a form of igneous particles.[71]

Having abandoned the mechanical explanatory model, Galileo saw no alternative but to hark back to the Aristotelian principles of condensation and rarefaction. In order to operationalize them, he had to find a way to explain, first, how changes of temperature can take place without assuming that heat occupies space and, second, how changes in temperature can take place before the Aristotelian principles take effect. Concerning the issue of space, the igneous particles do not seem to have any shape, and no more space is automatically created in the bowl when they leave the air. Concerning the processes of condensation and rarefaction, had Galileo postulated that changes of temperature are their consequence, he would have had

---

[68]The fragment shows how Galileo approached the idea that heat has a discrete nature. Since he published his conception of heat according to this approach in *Il Saggiatore* in 1623, and since the debate that ended with this publication began in 1619, it can be circumstantially inferred that this fragment was written between 1615 and 1623, perhaps around 1619.

[69]Galileo used the word *temperie*, which denotes not only "temperature," but rather the general situation of the weather or climate.

[70]Galileo's explanatory model, in this case, corresponds to the one used in his *De motu antiquiora* where he approached the phenomenon of free fall in terms of hydrostatic phenomena. For more details, see Van Dyck (2005, 868).

[71]For an introduction to the philosophical background of early modern atomism, see Lüthy (2003).

to address the problem the engineers were facing already, that is, that such processes involve changes in form, despite the fact that no change in form can be observed in the thermoscope, for instance, from air into water. In fact, introducing the *calidi* allowed Galileo to explain how changes of temperature can take place without appealing to the processes of condensation and rarefaction. These start operating only after the *calidi* have left the air and the latter has become colder. At this point the condensation process explains how air contracts and, since no vacuum can exist, the liquid ascends. From this perspective, Galileo's first formulation of the first atomistic conception of heat is a consequence of his integrating some aspects of Aristotelian doctrine into his new theory.

Yet the conclusion of Galileo's fragment that cold is nothing but a loss of heat does not seem to follow from this line of reasoning. This conclusion can be better understood through a mental experiment Galileo relates in his *Discorso intorno alle cose che stanno in su l'acqua* of 1612:

> [. . .] and one sees, thanks to the experiments, this air ascending faster through the water than the igneous exhalations through the air: hence one necessarily concludes that the same emanations ascend much faster through the water than through the air, and that, consequently, they are moved because they are pushed away by the environmental medium, and not because of an intrinsic principle, which is in them, [when the movement is] fleeing from the center, toward which the other heavy bodies tend. [...] in the elementary bodies there is only one intrinsic principle of movement, which is that toward the center of the Earth, and that the only reason for the upwards movement (speaking only about that [movement] which appears as a natural motion) is the pushing away by the medium [which is] fluid and heavier than the mobile; [...] (*EN*, IV:86).[72]

According to the Archiemedean framework, of which Galileo evidently made use (Bertoloni Meli 2006, 6), the only intrinsic motion of the bodies is the addressed motion downwards. Since such a downward motion took place in Galileo's model of the thermoscope when the temperature rose, the cooling process turns out to be only a loss of heat, because the ascending motion resulting from the cooling process is not an intrinsic motion, but rather, of course, one caused by the different specific gravities of the mobile and of the medium. Through the combination of the last fragment and this passage from *Floating Bodies*, it emerges that Galileo was on the verge of superseding the early distinction between degrees of cold and degrees of heat. But Galileo never actually took this step.

The explanation of the way the thermoscope worked is therefore based on the following five principles: (1) the principle of rarefaction and condensation, altered such that they do not involve change of form; (2) nature's abhorrence of a void; (3) the natural inclination of bodies to move toward the center of the Earth; (4) the difference between the specific gravities of the media as the cause of those natural

[72]For the emergence of this model in Galileo's early text *De motu antiquiora*, see Bertoloni Meli (2006, 52–53).

movements which are not directed toward the center of the Earth;[73] (5) the discrete constitution of heat. The first three principles belonged to the Peripatetic school and were derived directly from Aristotelian doctrine. The fourth principle, namely the introduction of the concept of specific gravity, applied the Archimedean explanatory model for natural motion. The last of the principles, according to which igneous particles exist, seems to be a genuine result of Galileo's analysis of the pneumatic instrument invented to measure cold and heat.[74] Neither in 1612, when he wrote the *Floating Bodies*, nor in 1615 when he exposed to Sagredo his early opinion on the thermoscope, had Galileo yet embraced this view.

Thus, what can be defined as Galileo's first atomistic conception of heat, presumably formulated around 1619, was developed from his work on pneumatics and, in particular, from his attempt to explain the way those pneumatic devices which are powered by a heat source worked. Galileo further developed his conception and published it in *Il Saggiatore* in 1623.[75] Having achieved this final step, Galileo then went back to his roots and attempted to solve Bardi's problem in a new way.

Galileo's Doctrine of Heat

The dispute that led Galileo to the publication of *Il Saggiatore* took place between 1619 and 1623 and concerned the nature of comets, after three of these celestial bolides appeared in 1618.[76]

The dispute behind *Il Saggiatore* A *disputatio* concerning the appearance of the comets was held at the *Collegio romano* in 1619 and originally published anonymously in the same year under the title *De tribus cometis anni MDCXVIII* (*EN*, VI:21–35). The author was the Jesuit Orazio Grassi (1583–1654), against whose

[73]The idea that upward motion is caused by heavier medium might have been mutuated from Galileo's early collaboration with Jacopo Mazzoni during his stay in Pisa as a lecturer of mathematics between 1589 and 1592. In a later work (Mazzoni 1597), Mazzoni clearly described this principle. For more details, see also Bertoloni Meli (2006, 61).

[74]Galileo also attempted to apply his discrete theory of heat to some subjects related to the science of life. Another fragment reports Galileo's opinion about human numbing: "That human beings die numb with cold happens because the environmental cold consumes all those igneous atoms that it finds in the limbs and, therefore, since the natural calor is no more there, one dies" (*EN*, VIII:635).

[75]During the same period other scholars like Giuseppe Biancani and Francis Bacon, for example, tried to investigate the principles on the basis of which the thermoscope works. In his *Sphaera mundi*, Biancani stated that the functioning of the instrument can be explained by means of the principles of rarefaction and condensation, and through the fact that nature abhors a vacuum. For more details, see Biancani (1620, 111).

[76]As a consequence of this dispute, the equilibrium dominating relations between Galileo and the *Accademia dei Lincei* on one side and the Company of Jesus on the other broke down once and for all. This equilibrium had allowed Galileo and the Jesuit natural philosophers to work in a climate of friendly collaboration. The new cultural environment that emerged after this quarrel can be viewed as the source of the political activism which eventually led to Galileo's abjuration. For more details, see Redondi (2004).

text Mario Guiducci, a pupil of Galileo, published a rhetorically quite provocative response—*Discorso delle comete*—in the same year (*EN*, VI:37–108). In his text Guiducci fiercely criticized not only Grassi's opinions, but also those of Tycho Brahe, which were widely accepted among the Jesuits, and of Aristotle. Guiducci's criticisms are elaborated explicitly on the basis of what he claimed he learned from Galileo.[77] The order of the Jesuits, and especially Father Grassi himself, reacted negatively to Guiducci's affront. Father Grassi immediately prepared a severe reprimand entitled *Libra astronomica ac philosophica* and also published in 1619, under the pseudonym Lothario Sarsio Sigensano (*EN*, VI:111–180). Believing Galileo to be the spiritual father of Guiducci's work, Grassi implicated Galileo directly in the quarrel. While Guiducci attested to his paternity of the text in a public letter sent to the Jesuit Tarquinio Galluzzi (1574–1649), Galileo read and annotated Grassi's *Libra astronomica*. In 1622 he sent the elaborated response to the *Accademia dei Lincei* in Rome, whose members decided to publish it a short time later. In late 1623 Galileo's response was published under the title *Il Saggiatore* (*EN*, VI:197–372). *Il Saggiatore* is the last public word by Galileo on this quarrel.[78]

The conception of heat between 1619 and 1621 The point of departure for Grassi's 1619 discussion of heat is the scholastic assumption that heat is caused by motion (*EN*, VI:32). Through Guiducci Galileo criticized this view, objecting that not motion, but friction is the cause of heat. Reproposing Aristotle's famous experiment with the arrow that warms up while flying through the air, Galileo stated that it was erroneus and offered an example that illustrates how heat is caused by the friction between solid bodies. When heat is generated, moreover, at least one of the two bodies is consumed. If there is no consumption, neither can there be any heat (*EN*, VI:55–56). Grassi's reply in the *Libra astronomica* is less a critique than an attempt to reconcile Aristotle with Galileo. He accepted Guiducci's statement, but added that, since there is no friction without motion, then motion is, albeit secondarily, the cause for the generation of heat. Moreover, Grassi's explanation about the generation of heat added two further aspects. First, that the processes of rarefaction and condensation operate such that the consumption of the bodies can be explained as an effect of the combined actions of these principles, on the one hand, and of mechanical friction, on the other. Second, he also took into consideration the friction between bodies which are not solid, like air, for example, in order to redeem the case of an arrow that warms up in flight (*EN*, VI:160–161).

---

[77]The *Collegio romano* protested that the true author of the text was Galileo himself. Although Guiducci and Galileo always denied that the latter was involved in the compilation of the work, today it has been confirmed that they were lying. In fact, a manuscript by Galileo is preserved in Florence, which makes clearly evident that he prepared the first draft of Guiducci's *Discorso delle comete* (*EN*, VI:672–680).

[78]When *Il Saggiatore* was published, Grassi's superiors prohibited his scholars from publishing any further texts on the same subject, probably after having decided that Galileo's response had surpassed the boundaries of tolerance, and that any action against his work must be of a political nature. Grassi, however, did write another work, published in Paris in 1626 (*EN*, VI:375–500). Galileo also began annotating Grassi's works of 1626 but he ultimately chose to leave his notes in the drawer. Galileo's notes were published by Favaro together with Grassi's work of 1626.

Galileo's admission of failure In 1623 Galileo challenged all of these aspects of Grassi's reply, though in different ways. First he rhetorically rejected the use of secondary causes, stating that, if Aristotle said motion was the cause of the generation of heat, he was not secondarily right, but merely wrong. Second, he approached the principles of rarefaction and condensation with more delicacy. As has been shown above, Galileo applied a modified form of these principles in order to explain the way the thermoscope worked, combining them with the conception of *calidi*. Probably because of the resulting uncertainty about extending these principles' applicability, Galileo admitted to Grassi in *Il Saggiatore*:

> [...] how this business of rarefaction and condensation works, about which it seems to me that Sarsi speaks with great confidence, I would have gladly seen explained in a clearer way, since, to myself, it is one of the most recondite and difficult questions of nature. (*EN*, VI:331).

Faced with these principles Galileo balked. As he admitted, he could not accept the way the philosophical schools used them and when he applied them himself, he combined them with other non-traditional assumptions. The fact that Galileo never published a word about the way the thermoscope worked or about his newly formulated pneumatic principle, and his admission that he was not able to understand the way the processes of condensation and rarefaction work, clearly testify to Galileo's dissatisfaction with his own theory about the functioning of pneumatic devices powered by a heat source.

Galileo's 1623 conception of heat The third aspect of Grassi's conception, according to which friction could occur even among bodies that are not solid, finally gave Galileo the opportunity to disclose the core idea of his new doctrine of heat. Galileo first rejected the rough idea that a stone which, thrown vigorously into the air, can warm up (*EN*, VI:330–331). Then, Galileo rejected the Aristotelian description of heat:

> [...] a true accident, affection and quality which really resides in the matter, from which we feel warm ourselves. (*EN*, VI:347)

The scientific investigation of the temperature of bodies would, according to the Aristotelian doctrine, be based on the relationship between an inherent quality of the matter and a sense, inherent to the perceiving body. Galileo objected against this view, offering a conception of heat as a factor independent of the singular perceiving bodies and, taking advantage of what he had achieved previously, stating that:

> [...] the operation of fire taken alone is nothing but, since it moves, penetrating all the bodies thanks to its greatest thinness, and dissolving them faster or slower according to the multitude and velocity of the igneous particles and to the density or rarity of the matter of those bodies; and of those bodies there are many, which, while dissolving, change into other smaller igneous particles, and it continues the dissolution until it finds dissolvable matters. But that besides the figure, the multitude, the motion, the penetration and the touch, there is in the fire another quality, and that this is the heat, I do not absolutely believe; [...] (*EN*, VI:350–351).

One of the principles Galileo used to explain the way the thermoscope works, namely the discrete nature of heat, became the core idea of his conception of heat:

The action of fire depends solely on the igneous particles of which it consists and on their motions. Fire, which was considered the hottest "body," does not contain heat. The sensation of heat is then easily explained by means of contacts between the perceiving body and the igneous particles:

> [...] those matters which produce and make us feel heat in ourselves, which we call with the general name of fire, are a multitude of smallest bodies, shaped in this way and that, moved with much and much velocity; which, meeting our body, penetrate it thanks to their greatest thinness, and their touch, made during the passage into our substance and felt by ourselves, is the affection which we call heat, glad or troublesome according to the multitude and the smaller or greater velocity of those smallest [particles] which sting and penetrate us, glad is that penetration, thanks to which our necessary and imperceptible transpiration is facilitated, troublesome that, when too great a division and dissolution of our substance takes place: [. . .] (*EN*, VI:350).

The new heat particles are no longer *calidi*, which is a concept suggesting that they are somehow ontologically defined on the basis of the quality "heat" (*caldo*). Instead, the new heat particles are defined only in terms of shape and motion. It is the effect of their encountering and penetrating the body that is called heat. The last Aristotelian residues were thus abandoned, paving the way for Galileo's general atomistic conception of matter, as expressed in the First Day of the *Discorsi* (*EN*, VIII:65–138).[79]

The final solution to Bardi's problem On the basis of the newly achieved conception of matter and heat, Galileo was finally able to furnish a new answer to the old problem of Bardi. A final fragment, undated but attributed here to after 1622, relates Bardi's problem and its solution in an updated form:

> The water placed in a room has the same temperature of the room where it is, since both equally partake of the igneous atoms. The reason why a hand, which is kept in the air and seems hot to you, then cools down when it is put into the water, is the following: if one considers both the external and internal heat [of the hand], while it remains in the air, its own igneous atoms can exit and these cause the heat; but put into the water, its [of the water] particles fill and close the entrances through which the mentioned atoms exit, because the parts of the water are bigger than the porosities through which they escape; and this does not happen in the air, since they find a free field, because they are not kept by the parts of water, since they are smaller than the pores through which they escape: because heat is nothing other than the contact and the tickling of those igneous atoms, which, when they escape, touch the limbs of the body (*EN*, VIII:635).

Galileo's final solution to Bardi's problem is based completely on his general atomistic conception of heat. All of the kinds of matter involved in the problem—heat, air, water—are conceived as constituted of particles. The mechanical interaction among these is the final key to explain pneumatic phenomena and, in general, all of those phenomena concerned with heat.

[79] In particular, for the analysis of the First Day of Galileo's *Discorsi* and, especially, for an interpretation according to which Galileo's atomistic conception of matter should be intended as mathematical rather than physical, see Biener (2004).

The Generation of a Heat Doctrine

During his apprenticeship in Florence, Galileo partook of practical knowledge in the field of pneumatics. The available sources do not allow the precise determination of the extent of Galileo's involvement, for instance, with colossal projects such as the hydraulic and pneumatic engineering for the garden of Pratolino. Certainly, however, his knowledge of the field allowed him to approach pneumatic phenomena as subjects for his teaching activity in Padova. Moreover, Galileo shared theoretical knowledge concerned with pneumatics and developed by hydraulic engineers at the end of the sixteenth century. At this time, in fact, and for certain specific fields of activity, engineers, too, began developing reflective knowledge, basing their investigations on their own experience. Such a process of generation of knowledge followed the same model as, say, the one of the Aristotelian commentators of the *Mechanical Questions*, seen in the previous chapter. Indeed, the hydraulic engineers formulated their theoretical knowledge within the framework of a new, enlarged and commented edition of another relevant work from antiquity, namely Hero's *Pneumatics* (Valleriani 2007). In particular, Galileo shared the then widely circulating view formulated by Giovanni Battista Aleotti and published in his 1589 edition.

Pneumatic devices challenged the Aristotelian doctrine and specifically Aristotle's processes of condensation and rarefaction, and this is the reason why engineers like Aleotti developed a new theoretical approach. From the perspective of Galileo, however, the appearance and especially the advent of a new instrument for use in science, namely a pneumatic device to measure degrees of cold and heat, represented a new challenge to the very theories developed by engineers at the end of the sixteenth century. Confronted with this challenge, Galileo decided to prove his first view by accumulating relevant experience and then to try to create a new theoretical explanation of how the thermoscope worked by resorting to the Aristotelian doctrine and eventually transforming it appropriately.

However, this intellectual undertaking by Galileo was a failure that created the background for developments that went much further: a new conception of heat completely structured on the basis of a corpuscolar and mechanical vision. Once more, therefore, the generation of new knowledge was the result of a process of integrating practical knowledge, directly shared and indirectly assumed via theoretical formulations of sixteenth-century engineers, with the fundamental Aristotelian scientific categories of his time. Galileo's atomistic conception of heat, as published in his *Il Saggiatore* of 1623, is therefore the result of the work of an Aristotelian engineer.

Part III
The Engineer and the Scientist

Chapter 6
Was Galileo an Engineer?

The most accepted view on Galileo portrays him as the great theoretician, the genius, the lone thinker who founded the modern science that changed the world dramatically through the scientific revolution. But Galileo was also an engineer, even a craftsman, who spent his life working with engineers, masters, and mathematicians. He was neither a genius nor a lonely thinker: His science is rooted in the practical knowledge of his time, and the paths of his speculations can be understood perfectly if the context of his work and of the problems considered urgently in need of a solution are taken into account.

Starting from the historical background against which the contours of Galieo's profile as a military engineer become evident, this last chapter integrates the results of both of the previous sections to construct a general vision of Galileo's scientific practice. It will be shown how Galileo can be seen as an Aristotelian engineer, and according to which epistemological model he succeeded in generating new scientific knowledge. The chapter concludes with a general definition of the early modern engineer-scientist based on the results presented in this work.[1]

Revolution of the Art of War

The historical background against which Galileo's practical activities should be regarded in their entirety is represented by the dynamic triggered above all by the revolution in warfare, a consequence of the introduction of firearms, and especially of mobile heavy artillery. The period from Charles VIII's descent to Italy, 1492–1494, to the mid–sixteenth century is characterized by a massive diffusion of firearms carried by single men, such as the harquebus, and also and especially by the creation and diffusion of a new kind of heavy artillery, namely mobile, heavy artillery powered by gunpowder. The diffusion of such weapons was due mainly to major and intensive developments in the field of metallurgy. Works such as

[1] While presenting the achievements of this work in summarized form, this last chapter only cross-references, either in the text or in the footnotes, the passages in which the issues are discussed. To avoid duplications, no further reference will be made to the sources already discussed.

M. Valleriani, *Galileo Engineer*, Boston Studies in the Philosophy of Science 269,
DOI 10.1007/978-90-481-8645-7_6,

De re metallica by Georg Agricola (1494–1555), first published in 1530 and again in 1541, and *De la pirotechnia* by Vannoccio Biringuccio, published in 1540, represent fundamental historical milestones for the development of this kind of art and it is no coincidence that they were published at this time.

After the first military campaigns using heavy artillery, it immediately became evident that no city or fortress could defend itself against these new weapons. Italian artist-engineers therefore were asked to find a solution. First, Italian cities, duchies, republics, and principalities had to equip their militaries with firearms. Second, fortresses and defense elements in general had to be rearranged, rebuilt or newly built in order to make them able to withstand artillery for a reasonable length of time.

Many sorts of new specializations were required to build and use firearms: Melting and casting processes, products of fine mechanics, gunpowder production, artillerist training, the development of ballistics, the construction of lifting machines, the building of larger and more robust ships, and the development of new mathematical instruments, were all necessary consequences of the need to have and use artillery. The use of artillery in the context of an attack on or a defense of a fortress, moreover, implied the development of new military strategies, which were necessarily based on the firepower available.

The power and functioning of the new artillery was codified in a geometric way with respect to the directions of possible trajectories. Comprehensive attack strategies were increasingly described in geometrical language, the position and range of light and heavy artillery, shapes of groups of combatants and their movements over the battlefield towards the fortresses were described in terms of geometrical shapes and their characteristics. Soon the bags of military officers and artillerists filled up with a number of mathematical instruments. The quadrant, caliber, compass, proportional compass and ruler were all instruments used at this time, and without which no war activity was conceivable. Practical arithmetic, geometry and drawing skills were part of the knowledge required to operate on the battlefield. Positioning artillery, and establishing the direction and velocity of the movements of soldiers, developed into an art for itself, and was based on geometry. Thus a new kind of officer skilled in this art was required, as well as people to train such cadets.

The diffusion of these new weapons and the rapid increase in their destructive power washed over the land like a tidal wave. Every village, city and fortress was in danger. Most of the fortified structures existing at the time were unable to cope with the intense cannon fire, simply because at the time they were built such weapons did not exist. The equilibrium between the attack and defense strategies of the medieval tradition had been thoroughly destroyed by developments in the field of metallurgy. A new equilibrium was later established with the emergence of the so-called "new method" (or geometric method) of fortification. In fact, the answer to the developments in metallurgy came from the field of architecture during the period that laid the foundations for the emergence of the discipline of "military architecture."

On the defense side, military architects planned the position, height and shape of protection walls and of fortresses on the basis of the same practical knowledge of geometry. The fortresses became ever more heavily equipped with artillery, and specialized artillerists and officers, with their instruments and practical knowledge, became the backbone of the defense. The geometrical shapes of fortresses were designed in order to deter the greatest possible number of attack strategies.

The knowledge related to the new art of war gradually became increasingly codified and its codification increasingly standardized. Such a process had begun back during the fifteenth century when the method of *fortificar di terra* (earthworks) came into use, particularly in Tuscany. This was a method of building basic blocks made of earth and wood and assembling them to obtain relatively stable but temporary architectural defense elements. These elements were built fairly quickly, for example, in front of old fortresses or close to the old walls of a city, and functioned as additional defense structures.[2] Due to their provisional nature they could resist attack for only a relatively short period of time and were thus conceived as a temporary reaction to immediate danger. They were not architectural structures built for the safety and security of a city as they lacked a stable character. However, this was certainly the best known solution until the 1530s. At this time, the bastion was conceived—the key architectural element of the new fortress.[3]

The conception of the bastion was the most important reaction to the appearance of firearms. According to the functions of such an architectural element, a fortress became the output of an elaborate, predefined geometrical computation. For certain architectural components of fortresses, it was even necessary to conceive a new kind of masonry. The bastion was immediately and widely employed with great success as a fundamental element of defense against the new attack strategies. A new sort of architect was therefore required. Cities, moreover, had to be defended by means of walls and, when cities had to be enlarged as a consequence of increasing urbanization, new sectors were planned according to military needs. This aspect, too, required new abilities and skills in the field of architecture.

Two further aspects make this process of change particularly relevant. First, it was imperative. Without undertaking such changes, a principality or town would have been occupied in short shrift, and its political body deprived of power. Thus the process of change was inevitable. Second, this process of change is better described as a tremendously powerful wave, which completely changed the Italian landscape and established new habits and needs. In the timespan of around half a century changes of this kind transformed not only whole territories, but also their economies and political systems. Military requirements and the related fundamental knowledge were recognized as vital for states, municipalities, republics and duchies, so that engineers, architects and all those possessing any of the practical knowledge required found a natural path for their work and pursued successful careers in the field of the art of war. According to Mario Biagioli, such changes caused those practical mathematicians of the sixteenth century whose work was somehow

---

[2]After the new method of fortification was introduced, the earthwork method was not abandoned. Since it could be useful in any number of situations, it was taught to military officers up until the seventeenth century.

[3]Two of the most illuminating scholarly works concerning the first developments of the art of war in Italy during the fifteenth and early sixteenth centuries are Lamberini (1990, 2007).

related to the field of the art of war to enjoy a higher social status than was possible up until that time, especially as compared to colleagues working in other fields (Biagioli 1989).

The increased relevance of practical knowledge, the growing need for its codification, and finally, the way in which it challenged the existing cultural and scientific order, derived not solely from the fields of metallurgy, military architecture and the practical geometry related to new military mathematical instruments. Other activities became central as well, beginning with the science of machines. Within the field of the art of war, no building site, fortress, large ship or harbor was conceivable without the massive deployment of a great number of large and efficient machines. Machine makers, with their knowledge of complex compositions of movements, of efficiency and of the resistance of materials, were behind all kinds of practical endeavors. To give an example, one of the issues at the end of the sixteenth century most relevant to the art of war concerned the preparation of fortress artillery in case of attack. This required artillery to be loaded onto their specific wagons and moved to an appropriate location, usually to the fortresses' strongholds. Fortress combatants were continuously trained for such events and, according to the military engineers of the sixteenth century, the artillery of an average fortress needed at least one week to be set up properly, and this was possible only with trained combatants, the precise orders of the officers and in particular, efficient machines that were able to raise heavy cannons quickly. Thus, one of the most relevant issues for military engineers was the use of machines to lift cannons onto the fortresses. If this was not done efficiently, even the best conceived fortress (and the greatest financial investment) would have been completely redundant.

Extremely relevant, but often forgotten among the early modern skills, are practical hydraulics and pneumatics. Water supply systems, the redirection of rivers for communication or irrigation purposes, the activation of water streaming for water replacement purposes, and water control systems to activate and deactivate mechanical and pneumatic devices were typical projects to which skilled hydraulic engineers were required to apply their knowledge. They were as vital to the states as the experts on metallurgy, the military architects and the machine makers.

Starting from 1609, finally, practical optics, too, appeared in the sights of military officers. Telescopes and binoculars of different shapes, eventually compounded with measurement equipment, immediately became instruments that no officer could do without.

The foundation of the *Accademia del disegno* in 1563 by the Grand Duke of Tuscany, Cosimo I (1519–1574), is one of the most significant demonstrations of the recognition of the increased relevance of practical knowledge in the sixteenth century. The most relevant purpose and consequence of the appearance of the *Accademia* was to encourage (and control) efficient teaching, as well as the diffusion of the practical knowledge required by the duchy. The enormous success of the model of the *Accademia*, which was imitated all over the peninsula in the space of a few years, demonstrates that the needs perceived and expressed by Cosimo I were not exceptional, and that he was creating a way to institutionalize and codify the knowledge required by all modern political authorities at the end of the sixteenth century.

Attending lessons at the *Accademia* or, to a certain extent those of the Abaco schools, or even completing an apprenticeship at an artisan's workshop, gave pupils the chance to obtain the basic knowledge that later allowed them to approach many different kinds of projects. Such pupils did not receive the teaching required to become, for example, a military architect. Instead they learned the basics of military architecture, drawing techniques, the composition of colors, the science of machines, sculpture, civil architecture, hydraulics, practical arithmetic and geometry. The aptitude they displayed for certain skills while working on specific projects determined whether they would work as artists, or as architects or machine makers. As mentioned, however, the greatest chance to improve their own situation towards the end of the sixteenth century was to work in the field of the art of war. This could take the form of designing or constructing a gunpowder machine, or devising a method to hollow out a cannon, or simply accumluating enough experience on building sites to coordinate the construction of standard fortresses on the boundaries of a state. Working on projects that were useful in the art of war, or of general use during wartime, also corresponded to the social expectations of those who received the typical artist-engineer education. They were expected to be experts on and, in the best of cases, to be able to improve military technology. This was their normal social and scientific behavior.

The art of war defined the predominant cultural aspects, and this was still the situation in Galileo's day. Artist-engineers were in demand to implement these changes, and as long as they bore fruit for the military art they became increasingly crucial for policy, security and the economy.

Galilei in the Current of Warfare

Galileo's apprenticeship, as described in the first chapter, focused especially on practical geometry. His teacher Ostilio Ricci certainly did not fail to demonstrate in practice what could be accomplished on its basis. The basics of geodesy, dioptrics, catoptrics, statics, and, moreover, practical arithmetic, were all imparted through practical problems and their solutions. During those years Galileo also shared the knowledge of practical pneumatics and the practice of spectacle makers, eventually visiting Buontalenti's workshop. Certainly he also became familiar with technical drawing, stereometry and perspective, as these were the basics for anyone entering a workshop for an apprenticeship, who was thus a potential artist. Galileo's training was more or less standard for artist-engineers and, in several aspects, oriented especially to the military art.

When Galileo decided to open his own workshop, as described in the second and third chapters, it was in Padova in 1592, a perfect place for a flourishing engineer's business. Galileo decided to hone his profile as a military engineer and architect, in perfect keeping with the contents of his apprenticeship. He revealed himself to be a good workshop keeper as far as the organization of the enterprise was concerned, that is, in finding ways to minimize costs. The *Ricordi autografi* show how skilled Galileo was at the administrative tasks concerned with the running of the workshop. These were fundamental prerequisites for engineers and architects to keep their businesses alive.

First, Galileo designed an instrument, the military proportional compass, which had tremendous success all over Europe for a long time. His practical skill and accumulated experience certainly played a very important role in this undertaking. Yet practical skill was not enough to design such an instrument. Clearly, Galileo first had to know the real needs of the officers on the battlefields. To accomplish this task, it would not have been enough for Galileo to collect information simply by reading treatises on fortifications or speaking with a few experts. The tradition of military architecture and fortifications, with all of the sub-topics that such a field of knowledge included, was too well established and codified at the end of the sixteenth century for him to specialize in these fields without being an acknowledged expert. Galileo certainly shared the knowledge of the military architects, engineers, officers and artillerists beyond his training in Florence, and in such a way that he became recognized as an expert and not a mere pupil: the other military architects considered him to be one of their own.

The military compass was the pivot around which his whole business rotated. His activity as a private teacher on a variety of topics, all of which concerned the work of a military officer, yielded the key product of the compass, and this was the reason for the success of Galileo's enterprise. Besides the use of the military compass, most of the other subjects he taught, such as military architecture and practical astronomy, were the ones taught by most of the low-paid professors for mathematics at the universities and, sometimes, by the even worse-paid teachers of the Abaco schools. Galileo's analysis of the machines, his attention not only to practical hints related to building procedures, but also to the work of those for whom such machines were intended, like the master masons, and, finally, his explanations of those devices very clearly show that Galileo not only shared the knowledge possessed by the men who usually commissioned such machines, but also aspects of the knowledge of those who actually built them: the only ones who were ultimately capable of making a working device. Galileo's course was on military fortifications and most of his potential pupils were future military officers. Even his lessons on mechanics, apart from the analysis of the simple machines, concerned primarily those compound machines typical for the life of a fortress.

When the telescope appeared in Venice, Galileo had seventeen long years of experience running a workshop, with a business consisting of building mathematical instruments for military purposes and in giving private lessons to train future military officers. On the basis of this experience, Galileo immediately comprehended the military relevance of such an instrument. This is probably the main reason why he immediately undertook the effort to improve its magnifying power, on the basis of the practice of the spectacle makers he had learned during his apprenticeship in Florence. In fact Galileo was able to "sell" the telescope to the Venetian executive council as a militarily relevant device, for which he was rewarded with lifetime tenure of his university chair. An improved telescope, combined with the idea of directing it to the sky, ultimately succeeded in making Galileo the most famous "optical engineer" of his age.

As described in the second chapter, Galileo first completed his apprenticeship in 1609 and 1610 in the workshops of mirror makers and spectacle makers, the first craftsmen who were able to make lenses suitable for telescopes: Galileo then long ground, polished, and monitored lenses for telescopes, whose manufacture did not differ in principle from that of the curved mirrors made of glass. He was considered to be the best optical engineer all over Europe for at least twenty years, and demand for his lenses and telescopes was extremely high. Because of this fame he was also officially commissioned by the Grand Duke of Florence to work at the Arsenal of Livorno in order to design further optical devices and accessories intended to simplify the use of the telescope, or of optical instruments in general, in special situations, like on board a galley. This was the setting for Galileo's design of the binoculars and the *celatone*, under the condition that these new devices could be used exclusively by the Tuscan military fleet.

As a favored figure at court, Galileo also took pains to provide some of those fashionable spherical mirrors for kindling, and mirrors of other shapes for their amusing distortional effects. He was an expert on the procedures for constructing a variety of mirrors of several shapes. He could evaluate them and even act as a broker for their trade. Indirectly, Galileo shared the knowledge of practical mathematicians like Ausonio, who wrote a text on the functioning of spherical mirrors, which is an overview presenting the fundamental information for building mirrors, but not a work on optics.

Finally, the intensity of Galileo's activities, the enormous amount of time he invested in them, and the scientific economy[4] by means of which he established and increased them, disperse any lingering doubts that he behaved and was regarded within the social context of the military engineer of the end of the sixteenth century. Galileo worked as an engineer, a term which, at the time, was also applied to architects, and as a practical mathematician, starting with the training in his youth and ending toward the end of the 1630s, when he was about seventy-five years old. Most of his activities, moreover, were particularly oriented to contemporary military needs.

Beyond Engineering

Why then is Galileo best known as the ingenious theoretician who founded modern mechanics, the lonely thinker who almost single-handedly launched the scientific revolution of the early modern age? Looking at the outline of Galileo's work as an engineer from a different perspective might be helpful in seeking an explanation to this historiographical oddity.

[4]Mario Biagioli (2006) clearly showed how Galileo's scientific economy up to 1610 corresponds to that of the engineers and practical mathematicians of the period. Although Galileo's economy of discourse changed dramatically after the publication of the *Sidereus nuncius*, aspects of this previous career remained unaltered throughout his whole life, as most of his technical developments in practical optics and the way he disclosed this knowledge clearly show.

During Galileo's time in Padova, he not only ran a workshop, but also earned a pittance as a professor of mathematics at the university. It was actually common practice for anyone in such a situation to augment his salary by giving private lessons on machines, for example, or on the use of the quadrant for artillerists.[5] The knowledge of artillerists, which he presumably partook of to accomplish his lessons, became the basis for his emerging new science of motion, published in the *Discorsi* in 1638 (Renn et al. 2001). It was this fundamental knowledge that allowed Galileo and Guidobaldo del Monte to set up the experiment to demonstrate that the trajectory of a projectile follows a parabolic path: Galileo's first step toward formulating the law of fall (Damerow et al. 2004), for which he today is often considered to be the father of modern mechanics. For his lessons on simple and compound machines, Galileo used his text *Delle macchine*, compiled in 1592, which is particularly oriented toward machine building. But after just eight years or less, Galileo superseded his first work on this subject by preparing a new text, *Le mechaniche*, which is a more comprehensive work on the science of machines, intended as an attempt to build a deductive structure, according to whose principles the work of machines is explained theoretically in terms of force, weight and velocity. Galileo eventually experienced a change of attitude, which can be paradigmatically described through the analysis of the different introductions that Galileo prepared for the two texts.

Both of the texts have introductions that are directly reminiscent of Aristotle's introduction to the *Mechanical Questions*. In his introduction Aristotle used two concepts—art and nature—in such a way that their relations to each other can be interpreted in three different ways. With the concepts of art and nature, emphasis is placed on their production: Nature is considered in its becoming and therefore in its producing phenomena; art is considered as *techné*, that is, "an art or science, involving a combination of various kinds of theoretical and practical knowledge" (Schiefsky 2007), and the emphasis is on the production of art and therefore on human beings. The art-nature relationship is therefore a nature-human being relationship, both seen as producers of objects and phenomena. Following the history of the reception of this puzzling problem, upon which the positioning of the entire discipline of mechanics depends, one can find three different interpretations of such a relationship. If art and nature are not considered to be on the same level, that is, if the laws of nature and those of mechanics belong to two different realms, then either nature or art can be conceived as predominant. A conception of nature dominating over *techné* was formulated, for example, by scholastic philosophers of the thirteenth century (Paré 1947, 53–80; Daston and Park 1998, 264). Mark Schiefsky showed (Schiefsky 2007) that many early modern Aristotelian commentators, who were well-versed not only in the *Mechanical Questions* but also in the

[5]Thanks the popularity achieved through Galileo's discovery of the Medicean Planets, when he wrote to Florence to negotiate about a new position, Galileo justified his request by complaining that his work in Padova did not leave him enough time to work on his major scientific interests. For more details, see Galileo to A. de Medici, February 11, 1609, in *EN*, X:228–230. For a translation of the entire letter, see pp. 223ff.

main works of Aristotle, were able to demonstrate that Aristotle himself introduced no contraposition between the laws of mechanics and those of nature, and, therefore, that the science of mechanics was worthy of being granted a place within the field of natural philosophy. The third position, according to which art tends to dominate over nature, started emerging in codified form during the second half of the sixteenth century and was supported and proclaimed mostly by educated engineers (Bredekamp 2007a, 41; Valleriani Forthcoming a).

In 1567 the renowned engineer Giuseppe Ceredi wrote about machines as "such impetuous and superhuman instruments" (Ceredi 1567, 41). At the *Biblioteca Nazionale Centrale* of Florence, moreover, an interesting copy of Domenico Mellini's *Discorso*, which belonged to the Florentine engineer Bernardo Buontalenti, is preserved, arguing against the belief that perpetual motion is possible. Buontalenti wrote his notes in the margins of this book (Mellini 1588).[6] In particular, one relevant passage of Mellini runs as follows:

> And although Aristotle said that the Art makes and leads to perfection some things, which are not made and reduced to perfection by Nature, of which Art is imitator, [however] as the same Philosopher said in the same passage of the second book of the Physics, Art is rather daughter of Nature, [...]; therefore it is not absolutely so, that Art can do more than Nature and is more than that.[7]

In the margin next to Mellini's sentence, Buontalenti polemically noted, in capitals, "L'ARTE VINSE LA NATURA," art superseded nature (Fig. 6.1).

Ceredi and Buontalenti are only two of many possible examples that could be cited here. Figures like Buontalenti, who in 1582 asked Oreste Vannocci Biringucci and Bernardo Davanzati to translate Hero's *Pneumatics* into Italian because he desired to read it,[8] and the many anonymous craftsmen who left no writings behind and never became famous in their day, in short: the tacit bearers of the practical knowledge were the ones who approached the construction of machines and the knowledge and experience related to them as a kind of battle against nature. It is beyond doubt that in many cases the contraposition between art and nature was instrumentalized and mentioned by artist-engineers in the manner of a literary *topos*, but, as Buontalenti's notes clearly show, there was also a kind of conviction among artist-engineers that the battle against nature raged on. It probably has to do not only with Aristotle's doctrine, but also with the fact that these artist-engineers were engaged in great enterprises like changing the course of a river, for example, and sometimes not only oversaw their execution but were also involved personally

---

[6]The collocation of Buontalenti's copy of Mellini's treatise is Palat. Serie Targioni 86, Biblioteca Nazionale Centrale di Firenze. For more details about Buontalenti's notes, see Fara (1988, 206–207).

[7]"Et se bene Aristotile disse, che l'Arte fa & conduce à perfezzione alcune cose, le quali non fà, & non riduce al perfetto la Natura, della quale ella è imitatrice, come disse il medesimo filosofo nello stesso luogo del secondo libro della Fisica, anzi è figliuola l'Arte della Natura, [...]; non è però che l'Arte assolutamente possa piu della Natura, & sia piu di quella" (Mellini 1588, 10).

[8]The reception of Hero's *Pneumatics* is discussed at length in Chapter 5, on pp. 172ff.

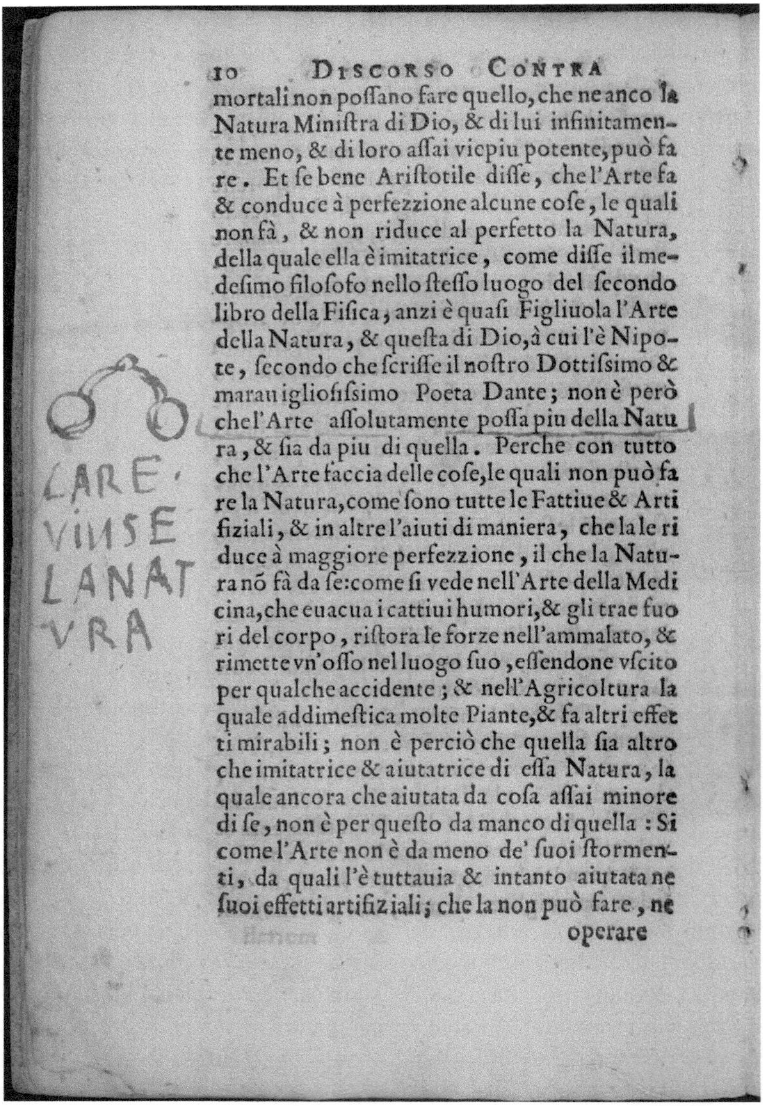

10 DISCORSO CONTRA
mortali non poſſano fare quello, che neanco la
Natura Miniſtra di Dio, & di lui infinitamen-
te meno, & di loro aſſai viepiu potente, può fa
re. Et ſe bene Ariſtotile diſſe, che l'Arte fa
& conduce à perfezzione alcune coſe, le quali
non fà, & non riduce al perfetto la Natura,
della quale ella è imitatrice, come diſſe il me-
deſimo filoſofo nello ſteſſo luogo del ſecondo
libro della Fiſica; anzi è quaſi Figliuola l'Arte
della Natura, & queſta di Dio, à cui l'è Nipo-
te, ſecondo che ſcriſſe il noſtro Dottiſsimo &
marauiglioſiſsimo Poeta Dante; non è però
che l'Arte aſſolutamente poſſa piu della Natu
ra, & ſia da piu di quella. Perche con tutto
che l'Arte faccia delle coſe, le quali non può fa
re la Natura, come ſono tutte le Fattiue & Arti
fiziali, & in altre l'aiuti di maniera, che la le ri
duce à maggiore perfezzione, il che la Natu-
ra nō fà da ſe: come ſi vede nell'Arte della Medi
cina, che euacua i cattiui humori, & gli trae fuo
ri del corpo, riſtora le forze nell'ammalato, &
rimette vn'oſſo nel luogo ſuo, eſſendone vſcito
per qualche accidente; & nell'Agricoltura la
quale addimeſtica molte Piante, & fa altri effet
ti mirabili; non è perciò che quella ſia altro
che imitatrice & aiutatrice di eſſa Natura, la
quale ancora che aiutata da coſa aſſai minore
di ſe, non è per queſto da manco di quella: Si
come l'Arte non è da meno de' ſuoi ſtormen-
ti, da quali l'è tuttauia & intanto aiutata ne
ſuoi effetti artifiziali; che la non può fare, ne
operare

LARE·
VINSE
LANAT
VRA

Fig. 6.1 Bernardo Buontalenti's note "L'arte vinse la natura" (Art superseded Nature). (Note at p. 10 of *Discorso di Domenico Mellini Nel quale si prova contra l'oppenione di alcuni non si potere artifizialmente ritrovare, ne dare ad un corpo composto di Materia corrottibile un Movimento, che sia continovo e perpetuo*, Appresso Bartolomeo Sermatelli, Fiorenza. Fondo Targioni Tozzetti 86, BNCF, Florence)

and, so to speak, even materially. Such enterprises probably really did give those craftsmen and artist-engineers the impression of being deployed against nature.[9]

Mentioning the battle between art and nature probably also corresponded to a particular social behavior, which expressed itself by rejecting nature (and natural philosophy), and therefore contacts with the academic world, in order to establish a sort of authority for the representatives of practical knowledge in line with their improving social status.

If the contraposition between art and nature in reference to mechanical science, especially when interpreted to imply art's domination over nature, is a sign of the way artist-engineers looked at their activities, it must be noted that this motif is found in Galileo's *Delle macchine*, but not in *Le mechaniche*. In the latter text, Galileo supports the idea that because no contraposition between art and nature exists, laws of nature and laws of mechanics belong to the same domain. This means, first, that there is yet another good reason to believe that *Delle macchine* is an expression of the knowledge of artist-engineers and, therefore, that the starting point of Galileo's reflections about the mechanical sciences was the practical knowledge of his day. Second, however, it suggests that Galileo experienced a change of attitude at the turn of the century, which directed his attention toward theoretical speculations rather than practical procedures, thus bearing a closer resemblance to the efforts of an Aristotelian commentator than a workshop-trained engineer.

Galileo as a designer of optical instruments and of their accessories, as well as Galileo the lens maker, are personae that found their deepest meaning in combination with Galileo the astronomer and mathematician. Galileo's celestial discoveries constituted the most sensational scientific news of those years. Accessories like the "ship over the ship" were practical contrivances designed to realize a method for calculating the longitude. The same method, however, included the compilation of the ephemerides of the eclipses of the four biggest moons of Jupiter, a task which could be accomplished only thanks to a very high level of computational and mathematical work. The lack of accuracy of Galileo's partial ephemerides does not seem due to an incorrect computational algorithm.[10]

The Aristotelian Engineer

When Galileo arrived in Padova, he was still lacking an important measure of the experience in doing, that is, that experience which allowed the discourse to be shifted from ideal machines to real ones. Most of the good machine makers and

---

[9]When the Grand Duke Francesco I selected the location on which the garden of Pratolino were to be built during the second half of the sixteenth century, he apparently selected a site that was particulary inconvenient and arid in order to display his power over nature (Montaigne 1929, 105; Valleriani Forthcoming a).

[10]The analysis of Galileo's observations of the Medicean Planets and of his attempts to calculate ephemerides for short periods has not yet been performed, although historical evidence for such research is abundant.

engineers, although mentally armed with the wrong conception, according to which two machines of different but proportional sizes should have had the same resistance to fracture, were, nevertheless, though not always, able to build great, efficient machines. The machine makers could achieve this result because they possessed a wealth of accumulated experience with the materiality of the machines, which they first obtained through an apprenticeship. Moreover, there were masters like some of the staff of the Venetian Arsenal, who had already grasped the idea that by proportionally increasing the dimension of a solid body, its resistance to fracture decreases. As described in the fourth chapter, Galileo had the chance to share this aspect of their knowledge with the machine makers and with the workers at the Arsenal. And he used this chance. He first came to the Venetian Arsenal because of his interest in specific nautical issues, as his preserved textual fragments show. If these fragments were bundled together in ideal order to make a manuscript, the result would bear close resemblance to the sketchbook of any early modern architect interested in shipbuilding issues. In this manner he entered in contact with masters of the shipwrights, masters of oar makers, experienced admirals and maritime merchants. Finally, his pseudo-apprenticeship at the Venetian Arsenal, initiated through the mediation of the Commissioner of the Arsenal at the very beginning of his stay in Padova, became particularly intensive when he was involved in the official inquiry of the Navy Committee of Venice concerned with the propulsion of great galleys. This was a crucial and costly military issue for the Venetian Senate.

However, Galileo's first practical approach to shipbuilding issues was simply too abstract, so that the Commissioner of the Arsenal felt the need to teach Galileo the technical and practical complexity of the issues from a practical perspective that regarded an entire ship to be a big machine. As a nautical engineer, Galileo was not at all successful. His oar model, as described in the second Day of the *Discorsi* and in the fourth chapter of this work, does not take into account the materiality of the oar, regarding instead only a geometrical, ideal oar. Although Contarini certainly must have found Galileo's idea useless, the doors of the Arsenal remained open to Galileo even after this experience, who doubtlessly partook of aspects of the shipbuilding techniques and experience accumulated over centuries. The functioning of oars and rudders, the way sails work in navigation, were the main topics with which the *Proti* ultimately entertained Galileo. In this case too, however, the interesting points for Galileo were mainly the ones resulting from the study of Aristotle's *Mechanical Questions* and, in particular, his Nautical Questions in the updated form given by early modern commentators. Thus in general, Galileo's sharing the practical knowledge of the Venetian masters was accomplished first by means of the Aristotelian lines of thought.

Galileo's science of the strength of materials bore the general statement that two machines, of different sizes but proportional to each other, do not have the same resistance to fracture, as the larger one is weaker. This declaration amounted to an attempt to furnish a solution to one of the most relevant problems of craftsmen. Machine makers, for example, since they believed that two such machines should have the same resistance, were unable to find any satisfactory and precise answer to the question as to why many machines whose models worked perfectly collapsed once they had been built in actual size. The whole theory therefore was not only rooted in practical knowledge, but also addressed primarily to the engineers, architects and machine makers. The critiques by the engineer de Ville, a member of the reading group organized by Micanzio in Venice to amend Galileo's manuscript of the *Discorsi* before sending it to the Elzevirs, show how an engineer could remain impressed by Galileo's ideas.[11]

When, however, Galileo discussed this point in the *Discorsi*, he did so not in reference to the practice of the engineers or of the machine makers, but mainly to theoretical arguments on the same topics compiled by Aristotelian commentators. Galileo and the Aristotelian commentators had a common theoretical point of departure: Aristotle's cantilever model outlined in Question 16 of his *Mechanical Questions*. In the end Question 16 was the only theoretical "lantern" at his disposal concerning the strength of materials as he watched the work of craftsmen in their workshops and shipyards. More than an engineer in the Arsenal, therefore, Galileo can also be regarded as an Aristotelian commentator looking at the work performed in the shipyards.

At the beginning of the seventeenth century Galileo shifted his interests to different practical realizations, also concerning the subjects in which he had trained, as described in fifth chapter. The invention and diffusion of the thermoscope was based on a body of knowledge, pneumatics, much of which he acquired during his youth, since, while he was accomplishing his training, the great garden of Pratolino was being built under the supervision and responsibility of the man running the workshop, with which Galileo was involved. Pneumatics was mostly the prerogative of engineers at the end of the sixteenth century. Galileo also took part in the debate that erupted among engineers to furnish an explanation of how pneumatic devices work. Such a debate started when the ancient work on pneumatics by Hero of Alexandria was made available in printed form thanks to the Latin translation by Federico Commandino in 1575. In this work Hero of Alexandria had shown clearly how the Aristotelian principles, commonly used to explain the functioning of pneumatic devices even during the early modern period, were clearly insufficient to give an exact account of what could be easily observed by activating such devices. Those engineers like Bernardo Buontalenti, Giovambattista Aleotti and Salomon de Caus, personally engaged in the design, conception, and evaluation of such pneumatic devices, appropriated Hero's work and, through a series of new commented editions of his work, expressed their own theoretical approach based strictly on their experience but increasingly abstracting along the way to a formulation of general principles concerned with hydraulics and with the doctrine of the elements and of their nature (Valleriani 2007). Like the engineers, Galileo, too, felt the need to provide a modified theoretical explanation for those devices, which eventually brought both the hydraulic engineer Aleotti and Galileo to formulate a new conception of heat.

The diffusion of the thermoscope as a scientific instrument, however, took place within a genuinely Peripatetic theoretical framework. Although Galileo linked the theoretical debate among engineers concerned with the functioning of the thermoscope, it is a fact that only those engineers who were also experts on Aristotelian

[11] De Ville's reaction is discussed on pp. 125ff.

natural philosophy contributed to such a debate. Galileo, aided by Sagredo, the Venetian experimenter, took over Aleotti's program and tried to furnish an explanation of the instrument by combining Peripatetic scientific principles with a new conception of heat, according to which it was constituted of particles. Galileo made use of the Aristotelian principles like, for example, nature's abhorrence of a vacuum, and of his description of the processes of condensation and rarefaction, which are essential blocks in the Aristotelian buildings of *Physics* and of *Meteorology*. Galileo's work on the thermoscope was, in conclusion, the work of an Aristotelian engineer expert in pneumatics.

The engineer Galileo Galilei, therefore, revealed himself to be a very expert commentator of many aspects of the Aristotelian doctrines, especially concerning the *Mechanical Questions, Physics* and *Meteorology*. Thus, taking into consideration only the sources and the results analyzed and achieved in this work, one is obliged to consider Galileo not only as a military engineer but also as a complicated practical mathematician who was expert on Aristotelian natural philosophy. This view, moreover, is perfectly consistent with relevant results gleaned from the framework of the historical research concerned with Galileo's science of motion. From this perspective Galileo's work has long been interpreted as the beginning of classical mechanics. This interpretation was based on the fact that Galileo formulated the following theorems: (1) the proposition that all bodies fall with the same constant acceleration, (2) the law of fall, (3) the proposition that the trajectory of projectile motion is parabolic in shape. Such theorems are indeed fundamental in the framework of classical mechanics. But Galileo formulated them within a conceptual preclassical and profoundly Aristotelian framework, and as such they have not been immediately recognized as milestones in classical physics (Damerow et al. 2004, 135–141). In other terms Galileo was an Aristotelian engineer.

A relevant aspect of this result is that it does not underline any sort of chronological development in the work and profile of Galileo. When he entered the Venetian Arsenal for the first time, for example, at the very beginning of his stay in Padova, he did so using Aristotle's *Mechanical Questions* as a sort of plot to guide him through the shipyards. Just as Galileo worked mostly on the theory of the strength of materials during his stay in Padova, namely until 1610, he again directed his efforts to the theory behind the second of his two new sciences, namely to his science of motion (Damerow et al. 2004, 140). The phase of Galileo's most intensive involvement with practical knowledge and of his most intensive activity as a military engineer therefore coincides with the period of his most intensive work on theoretical mechanics. In conclusion, Galileo as a military engineer did not come before Galileo the Aristotelian engineer; rather, they existed at the same time.

Generation of Knowledge

The figure of the Aristotelian engineer allows the formulation of a model according to which new scientific knowledge was generated. Both case studies that make up the second section of this work show the gnoseological processes that led Galileo to formulate the core ideas of two of these major theoretical achievements. In both cases Galileo integrated practical with theoretical knowledge while focusing on the explanation of the behavior of a single object, typically belonging to the tools and instruments of artist-engineers. In the case of the science of the strength of materials, Galileo achieved his fundamental idea while focusing on the oar and its mechanical behavior; in the case of his atomistic conception of heat, Galileo was trying to explain the functioning of the thermoscope, which was a pneumatic device that had been used since antiquity for entertainment and built by machine and fountain makers. In both cases Galileo had practical knowledge at his disposal that he shared in different contexts, and which he filtered and reanalyzed in the light of Aristotelian theoretical conceptions. Both of these theoretical developments by Galileo show how, in focusing on those objects, Galileo conveyed theoretical Aristotelian concepts together with aspects of his practical knowledge. There is therefore no hint to support the naive idea that practical knowledge or practical realizations represented such a challenge for the mathematicians that they developed new revolutionary theories by simply observing and sharing those aspects of the work of the artist-engineers. New knowledge was instead generated on the basis of a long and articulated interaction between theories at their disposal and shared practical knowledge to which they eventually were applied. The key to understanding Galileo's scientific advances thus lies in the type and modality of the interaction between practical and theoretical knowledge. For it was such an interaction that took place when Galileo conveyed and integrated such knowledge while focusing on one material object.

Engineer-Scientists

Galileo's work can be finally reconsidered as the work of an Aristotelian engineer within the more general framework constituted by the description of the early modern engineer-scientist.

The figure of the engineer-scientist is historiographically not completely defined. In his book, Edgar Zilsel introduced the figure of the artist-engineer (Zilsel 2000). He based this connection on the similar cultural background of early modern engineers and artists, who are social actors almost completely without connection today. Engineer-scientists were not like Zilsel's artist-engineers. The engineer-scientist of the Renaissance was familiar with the practical knowledge of the artist-engineer, but not exclusively. As in Galileo's case, engineer-scientists accomplished, partially or completely, the typical training program of an artist-engineer. Like artist-engineers, engineer-scientists, too, cannot be unequivocally identified with a training institution or an activity. Examples of engineer-scientists include military engineers at court, mathematicians at the university, instrument designers, astronomers, philosophers, and theoretical mechanicians, as Galileo was all of these things. An engineer-scientist was, however, also a person who generally enjoyed a relatively high level of education, especially in mathematics (Dijksterhuis 1983,

271–272; Bennett 1991), having studied at least mathematics and geometry and in certain cases also philosophy. Except for the period of the apprenticeship, an engineer-scientist was almost never personally employed in workshops or building sites, but he was aware of the work procedures followed in these locations and was able therefore either to commission craftsmen or other persons involved with practical activities, to supervise or teach them, or simply be consulted to evaluate their works.[12]

In all of Galileo's activities there is a common denominator: his network of friends, pupils, engineers, practical mathematicians and craftsmen. Galileo constantly enlarged his network, maintaining contact with all of its members, even with those who entered it very early, like for example, Ludovico Cardi da Cigoli, his "classmate." Some other important examples among his friends in Venice are Giovan Francesco Sagredo, Paolo Sarpi, Fulgenzio Micanzio, Agostino da Mula, Giovan Vincenzo Pinelli, Sanctorius Santcorius and Giacomo Contarini, together with the craftsmen with whom they normally worked, like, for example, the lens makers Master Antonio and Master Bacci, the producers of most of "Galileo's lenses" between 1613 and 1616, or the workers of Murano, at the same time servants of Sagredo to whom Galileo and Sagredo entrusted experiments intended to improve the quality of the crystal for lens making.[13] The former pupils most important to Galileo's network include Benedetto Castelli, Bonaventura Cavalieri, Paolo Aproino, Daniello Antonini, Niccolò Aggiunti, and Filippo Salviati (1582–1614). Galileo's masters who belonged to the network were the brothers dalla Luna and Ippolito Francini, all of them also lens makers, and Marcantonio Mazzoleni, the compass maker. Mathematicians, instrument designers and other engineer-scientists like Giovanni Battista Baliani, Luca Valerio (1552–1618), Antonio Magini, Cesare Marsili and Guidobaldo del Monte also belonged to his network. Galileo was also a member of academies like the *Accademia dei Lincei*, itself an international network. After 1610 the source of power for Galileo's network was the court. It was, finally, also thanks to the wealth of the Grand Dukes behind Galileo, or their patronage, that Galileo became a point of reference for his contemporaries.

Between 1613 and 1615 the activity of Galileo's network was exceptionally intensive. Just having published the second version of his *Floating Bodies*, thanks to his network he began working (1) with Sagredo on the way the thermoscope works, and also performing pneumatic experiments that led him to abandon his first conception of heat, (2) on the improvement of the telescope, using biconvex lenses, also

---

[12]In their paper entitled *Hunting the White Elephant*, Jürgen Renn, Peter Damerow and Simone Rieger analyzed the figure of the early modern engineer-scientist for the first time (Renn et al. 2001, 66–68). Accordingly they furnish the following definition: "a new category of intellectuals [...] who were no longer necessarily and, in any case, not completely involved in technical practice in the same way as the engineers themselves, but who rather specialized in the reflection of the new type of knowledge produced by this practice and, of course, in the attempt to make that reflection useful again for practical purposes."

[13]Galileo's and Sagredo's experiments to improve the quality of the material for telescope lenses are discussed on pp. 48ff.

with Sagredo, (3) on the method to calculate longitude, in order to "sell" it to the Spanish Crown, thanks to the diplomatic network of the Grand Duke. During the same period he was involved in the debate with the Jesuits concerning the discovery of sunspots. Finally, he also wrote his four famous Copernican letters, open letters written to avoid a conflict between the Copernican view about the structure of the solar system (actually, of the world) and the Biblical view, as professed by the Holy See at that time.[14]

In Galileo's day, moreover, one of the most relevant perceived differences between engineers and engineer-scientists was probably the latter's knowledge of Latin. This situation suggests that engineer-scientists were generally extremely active as bridges for communication. Although from the end of the sixteenth century engineer-scientists expressed themselves in the vernacular, their knowledge of Latin certainly allowed them to act as scientific intermediaries. James Bennett, moreover, showed how producers of mathematical instruments played an intermediary role between the world of practical realizations and that of theoretical speculations in seventeenth-century England (Bennett 1986). Galileo was also the designer and producer of mathematical instruments, like, for example, the military compass.

This intermediary function ultimately could be the key to determining historiographically what an early modern engineer-scientist was. He was a point of contact between the world of craftsmen and foremen, and that of the cultural authorities. He was the center of a network ramified over different social levels. From the perspective of a person like Galileo, the cultural authorities were the Peripatetic natural philosophers, the humanists, the theologians, all of whom expressed themselves in Latin, and the political and executive bodies, like, for example, the Commissioners of the Arsenal, the state secretaries of the Grand Duke, and the Tuscan ambassadors.

Peripatetic knowledge, transmitted in Latin, established its status by gaining credibility through strict relations between "Christian" interpretations of Aristotle's work and the doctrine of the Roman church, depository of the unconditionally true Holy Writ. Peripatetic philosophers often held university chairs for philosophy and, in certain Italian cities, these chairs were under the direct control of the Jesuits.

The figure of the humanist emerged primarily within the institutional bodies of cities, which, after the experience of the communes, slowly changed their nature in favor of bodies that were often more stable politically: duchies, principalities, and republics, besides the state of the Church. Humanists were originally people who worked in administrative bodies, especially in those in charge of cultivating relations abroad, as curators of the political and economical contacts of their lords. In particular, they had to know Latin well and, especially, have good rhetorical skills in written composition.[15] They improved their social relevance over the period to

---

[14]Galileo's Copernican letters are: Galileo to Benedetto Castelli, December 21, 1613, in *EN*, V:281–288, Galileo to Piero Dini, February 16, 1615, in *EN*, V:291–295, Galileo to Piero Dini, March 23, 1615, in *EN*, V:297–305, Galileo to Cristina di Lorena, 1615, in *EN*, V:309–348.

[15]For a general view on the social position and education of the humanists, see Zilsel (2000, 48–64). Further details are in Olschki (1919–1927, III:26–30).

the extent that they attained a status equal to that of the natural philosophers. The humanists became literate, devoted to the ancient languages, and editors of annotated publications of classical works dating back to the Greek and Roman periods. The work of the humanists specialized to such an extent that philology became established as its own discipline.

The vernacular, on the other hand, sustained by the works of Dante Alighieri (1265–1321), Giovanni Boccaccio (1313–1375), Pietro Bembo (1470–1547), Francesco Petrarca (1304–1374), Torquato Tasso (1544–1595), Ludovico Ariosto (1474–1533) and Ruzzante (1496 ca.–1542), was the means of expression for institutions or centers of production like, for example, the Abaco schools, most of the newly founded academies, the workshops for painters, sculptors, architects, cartographers, machine makers and instrument makers, music instrument makers, shipwrights, founders, smelters, smiths and carpenters. In other words, the vernacular was the language of those dedicated to the arts and mechanical production, who were attaining a new and more important social status during the Renaissance.

The development of commerce and, in general, of contacts between cities, regions and countries for political, cultural and economical reasons, created a communication network which allowed all of the figures involved in this social dynamic to increase their cultural exchange tremendously. The crusades had already led to a shocking interaction between the European countries and the Near East. As a consequence of this interaction, not only Arabic numbers and new mechanical milling devices were introduced into the European countries (Popplow and Renn 2002), but especially the awareness that such exchanges, also accomplished with ad hoc expeditions, were politically, economically and culturally fruitful. These were the principles that led to many expeditions to the Near and Far East at the end of the Middle Ages and during the Humanist period. Some of them were searching for the legacies of ancient culture, preserved in those countries thanks to the long life of the Byzantine Empire and because of the process of transmission of knowledge into Arabic accomplished during the Middle Ages. Most of the expeditions were commissioned by dukes, princes, doges and the Pope.

The rediscovery of ancient knowledge was presented as the result of the patronage of the protagonists of the political reorganization of those years. Among the enormous amount of ancient Greek and Latin literary works, the expeditions of the fifteenth and sixteenth century brought material that also breathed new life into disciplines like geometry, arithmetic, algebra and mechanics. The contemporary diffusion of printing thus played a crucial role in the frantic work of the humanists in reading, philologically comparing, translating and commenting the ancient works. Aristotle, Ptolemy, Archimedes, Pappus, Hero, Euclid, Pythagoras, Plato, Vitruvius are only a few of the ancient authors to whom the humanists and the intellectuals of the Renaissance drew their attention. The new ancient knowledge was eventually re-elaborated in light of the knowledge achieved by that time thanks to the efforts of earlier engineers, scientists and craftsmen like, for example, Jordanus de Nemore and Maurolico. The patrons improved their images by supporting, first, these expeditions, and second, those who worked on the ancient writings. The practice of dedicating the printed books to important personalities like dukes and princes also began at this time.[16]

The spectrum of accumulated experience that characterized the practical knowledge of the Renaissance was very broad. Thus oral transmission, the usual method used for initial training, eventually turned out to be insufficient for such *factotums*. From the fourteenth century on, but especially during the fifteenth, engineers began recording their knowledge. In short, more and more treatises on fortifications, architecture, surveying, artillery, metalworking, clocks, shipbuilding, and machine building found their way into circulation, first in the form of manuscripts and then as printed books. According to Bert S. Hall (Hall 1979), treatises displaying practical knowledge were very rare before 1400. Their number, however, increased enormously and continuously after this date. After 1530, moreover, such treatises were printed, so that their value became ever less dependent on the actual technical realizations of their authors. This process of codification, as well as the rigorous work of the humanists, are certainly at the root of the emergence of the figure of the engineer-scientist as a bridge between these two frameworks.

During Galileo's lifetime urbanization, wealth, competition and the development of the art of war explain the great impulse for arts and mechanics. Craftsmen, engineers and architects of every sort were not the only protagonists of the great economical and social changes of the early modern period. Engineer-scientists were at the real center of this cultural shift. They were surrounded by those active in the political bodies, business people, administrative boards, humanists, natural philosophers, artists, foremen, engineers and mathematicians, mechanicians, officers, artillerists and shipwrights. The point of contact among all of these was the science of mechanics, in its "low" and "high" forms (Lefèvre 2001, 23–24).

Galileo was not a genius, but he certainly was a quite exceptional person. His apprenticeship and lifelong activity as an artist-engineer, his curiosity for the technical and artistic creations as well as for natural phenomena, his obstinacy apparent in the theoretical speculations—cultivated in the case of the two new sciences, for example, for around forty years—his good nose for business, and his excellent communication skills, made him the most popular engineer-scientist of his day. Galileo was not an ingenious, lonely thinker, but today he appears to have been a great scientist. The greatness of his science, however, was a consequence of its links to practical knowledge. Galileo was neither a lonely theoretician nor a pure engineer. He was a skilled and intelligent engineer-scientist.

[16]Along with the practice of dedication, the patronage system was born, which later revealed its epistemological function. According to Mario Biagioli, scientific credibility was, in fact, strictly connected to the patron to whom the work was addressed. In this period and even later, scientific production was, in other words, a production of the courts, regarded as printed output of discussions held around the prince (Biagioli 1993).

Sources: Galileo's Correspondence

Notes on the Translations

The following collection of letters is the result of a selection made by the author from the correspondence of Galileo published by Antonio Favaro in his *Le opere di Galileo Galilei*, the *Edizione Nazionale* (*EN*), the second edition of which was published in 1968. These letters have been selected for their relevance to the investigation of Galileo's practical activities.[1] The information they contain, moreover, often refers to subjects that are completely absent in Galileo's publications. All of the letters selected are quoted in the work. The passages of the letters, which are quoted in the work, are set in italics here.

Given the particular relevance of these letters, they have been translated into English for the first time by the author. This will provide the international reader with the opportunity to achieve a deeper comprehension of the work on the basis of the sources. The translation in itself, however, does not aim to produce a text that is easily read by a modern reader. The aim is to present an understandable English text that remains as close as possible to the original. The hope is that the evident disadvantage of having, for example, long and involute sentences using obsolete words is compensated by the fact that this sort of translation reduces to a minimum the integration of the interpretation of the translator into the English text.

[1] Another series of letters selected from Galileo's correspondence and relevant to Galileo's practical activities and, in particular, as a bell caster is published appended to Valleriani (2008).

M. Valleriani, *Galileo Engineer*, Boston Studies in the Philosophy of Science 269, DOI 10.1007/978-90-481-8645-7, © Springer Science+Business Media B.V. 2010

Galileo to G. Contarini in Venice. Padova, March 22, 1593[2]

Most Illustrious Lord,

I heard from the Illustrious Lord Gianvincenzo Pinelli the problem of Your Most Illustrious Lordship, about which I will tell you what I believe is true: and it is this.

Concerning the need to apply more or less force in propelling the vessel forward, it does not make any difference if the oar lies on the live or dead part of the deck, since all other circumstances are the same. And the reason is that, since the oar is practically a lever, as long as force, support and resistance divide it with the same proportion, it will operate with the same vigor, and this is a universal and invariable proposition. And I do not believe that making the wings in the galley will achieve anything but the ease of having more space for the soldiers and convicts, who otherwise could not be seated in rows of four or five per oar, especially toward stern and bow, if there were no wings. But if they could sit and row both in one way and in the other way, I do not necessarily believe that placing the protection inside or outside the live part of the galley would make any difference if, however, the oar is divided with the same proportion. And I do not see anything that could hinder or facilitate the rowing other than placing the protection further away from or closer to the handle: the closer it is, the more one can apply force. And the reason is the following, a reason that has perhaps not been investigated by anyone else: The oar is not a simple lever like any other one, indeed, there is a great difference for the following reason. Ordinarily the lever should have a mobile force and a mobile resistance and a support at rest, but in a galley support, force and resistance move. It follows from this that support and resistance are the same because when the blade of the oar is placed in the water, the water becomes the support, and the protection becomes resistance. But when the oar moves the water, in this case it becomes the resistance, and the protection is the support. And since, when the support is fixed, the whole force is applied to move the resistance, if the oar is immersed so that the water becomes almost immovable, then most of the force is employed to propel the vessel. On the contrary, if the oar is immersed so that the water is moved easily by the blade, then one is not able to apply the force to move the boat. And since the greater the length of the part of the lever is toward the force, the more easily one can move the resistance, when the part of the handle is very long, the water will be moved more easily, and hence its support will be weaker and one will propel the vessel less. On the contrary, when the same part between the protection and the force is shorter, then it will be more difficult to move the water with the blade and consequently, since it is needed as support, it is more solid and one is able to propel the vessel with more force. And one concludes that, the closer the protections are to the handle, the stronger the force can be applied in propelling the vessel, as the water is not able to be moved so easily with a blade very distant from the protection by a force close to the same protection. Hence, in such a case, the water functions

[2]*EN*, X:55–57.

more as support than resistance. All of this is very evident from experience. Hence, since only the larger or smaller distance between the protection and the force can give ease or discomfort to the stroke, I absolutely do not believe that placing the protection on the live or dead part of the deck makes any difference.

This is what occurs to me up to now in answer to your doubts and I believe that Your Most Illustrious Lordship has spoken more clearly about this. But if you would like to share your thoughts on this business with me, I would remain infinitely obligated to you since I am sure I would learn much and perhaps your arguments would induce some other things to occur to me. I beg you that when similar problems are in the air, you condescend to share them with me, since I take great pleasure in thinking of curious things.

I sent the letter of Your Most Illustrious Lordship to my friend the sculptor but have not yet received an answer. With this, I bow very modestly to you and beg you to summon me.

From Padova, March 22, 1593.

Very Obligated Servant of Your Most Illustrious Lordship

Galileo Galilei

Addressed to: Most Illustrious Lord and My Very Cultivated Master

Lord Iacomo Contarino

Venice

G. Contarini to Galileo in Padova. Venice, March 28, 1593[3]

My Most Magnificent and Excellent Lord,

To my great pleasure I saw what Your Most Excellent Lordship wrote to me in reference to the oars of the galleys. Although I do not have much time, since you ask me to write to you about some of the things related to this business that are on my mind, I tell you that, for the observations I did, the oars normally used are not proportional to the body of the vessel. But one should put some thought into proportioning these two things together so as to obtain what is required, that is, agility and velocity. And in my opinion one can obtain the mentioned proportion from three things: from the width of the live body of the vessel, from the height between the water and the position of the oar, and from the motion that the galley-slave communicates [to the oar] while pulling it.

Speaking first about the last issue, which is the moving force, I say that when the galley-slave begins to row, one should consider that he is obliged to perform one of three motions: either to pull the oar from the bottom upwards, since it is low; or from the top downwards, since it is high; or straight to the breast, since it is placed between these two extremities. Now one can understand very well that those two above-mentioned extreme movements, being violent, tend to lack force and are very difficult to endure. Thus, one should place the oar in such a way that, by pulling it, it comes to the chest when the man is standing with the advantage that, when the galley-slave loses his balance due to the weight of his body, he moves the oar more easily than when just using ordinary force. In supposing this, which is not difficult, it is necessary to understand what is the right position and the right height to position the oar in reference to the protection above water-level. And one can find this easily if the height of the man who rows and the length of the oar are considered. Thus, the oar should be placed at a height where it can touch the water without the galley-slave performing one of the two extreme movements mentioned above. To do this, the oar should be long and touch the water at a great distance from the ship so that the galley-slave does not have to row with raised arms. But another need arises because, having a long oar, a greater force is needed to move it, to row with it, as well as to raise and to lower it. But at this point, the width of the ship becomes relevant because it is from this width that one obtains the length of the oar's handle and the width of the apostis; if the ship is not able to contain the length of the handle on its live body, one may solve the problem with the apostis. Since the handle is made not only to move the pole of the oar that is outside of the protection, but also to act as a counterweight to the weight of the mentioned pole, the weight of the latter is so great that, even if the handle is made long and thick, it is not sufficient and it is necessary to add to it 50, 60 or 100 pounds of lead, so that, keeping his balance, the galley-slave has only the effort of pulling it. Besides this thickness, one has to consider the force of those who must move the vessel, which

[3] *EN*, X:57–60.

has to be of a lot of men, as it is said, because of the heaviness of the ship and the length of the oar.

By professionals of the art, the oar is divided into three parts: two parts are from the protection to the blade, and one is reserved for the handle. This third part is so long that, because of the above-mentioned reasons, it cannot be contained inside the live body of the galley. It is therefore necessary to enlarge the apostisses so much that from the gangway to the protection one can contain the handle, therefore one uses the weak side of the ship and the force is thus placed in the dead part of the galley. Now this space does not have to be completely occupied by the men who pull the oars because if one puts 4, 6 and more men to this oar, one must take care that the last man is sufficiently distanced from the protection so that he can apply force to it and add his own force to that of his other companions, who are nearer the gangway; if he were too close, he would be useless. Moreover, the combatants should remain between the last man and the apostis, without hindering the rowers. The third part of the oar must therefore be between the gangway and the protection. This third part should be so long that it enables the above-mentioned things. And on the basis of this third part within the ship, one can deduce the two parts outside the ship. The movements made by the galley-slaves are very different because the first one exerts more force than the second one and the second more than the third, and so successively until the last one. And one deduces their force theoretically. Since the motion of the oar at its extremities is circular, let us compare the circles made by the blade to those made by the handle, and one sees the proportion of the force that moves to that which is moved. And let the circles of the first galley-man be compared to those of the second, and one also sees the proportion of force between them. *Therefore what one says cannot happen, that the longer the handle, the easier it is to move the water. And therefore its support will be weaker and the vessel will be propelled less*, since *it is certain that with a short handle, one will never have force both to steer the oar and to row it.* It is very true that if it is necessary to make the oar outside longer than that which supports the third part, which has to stay inside, one has to accept this situation. But if one wants to use it, one is forced to apply a greater moving force, that is, a greater quantity of men, and by using more men, the outside of the live body of the ship needs to be widened even more. In my opinion, this is not bad, since the solution to the second doubt originates from this, that is, which is the greater force: the one that remains on the live body of the ship, or the other performed on the apostis outside the live body. The force that remains on the live body is closer to the center and therefore moves with greater difficulty, and that which is distant from the centre forces the ship to be propelled more easily. This happens because the closer force has two opposite motions, one of pushing and the other of bringing the oar under the ship, because it is better to immerse it as much as possible thus bringing the blade closer to the ship. And if one wants to operate it, it is better to go deeper under water; if in this way it goes under the ship, the resistance of the water becomes greater, which, resists moving all the more. Hence, it is more useful if the protection is far from the center, though it suffers the imperfection of the weakness that experienced men will always be able to remedy thanks to their art.

I thought I would have filled less than a half side of a folio and nevertheless I allowed myself to carry on until now and, going further, it would disturb you and one could object that I seek to *instruere Minervam.* Your Most Excellent Lordship who loves and commands me.

From Venice, March 28, 1593.

Very Devoted Brother, to serve You always, of Your Most Excellent Lordship

Giacomo Contarini

Galileo to A. Mocenigo in Venice. Padova, January 11, 1594[4]

Most Distinguished and Illustrious Lord,

From the words of Your Most Distinguished Lordship and from the very confused description of the construction written by Hero at nr. 7[5] I recognize that lamp, whose building method Your Most Distinguished Lordship wishes to know. And I read it several times and in conclusion am unable to deduce from those words a clear understanding. But since we are not obliged to remain close to his words, it seems to me that he wants to describe a construction method similar to the following (Fig. 1).

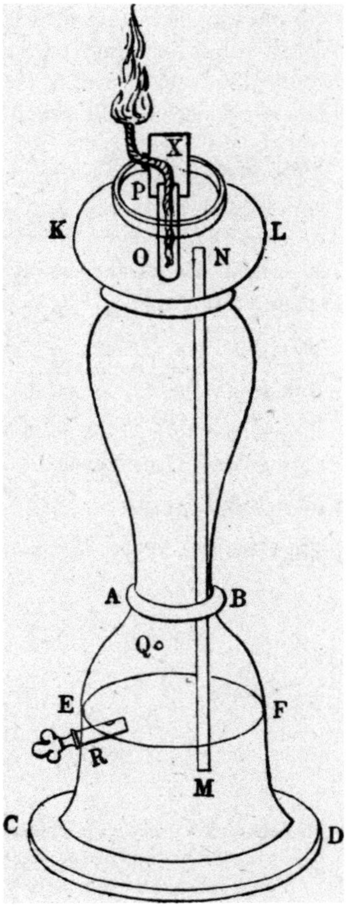

Fig. 1

[4]*EN*, X:64–65.

[5]It is the description nr. 72 of Commandino (1575).

Let a lamp be constructed which has the concave base ACBD, cut by the bar EF. Let KL be the vessel containing oil and, from the bar EF, let a pipe MN be erected and inserted into the bar distant enough from the opening of the vessel to allow the air to escape. Moreover let another pipe XO be placed through the opening, at a distance from the bottom of the vessel, to allow the oil to run and which extends a little from the opening. To this extending part another pipe P is adapted, whose opening at the top is closed, and to whose end another pipe is glued together, similarly perforated, and through which the lamp-wick passes. The key R is integrated in the place CDEF under the bar EF so that if opened, the water passes from AEFB to the same CDEF. Let there also be a small hole Q at the covered part AB, through which the place AEFB is filled with water. Now, taking out the lamp-wick, we fill the vessel with the oil through the pipe XO and the air exits through the pipe MN toward the bottom, thanks to the open key, and then through the hole. Once the vessel is filled with oil, we place the lamp-wick into the pipe X. Closing the key, we fill the place AEFB with water. Once the lamp-wick is soaked with oil, and once the key R is opened, the water passes to the space ECDF, and the air, pushed through the pipe MN, forces out the oil through the pipe OX toward the lamp-wick. And to avoid a great stream, we close the key.

This is what I am able to understand until now from the words of Hero, which, as I mentioned, are very confused. And I decided to send it to Your Most Distinguished Lordship so that, based on your experience, you could eventually take something better from this, if it is really so that the described method results in what this proposal promises.

With this, reverently kissing the hands of Yours, I remain your very devoted servant. Our Lord bless you.

From Padova, January 11, 1594.

Of Your Most Distinguished Lordship

Very Ready Servant

Galileo Galilei

G. Sagredo to Galileo in Padova. Venice, January 17, 1602[6]

I send to Your Most Excellent Lordship two instruments to make screws, which need to be repaired. I have made the small one myself some months ago, but since it seems to me that it does not have the good grace I would like, I wish to repair it but do not want to do this myself because, in fact, I am not longing to work. Hence, since I would like to make a very small machine, I beg Your Most Excellent Lordship to arrange that Master Fait repair it for me immediately. I would also like him to repair the other one for me as well, in such a way that it works properly. And excuse me for the chore.

I have had another small machine made, with a wheel of ivory and a concave perpetual screw, like those of Master Fait, but without using the lathe, only the simple chisel. It turned out really beautifully, although it has been made by a novel master. I also ordered construction to begin on another one for dragging, almost completely of iron, with the form of those of Master Fait, but with one more wheel. But I do not know whether it will turn out well, since the boy behaves like a smith and does not want to obey. And in doing such things more for gallantry than for something else, I am sorry because the stubbornness of that person takes away that bit of pleasantness one could lend to it by making some ornament. And finally, I recommend myself to you.

In Venice, January 17, 1602

[6] *EN*, X:86

G. Sagredo to Galileo in Padova. Venice, August 23, 1602[7]

My Most Magnificent and Excellent, Very Respected Lord,

Your Most Excellent Lordship has heard from Monsignor Gasparo about the store of machines I have made and the wish I have to take from the hands of Master Fait one of those drills with which he hollows out the gears of the wheels for the perpetual screws. I do not know whether one would be able to obtain it so easily if he realizes it will show up in my hands, since he is really annoyed that I have taught my boy how to make his machines. But I beg you to manage to do this business in whatever way seems possible to you.

I certainly want to come there soon. Our Lord Veniero and I wish to make a short trip to Cadore and to some other places close by during the next month of October. But since our trip through fantastic places would be really dull without the company of Your Most Excellent Lordship, I would like to advise you of it in good time so that, in order to favor us both, you may prepare yourself to do us this favor. And the greater your unease in doing us this favor, the greater effort we oblige ourselves to make in your favor at the time of your reconfirmation and so I would like to know when it will be. This is the end of this letter, praying to Our God for your every happiness.

Venice, August 23, 1602

Of Your Most Excellent Lordship

Lord Galileo

Very Devoted to serving you

Giovanni Francesco Sagredo

Addressed to: Most Magnificent and Excellent Very Respected Lord

Lord Galileo Galilei, Mathematician

Padova

---

[7] *EN*, X:90–91.

Galileo to A. de' Medici in Florence. Padova, February 11, 1609[8]

Most Illustrious and Most Excellent Lordship and Most Cultivated Master,

I understood in detail thanks to Sir Benedetto Landucci, my brother-in-law, the most courteous affection with which Your Most Illustrious Excellency has shown himself to be well disposed regarding the advancement of favor that he requested and finally obtained by means of your favor. Hence I thank you endlessly and assure you that Your Most Illustrious Excellency does not have to subordinate me to all your readier and more faithful servants, but for the weakness of my forces.

Moreover, my brother-in-law commands me to write to Your Excellency something new about my studies and that this is your wish, which I receive as a very great favor and which represents an incentive to inquire more than usual. Hence, I inform Your Excellency that I was busy in reference to some reflections and to different experiments related to my treatise on mechanics, in which I hope there are more new things than old ones, things about which no one has speculated in the past. *And recently I also finished finding all the conclusions, with their demonstrations, related to the forces and resistances of woods of different lengths, thicknesses and shapes: how they are weaker in the middle than at their ends, and how they will support greater weight if this is distributed all along the wood rather than at just one point, and the shape this should have so that it is equally resistant. This science is imperative to build machines and every kind of engine, and yet no one has performed this study to date.* I am working now on some problems that I still have to solve and that are related to the motion of projectiles. Among them, there are many connected with shots of artillery. And lately I found this too: if one places a weapon on some site that is raised in reference to the plane of countryside and, if the weapon is well aimed, then a ball exiting the weapon always declines and falls with the same velocity in such a way that with all the aimed shots, the ball, no matter how much gun powder forced it, reaches the ground in equal time. And this remains true if shots are either very distant or very close, and also if the ball just exits the weapon and falls perpendicularly onto the plane of the countryside. And the same happens in reference to elevated shots, which are all shot in equal time as long as they reach the same perpendicular height. For example, the shots AEF, AGH, AIK, ALB, between the same parallels CD and AB, are all shot in equal times (Fig. 2), and the ball covering the line AEF takes the same time as along the line AIK and any other. Hence, their half, that is, the parts EF, GH, IK, LB, are covered in the same time and correspond to aimed shots. On the subject of water and other liquids, which is also a relevant issue, I have likewise discovered very great qualities about nature. But due to lack of time, I cannot write about them now since I have to send many other letters. Hence, when I have a better opportunity, I intend to tell Your Excellency about three or four conclusions and effects that I saw and proved, which are probably more magnificent

[8] *EN*, X:228–230.

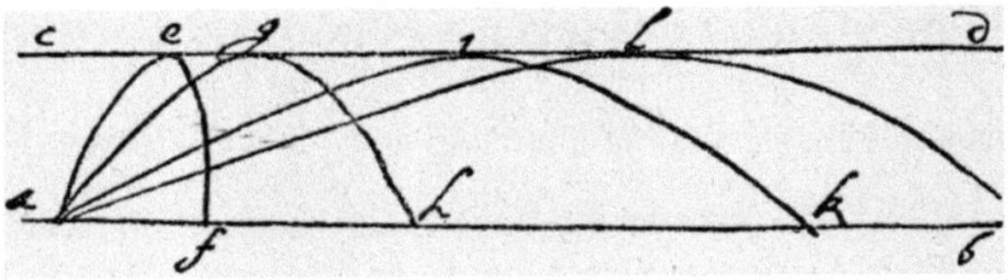

Fig. 2

than the greater curiosities investigated by men until now. But this is sufficient at this point.

I must still beg Your Most Illustrious Excellency to keep me in that place of your favor, which your great goodness has permitted me until now. And be certain that you have a servant, who has the most possible devotion. And in order to end, bowing respectfully, I kiss your hands, and [...] God great felicity.

From Padova, February 11, 1609

Most Devoted and Obligated Servant of Your Most Illustrious Excellency

Galileo Galilei

G. Bartoli to B. Vinta in Florence. Venice, September 26, 1609[9]

[...] Concerning the secret or the cannon with long sight, I have to say that they really sell it in many places and that every spectacle-maker declares himself to be the inventor, and produces and sells them. A French man in particular, who makes them secretly, sells them for 3 and 4 sequins, or even 2 and less I believe, depending on their perfection, since one kind is made using rock crystal, and these are expensive, that is 10 or 12 scudi just for the lenses, and then use Murano crystal and ordinary glass. And this French man declares that he possesses the true secret, similar to or better than that of Galilei. Concerning me, since I have tried some of them and in particular one cannon sold for 3 sequins to the Master of the Post of Prague, I acknowledge that I am not entirely satisfied. Since it is longer than one braccio, one needs to try for long time to target with the eye the thing that one wants to see, and once it is targeted, one needs to keep the instrument very well at rest, because the smallest movement loses it. It is said that Galilei's does not have such imperfection (only a little bit) but, since he offered it as a secret and since he has to make 12 of them for the Seignory, he has been ordered not to teach this to anyone. I could not speak with him because he is in Padova. But I heard that soon it will be found also by other people because the secret lies in the quality of the matter of the lens and of the way this is applied to the cannon. I will try to find one that is appropriate and will send it [...].

[9] *EN*, X:259–260.

M. Hastal to Galileo in Florence. Prague, August 24, 1610[10]

Most Illustrious and Excellent Lord, Very Cultivated Master,

Because of the short time, I resolved to use the laconic tune.

His Caesarean Majesty, since he received from me the summary of the last letter written by Your Lordship to the Lord Ambassador, wanted to see the original text, which I found and gave him, and then also gave back. Thus, His Majesty's mouth is watering to know the meaning of the string of letters with changed positions [anagrams] and which contains what you recently discovered.

Moreover, *he ordered me to ask Your Lordship whether by chance you possess the secret of the Archimedean parabola, which kindles far away, and how far away. I answered him that I will write and also that I know with certainty that a very close friend of yours has it [the secret], but that he holds it in such value he does not want to sell it to the Grand Duke Francesco for many thousands of S[cudi]. I want to say Master Paulo. I was told about this in my hometown by a great mathematician, a friend of yours.*

Concerning the telescopes, His Majesty told me to tell the Most Illustrious of Tuscany that he should write to Your Lordship that you must take more care of those instruments that you make by your own hand, since he was annoyed that the priests took from your hands the one first destined for His Majesty. So, he is waiting for very perfect ones as soon as possible, at least one. With this close, I kiss your hands.

From Prague, August 24

Of Your Most Excellent

Zugmesser arrived during the night: I did not meet him yet. I will not fail

Very Devoted Servant

M. Hasdale

Addressed to: Most Illustrious and Excellent Lord

The Lord Galileo Galilei

Florence

[10] *EN*, X:426–427.

D. Antonini to Galileo in Florence. Brussels, February 4, 1612[11]

My Most Illustrious and Very Respected Lord and Master,

I have not received news from Your Most Illustrious Lordship for such a long time that I begin to doubt with great pain whether I am still favored by you. I would doubt even more if my own fault [...] of being so negligent in writing to and revering Your Lordship did not help me. But since I know I have committed a sin, it is also better I regard myself as being obliged to do penance. Although I have not been so negligent as to forget the infinite obligations and debts I have with regard to you, because this has been deeply impressed on my mind by your courtesy and it will not be forgotten. Having to come to you with vain words, however, without anything worthy of your ears, fearing it could be only disturbing for your studies, this has made me not very diligent in reminding you that I always live as your very affectionate servant.

These past days I have spent showing to these incredulous persons some of those things that Your Lordship discovered in the sky, and I made them confess that everything is true. Then I have a great disputation in reference to another thing, and it comes from this: Many days ago I learned that the King of England has a perpetuum mobile, within which certain water moves in a glass pipe, sometimes rising, sometimes sinking, like the ebb and flood of the sea. Thinking of this, it occurred to me that this is not ebb and flood, but that it is said in this way to cover the true cause. And the truth is that this motion is produced by the mutation of the air, that is, it is caused by warm and cold. This I gained from the speculations about those experiments made with a welcome-glass[12] that Your Lordship knows. And hence I also started to make one of these mobiles. I did not make it like the one in England had been drawn for me, which has a round pipe like a ring, but with a straight pipe, as Your Lordship is able to see in the sketch I send to you (Fig. 3). You must consider the small pipe AB of glass and the rest of metal that is well closed. In the vase B there is the liquid, which, when the enclosed air contracts, rises up along the pipe of glass, and when that air rarefies, it sinks. Then behind the pipe I placed a tablet marked with equally distant thick lines and their numbers in such a way that it is possible to make notes of the motion. The opening C marks a small hole that allows

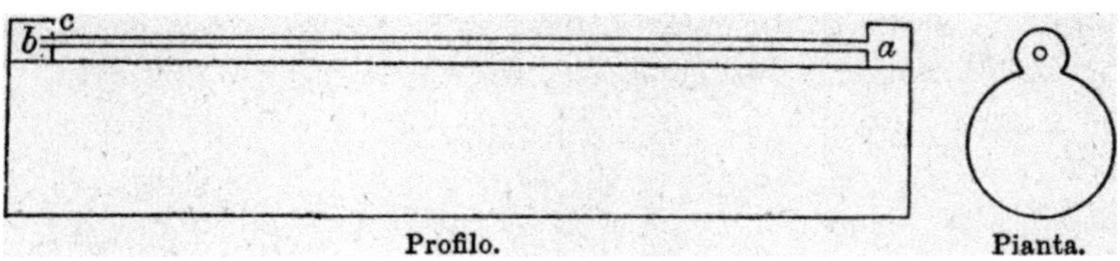

Fig. 3

[11] *EN*, XI:269–270.

[12] Thermoscope.

the air to replace the liquid when this rises up along the pipe from the vase B. I made it, as I tell Your Lordship, on a whim. But then, coming to the ears of this Prince, he wanted to see it and I have not only shown it to him, but also given it to him as a present. Now the quarrel I have in reference to this is ridiculous. Because these good Italian spirits absolutely do not accept this story, saying these formal words: How is it possible that what many great men have not been able to do has now been done by this bad youngster, who moreover has never seen and experienced the war. Now Your Lordship sees that I have more occasions to laugh than to discuss. But we can disregard these problems since, if they had spoken honestly, I would admit they are right *because I know well that there is no difference between this motion and that of a water mill apart from the cause of motion, which is seen by everybody, whereas in this case it is not.* I also had the possibility, required by this Highness, to apply this irregular motion to a regular one in order to run a clock. I am just starting to have it built. It will be a very artful machine and I hope it will turn out well. If it works, I will send Your Lordship its design. In the meantime keep me as your servant and favor me sometime with some of your new speculations. I kiss your hands.

From Brussels, February 4, 1612

Most Affectionate Servant of Your Most Illustrious Lordship

Daniello Antonino

Addressed to: Most Illustrious Lord, my Most Respected Master

Lord Galileo Galilei

Florence

[By other hand:] free as far as Mantova

G. Sagredo to Galileo in Florence. Venice, June 30, 1612[13]

Most Illustrious and Very Excellent Lord,

I give endless thanks to Your Most Excellent Lordship for the formations of the Medicean Planets that you kindly sent to me. I will share these with the Most Illustrious Mula and with some other friends as I did last time too. I am still waiting with great eagerness for your instructions about sight, and I would also ask that you do not forget to write to me about your opinion of the book entitled *De radiis visus et lucis* by the Archbishop of Spalatro, who, on folio 15, refutes with great confidence my opinion that I communicated him, that is, that sight is made within the eye thanks to refractions made by the species when they pass through the crystal humor. If Your Most Excellent Lordship pleases to make objections better founded than those of the Archbishop, I will receive them with great pleasure because I am greatly obsessed by this opinion and I wish to abandon it if it is wrong, once illuminated by those reasons of which you would not approve since you understand everything well. Father Master Paulo spoke with me very plainly about this issue, but only told me that sight, in his opinion, is not made in this way. His and my duties always hindered my speaking with him again about this subject.

The Most Illustrious Mula is greatly distracted by public duties, family matters and some other affects that keep him occupied with other thoughts. On his arrival in this city, however, he showed me a very great number of wooden tablets engraved with several demonstrations. These were to be used for one treatise, written in his own hand on paper, of about 100 folios. But he did not allow me to read anything, although he showed a great desire to share his thoughts with me in order to dispel some small doubts, which, as he said, he still had to solve in order to fully demonstrate the entire science of sight in a way completely opposite to what has been written by Vitellio and others until now. I told him my opinion and, as usual, he did not want to hear anything else and declared my thoughts to be wrong. After three months, however, once he secretly communicated to me the fundaments of his doctrine, he could not deny that, besides the three ways he told me, my own opinion could be added to them as a fourth way. Since that day, he has not spoken again with me about this issue, although he first wanted me to stay with him in order to show me his book.

None of the things I have read or heard from him and others ever generated doubts in my mind about my opinion, and I am only waiting for the opinion of Your Lordship, since I appoint you as the unappealable judge of this lawsuit.

The Lord Mula was at the patron fair and told me he has seen *an instrument by Lord Sanctorius with which one measures the cold and the heat with the divider. Finally he communicated to me that it is a large bowl of glass with a long neck* and *I immediately applied myself to producing some of them very exquisitely and beautifully. I make the ordinary ones with an expenditure of four Lira each, that is,*

[13] *EN*, XI:349–351.

a small water vase, a small decanter and a glass siphon. My method of production is such that I can assemble up to ten of them within one hour. The nicest one I made was produced by means of a small flame. It has the size and the shape of the one in the drawing included here, with all its parts. I am waiting to hear that you have made «mirabilia magna.»

From among my lenses, Bacci chose three, which I send to you. Two of them of six fourths are from my poor man; the other one of eight is from Bacci, who promised to give me a good one of four and a half, but then did not do so. If he is not able to do it next week, I will send you one of my two lenses of that size, which I have in my possession.

On my return I bought the *Diagnia*[14] in Milan and read it on the coach. I consider it to be very impressive buffoonery, which in my opinion is absolutely not worthy of an answer. If I have time to waste, I will also read that small book of Lagalla and that of Martino. I cannot write any longer. I kiss your hand.

Venice, June 30, 1612

Of Your Most Excellent Lordship

Yours

G. F. Sagredo

Addressed to: Most Illustrious and Most Honored Lord,

The Lord Galileo Galilei, Philosopher and Mathematician of His Highness

with a small box

Florence

[14]F. Sizzi, *Dianoia Astronomica, Optica, Physica*, Venetiae, 1611.

G. Sagredo to Galileo in Florence. Venice, May 9, 1613[15]

Most Illustrious and Most Excellent Lord,

Although I told Your Most Excellent Lordship in my previous letter that I received the four books on your solar letters, and what I decided to do about distributing them, and how I decided to retaliate with these letters, I would like to let you know that I executed the plan and would like to add my thanks to you because one of the books remained in my hands.

I never heard that you received the last map I sent to you, although I suppose that, since I delivered it to the Lords Guadagni, it cannot be lost.

Until now no mathematician has been chosen because no subject of great fame has applied. My Lord Father is a Reformer and has asked me to gather information about subjects worthy of that chair. Would Your Most Excellent Lordship therefore kindly do me the favor of writing to me about your opinion on this matter? Glorioso, among those who have applied, is beyond compare. However he is so cold *in agibilibus* that, since no one has ever seen a sign of the vivacity of his intellect, many believe that, besides the teaching of ordinary things, the University of Padova would not gain any splendor from him.

It seems that the stones sent in the small box by Your Most Excellent Lordship do not hold the light. I would like to know whether they are natural or artificial, and some other particulars.

Cremonino's work is not yet completely printed and one believes the printing will go on for other three months.

I saw what Lord Velser wrote to you about me in his printed letters. Now it seems to me that I have been lucky since I did not write to him (as it happens) about any rubbish that would then have been printed. In conclusion, I learnt that I have to be very reserved with you literates, who print yours and other people's things. I have a kind of friendship with that Velser, introduced by the Jesuits who are very close to him. Although we write to each other often, I believe nevertheless that he is not completely satisfied with me because I have not been generous with the title. But this is the Jesuits' fault. They instructed me to call him Very Illustrious because, since I am not looking for trouble, I would not have had any problem calling him Most Illustrious and receiving Magnificent and Sir. But I decided to keep the initial use so as not to display any irresponsible behavior.

The instrument to measure heat, invented by Your Most Illustrious Lordship, I reduced to several very pleasant and elegant forms, so that the difference in temperature between one room and the other can reach up to 100 degrees. *With these [instruments] I speculated about several marvelous things, like, for example, that the air in winter is colder than the ice and snow, that the water now seems to be colder than the air, that a small amount of water is colder than a large amount, and other such similar perceptions. Our Peripatetics cannot give any resolution for*

---

[15] *EN*, XI:505–506.

these, to such an extent that some of them (among whom is also our Gageo) are so far off track that they do not yet understand the cause of the first operation, since they believe that one should see an opposite effect, because, since heat (as they say) has an attracting virtue, it should be that, when the vase is warmed, it pulls the water toward itself. And such men claim the main chairs of Padova!

I cannot go on any longer and so I will stop boring you with my idle talk, and recommend myself to you as usual, etc.

From Venice, May 9, 1613

Of Your Most Excellent Lordship

Yours

G. F. Sagredo

Addressed to: Most Illustrious and Excellent Lord

The Lord Galileo Galilei

Florence

G. Sagredo to Galileo in Florence. Venice, July 27, 1613[16]

Most Illustrious and Excellent Lord,

I received the small box in very good condition, according to the notes in the letters of Your Most Excellent Lordship, and I also received some money of which you make no mention. I believe this corresponds to what I gave to Bacci, although you did not seem to know that I paid him. Hence, I believe that many letters were lost, and therefore I would also like to tell you that I did not hear if the last map arrived and I received no money for this one or for the first map I sent to you I tell you this so that I am able to get the money back in case you have already given it to the carrier or somebody else. Concerning our account, it is better for me that we do not make any final balance in order to avoid paying my debt.

Bacci tells me he has sent you some very good lenses but I was not able to see them to compare them with my own. If you order me to pay him for them, I will execute your order and any other business.

After the arrival of the very precious wine of Your Lordship, and with this heat, my intellectual purpose lies in measuring that heat while I have cold drinks. The instrument to measure heat has been almost perfected and I have also been making ephemerides with it for fifteen days. I will send a copy of it with the next post, since I have not enough time now. *I also found: a decanter that, when the wine passes through it, cools down immediately, and if needed, warms up; some glasses in order to drink it with ice, and another where, once the wine is introduced into it, one can see how many degrees of cold it has taken, and it can also be used to drink; an inkwell that preserves the ink in this hot weather so that it does not dry up, become thick, or make the pen too wet, which is cheap and lasts a long time. After drinking two glasses of the wine of Your Lordship, these inventions came to me so now I hope that, as soon as I have drunk only one of your flasks, I will have invented divine things*. My duties do not allow me to return your flasks with this postal dispatch, but I will do it next week. Although the quality of your wine took away my hope of finding something equivalent, I will nevertheless try not to give up. I cannot be longer and *more philosophico* I close in this way.

Venice, July 27, 1613

Of Your Most Excellent Lordship

Lord Galilei

Yours

G. F. Sagredo

Addressed to: The Most Illustrious and Respected

The Most Excellent Lord Galileo Galilei.

Florence

---

[16] *EN*, XI:544–545.

G. Sagredo to Galileo in Florence. Venice, August 24, 1613[17]

Most Illustrious and Excellent Lord,

I did not send you the large lens that I promised because (as I wrote to Your Most Excellent Lordship) it is only of mediocre quality. I believe I can better satisfy your desire and my debt if I left with you the first one I already sent, since it is better and more worthy and also because fortune has brought me another similar one. Now that I have received back this first one, and now that I know of your wish to receive another larger one, I will not send this second one to you, but will try to carry out some experiments in order to obtain one lens of the size you would like to have. My master is a poor man called Master Antonio, who works under the insignia of St. Lorenzo in Frezzaria, and I consider him more adequate and serviceable than Bacci. He has a die of mine that I gave to him as a present with which one can make very good lenses of six fourths. He has another die of mine too that he repaired himself, with which one works those [lenses] of thirteen and fourteen fourths. The lenses worked with this die on both sides are, as you know, of seven fourths and many of them are really good. The lenses that are worked with this die on one side and with the smaller one on the other are one braccio long. He can also make perfect lenses of three fourths, working them on both sides with the die of six fourths. [He can make] any kind you would like to have. You can always write to him about what you wish to have, because he would very much like to be your servant and professed himself to be obligated to you. In fact, though he is really poor, he has nevertheless improved his situation considerably thanks to this business of the lenses on which he works continuously. He has almost completely abandoned his usual activities, which was making mirrors and working stones of whichever sort.

I am really sorry that Your Lordship is affected by a severe scarlet fever, especially with this temperature. You truly arouse my compassion. For God's love do not go outside in the night and be sure that night air is very dangerous. Forget Jupiter and Mars and all the planets in the sky. Take care of your health and of your life. Consider your studies as a pastime and embrace the true philosophy, enemy of ambition and slave to good health and taste: the best thing in our lives. Thanks to God I have had a very happy life since my return from Syria. There is no one in my house who orders me around. With my father Lord I only have the business of greeting and of confabulation. I have freed myself completely from household duties. Concerning the other businesses, my brother has seven eighths of them since I made him owner of everything, because everything must go to his sons in any case. A very small part of the business is in my hands and I can accomplish this as well if I stay at the country house, and this depends only on my orders and on writing three of four lines a day. If I want to go to the country house, I am there in a couple of steps. If I do not want to move, I have six rooms here in the house where I can lodge the custodian of the country house too, and the entrance is separate. I am served by Lopotoppo. In

[17] *EN*, XI:552–556.

conclusion I only take care of my own conservation and taste, as if when I perish, the world would perish too. Business and ambition do not touch me. It also seems to me that this whole ease of mine, which I have without depending on anyone, will not disappear (if the world does not turn upside down). So I only take care of that continence necessary for the conservation of health, in reference to which I am truly diligent since I do not want that my present satisfaction takes away my taste for many satisfactions in the future. I have completely banned doctors, having decided that they can come to me only in very dangerous cases. My health rules are: leaving the table a little hungry, drinking with self-restraint, eating tender things, friable, which nourish well and taste good.

Great wines are ordinarily excluded but I sometimes drink some of the good ones after fruit, but without exaggeration. I like to share them with friends for whom I have a good reserve. I avoid the cold like a mortal enemy and also excessive heat, which can cause inflammation. I have banned hard work and my movements are very moderate and always made with comfort and taste. All of my business is voluntary. In fact, I am convinced that this world is made to serve me, and not me to serve it.

And I would like my Lord Galileo to do the same, for whose love I damn the courts and ambition a thousand times a day. Do not bother, please, to answer certain ignorant philosophers; do not waste your time reading their madness. Do not write any more demonstrative things in form of dialogue. If the preachers do not die for their obstinate sinners, why should you be the martyr to convince the ignorant people, who, finally, since they are not predestined or elected, deserve to fall into the fire of ignorance. And the more cheerfully [they fall], the more these good people, since they feed themselves in these flames without pain, believe to enjoy the sky of sapience and that the soul of Your Lordship is lost. And so they would be convinced as well that they could bring you into their ignorant paradise thanks to their orations? You should philosophize (as I do) whilst walking, promenading and sitting. Be your own master and pupil. Do not stick to the books and do not kill yourself writing; use (if you can) someone else's hand. Only write back to those who deserve it; to some people write laconically, only four lines, and apologize on the basis of your infirmity. And each time have written to me a quinternion of paper, since I am convinced that it will be tasteful and without effort. And then, absolutely do not forget these blessed mud baths of Padova because they will surely free you from that scarlatina. I offer myself as your archiater.

Already some weeks ago the Lords Guadagni sent to my home the money paid for the maps, and they apologized profusely for the delay. *More philosophico*: I considered it superfluous because I would have been sorry only had the money been lost.

Concerning the equation, I saw Your Most Excellent Lordship's answer, which however does not touch the point that is reason for the thousand disputes with the mathematicians of these parts, and especially with Master Paulo and with Mula, because what Your Lordship demonstrates is true for every meridian and in every instant. But my paradox is this: I believe that the rules known until now for using those equations to formulate a general equation work well in reference to all the meridians except for two, which are contiguous. Among these one finds the difference of one day and this difference cannot be avoided by human work and sapience.

Hence it follows that, given a Supreme Pontiff, spiritual and temporal king of the universe, who would like to confirm or constitute a new denomination of the days, establishing holidays and eves, even if he could instill and command his will over the whole world in one instant and with one gesture, he could not avoid this difference of one day that takes place between two of the meridians, so that along one meridian the eve would take place and along the other the holiday, and the difference mentioned would remain among the inhabitants of those meridians *perpetuis temporis*. And since this imagined thing is completely true, it must be that this place really exists in this world where Roman Christianity lives. Deductively I find that this difference takes place where the Portuguese and the Castilian Spaniards met on their navigations, which was between Manila in the Philippines and Malacca. And when the Castilians pass through that place they gain one day, whereas the Portuguese lose one. And if this meeting had taken place on land, in two places close and contiguous, the same would have happened.

This thought of mine, although very true and demonstrative, since it is also new, is considered not to be credible by our mathematicians here, who, equivocally basing their reasoning on the uniformity of the sphere, could not understand that this accident happens in one place and not in others. And that Saint Beatitudes, transformed into a monarchy, could not make *unum ovile* to celebrate the holidays because these close inhabitants experience a difference of one day. It seems very strange to them that this accident of experiencing a difference of one day, as happened to Magaglianes because he navigated around the world, could happen to those inhabitants too when they undertake a short trip to the mentioned place, and this would not happen to a person, who, starting from Malacca, would go to Manila going around the world (and not taking the shortest route).

I wait until you tell me, with ease, how you find my new speculation. I already had this thought four years ago and did not communicate it to anyone, considering it to be so clear a thing that everyone would know about it. But when I communicated it casually to Lord Mula, I received so many objections that I decided to appeal myself to the Father Master, who does not want to understand it. This is the reason why I divulged it as a higher speculation than I considered it at the beginning.

Since I see that you will not return, I will convince my Lord Father to provide for the chair of mathematics, which will probably be assigned to Lord Glorioso, who in fact is a very intelligent man, although very cold and, *in agibilibus*, does not completely satisfy me. I do not like Kepler at all and believe he is a Calvinist too. That Luca Valerio requires perhaps too high a salary. Therefore it is more convenient to elect the mentioned Glorioso.

Cremonini's treatise has not yet been completely printed, since it still lacks the table of content. I will have it next week and will send it to you. But please, Your Most Excellent Lordship should not pay attention to the jabber he writes. If his reasonings about the new appearances in the sky are not too long, I will read them and afterwards I will write to you what, in my opinion, you should answer him.

According to the little bit I studied, I see that, in reference to the mirrors, only the simple reflection has been written about, as happens with those of steel and others which do not have any transparent part. There is no mention of the refraction that takes place in those of glass. And from this it seems to me that it remains a great and new speculation to accomplish in reference to this point. Although it seems that this also happens in ordinary mirrors with flat surfaces. Nevertheless in other mirrors of regular shape one will see unexpected effects. I wrote about this to Lord Magini, who has an opinion completely contrary to the truth. At the beginning, Father Master also seemed to tend toward Magini's opinion, but then I was able to convince him. Please, Your Most Excellent Lordship, do me the favor of your thoughts on the effects of the figures attached here and indicate your opinion to me and I, although a very weak geometer, will tell you what my opinion is, but always keeping your opinion as true. It is late; I cannot go on any longer. I kiss your hand and pray to the Lord God for your health and happiness.

In Venice, August 24, 1613

Of Your Most Excellent Lordship

Yours

G. F. Sagredo

Each of the drawn forms can be considered [investigated] in two ways, since one can place the leaf [stratum of quicksilver and tin] on both of the surfaces, except for C, and D, which [whose surfaces] are equals (Fig. 4).

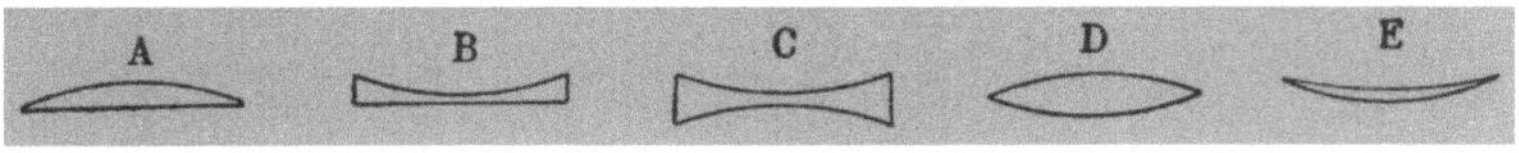

Fig. 4

G. B. Baliani to Galileo in Florence. Genoa, April 4, 1614[18]

My Very Excellent and Ready Lord,

The letter of Your Lordship gave me much pleasure as well as much displeasure to learn that you are not in good health, although persons like yourself should enjoy a very long life with good health in order to be able to bring to the world those benefits, by means of their efforts, as Your Lordship is doing the whole day.

I will briefly answer the very pleasant letter mentioned and this because I remain truly satisfied by the answers Your Lordship gives to my reasonings, which I wrote to learn something from your answers rather than because I had doubts that something Your Lordship had said in the Letters is not completely correct. Since the letters are very rich in doctrine and novelties, from the time I wrote to Your Lordship, I remained without them and neither do I have them now. And this because I have a lot to do in sending them to this or to that curious person who wants to see them, since there is no lack in Genoa of those who are curious about things on mathematics and, especially, about those of Your Lordship.

I noticed you do not say anything about what I wrote to you, that the changing of the sunspots may be accidentally due to changes of weather, and that in these last days of March the weather was colder and more turbulent than is usual for the season. And although I know one can think that the reason is the conjunction of Saturn with the Sun, I cannot however force myself to think that it may not be that there were and still are more and denser spots on these days than during the month of January.

Above all, I have been grateful about the ingenious way of finding out the weight of air. And because Your Lordship wishes, I will tell you how to cook without fire. I made an iron receptacle with a flat base, round, with a diameter of about one span, and another iron tool, also round and flat, with the same diameter. I let the last iron tool turn quickly, either by means of a large wheel or of running water. Above this iron tool I had the base of the mentioned receptacle placed in such a way that it stays very close to it. Thus, by rubbing each other, the iron tools mentioned warm themselves so much that one can also heat and cook what one puts into the vase. And I stop now and I kiss the hands of Your Lordship and pray for immediate and long good health; and I will give to Lord Pinelli your recommendations as soon as I see him.

From Genoa, April 4, 1614

Of Your Most Excellent Lordship

Very Devoted Servant

Giovan Battista Baliano

Addressed to: My Very Excellent and Ready Lord

Lord Galileo Galilei

Florence

---

[18] *EN*, XII:44–45.

G. F. Sagredo to Galileo in Florence. Venice, February 7, 1615[19]

Most Illustrious and Excellent Lord,

Today, in recording some of my writings, I found a letter from Your Most Excellent Lordship written on the 27th of September, to which I do not remember having given an answer, neither do I know how I didn't notice that it left the bundle. However, since this debt, transformed into money, has not been paid immediately, Your Most Excellent Lordship will be satisfied in receiving the payment in the term used among credit-merchants of four months.

At first I will say to you that if Your Lordship would like to exchange letters every week, you are also obliged to write each week, even if you notice that my answer is late. Because I actually promise not to fail to do so on my part, not so much to please you, but more to give me the pleasure of reading your letters, for which I am more grateful than for those of anyone else, as my affection toward you and the esteem in which I hold you is much greater than for any other.

Father Master Paolo is very well, thanks to God, and each time he meets me he wants to know about you. Lord Mula is *podestà* in Verona and I hope to meet him in the space of a few days. Lord Veniero is healthy and he likes you as ever. Lord Francesco Moresini, God willing, will return from Crete within two or three months. Indeed, the company is alive and healthy and wishes, more than anything, for the presence of Your Most Excellent Lordship, who, not being able to satisfy our wish personally, can at least console us with Your letters.

No long telescope, or at least ones better than the first, have turned up in my hands, either because the master was not skilled enough to make them or because I have not solicited him for a long time, neither am I using them. When the climate becomes agreeable and one is able to leave the windows open without unpleasantness, I plan to use it sometimes. And if some novel and good things occur to me, I will share them with you.

The reconfirmation of Lord Cremonino's chair has not yet been announced. The Lord Prosecutor, my father, has a bad opinion of this person believing him to have promoted atheism in many young people with his doctrine of the soul. It seems that this opinion is very widespread among the nobility. Hence, many people consider him to be a scandalous man, careless and unworthy of confirmation in the University of Padova. However, the Lord my father will leave the position in a few days and a new Reformer will be elected in his place.

Concerning new speculations, I have so many in mind that the matter of speculating is never lacking in me. But not being able to understand the old ones without the help of Your Most Excellent Lordship and without your presence, I manage rather to realize my comforts and some tastes than to speculate, seeming to me, in this way, not to lose time.

I have multiplied and refined the uses of the instrument to measure heat and cold, so much so that there is much to speculate on. But as I said above, without your help

[19] *EN*, XII:138–140.

I can hardly satisfy this need myself. *With these instruments I clearly see that the water in our wells is much colder during the winter than in the summer. For my part, I believe that the same thing happens in live fountains and subterranean locations, although our senses consider these in a different way.*

I wrote this letter at the end of last week but, because a comedy [of events] prevented me from closing and sending it, I kept it until today. I must tell you that *two days ago, when it snowed, my instrument displayed 130 degrees more heat in this room than what [it showed] two years ago during a time of very rigorous and extraordinary cold. The same instrument, immersed and buried in the snow, displayed 30 degrees less, that is, only 100. But then, immersing [the instrument] in snow mixed with salt, it displayed a further 100 [degrees] less. I believe that it really displayed even less, but one could not see it because of the hindrances [caused] by the snow and the salt. Since, at the hottest point of the summer, it had displayed 360 degrees, one can see that salt added to the snow increases the cold by as much as one third of the difference between the greatest heat of the summer and the greatest cold of the winter*. This is such an admirable thing that I cannot give any imaginable reason for it. I would be glad to hear Your Most Excellent Lordship's opinion and also what you have seen about the practice of cold caused by saltpeter. Although I have listened to many things about this, in effect I did not see anything.

I think it would be difficult to send instruments there for this intention so that you are able to see that experiment; it may be easier to manufacture them there. However, if you so indicate, I will serve you according to your wishes. And in closing, I kiss your hand.

In Venice, February 7, 1615

Of Your Most Excellent Lordship

All Yours

Giovan Francesco Sagredo

Pardon me, I do not have time to see these folios again

Addressed to: Most Illustrious and Ready Lord

Most Excellent Lord Galileo Galilei

Florence

G. F. Sagredo to Galileo in Florence. Venice, March 15, 1615[20]

Most Illustrious and Most Excellent Lord,

If it is true that the fact I consider you to be a friend has such value, then Your Most Excellent Lordship is owner of it and no further gift *inter vivos* is required. Or, to put it better, if it is true that from my favor you can obtain real happiness for a hundred years, as your letters tell, I would be satisfied to give it to you as a very solemn mortgage, charging you with the right to possess it completely, to press it and distill it so as to extract from it the fifth essence, if this can, as you write, add years and centuries to your life, and if it offers you perpetual happiness and satisfaction.

I am really sorry about your bad mental and bodily health. I am more worried about the bodily health than the mental one, as it is more difficult and recondite to remedy this. Whereas for mental health it seems to me that if one lives prudently and according to a founded and true philosophy, good health depends only upon our will, since there is no doubt that if one is generous and abandons certain opinions invented by human lightness, one is able to tolerate very easily any sinister encounter. But only if this is not shared with the body, which, by means of speculation, does not receive heat, cold, nourishment, taste and all other things necessary for the sustenance of the life and for the taste and the happiness of the senses. Therefore Your Most Excellent Lordship continue reading Berni and Ruzzante and leave aside Aristotle and Archimedes. Make your speculations in bed so that the mind shares the ease of the body. Contemplate tasteful things and take care of your health, not with medicines and diet, but with quiet and prudent sobriety. Avoid those dishes that you know can disturb you and choose those which are healthier and more tasteful to your senses, retaining a little appetite at each meal so as to better taste the next one. Be sure that you will get over every indisposition with this rule, because, thanks to God, you still have many years before you enter into old age.

Thanks to divine clemency, I am doing well by means of this same rule, healthier and more vigorous than two years ago. Concerning the soul, I live very happily far away from all work and trouble. No accident seems to me to be new and unexpected. All my wishes are very limited and moderate. I happily receive every good thing that happens to me and, to make it even more pleasurable, I consider it as being inconvenient or as not belonging to me. If I consider it to be a gift or a loan of fortune, and not as an ordinary and due revenue, I enjoy it even more. And for the same reason, I can easily give it away if the situation should require it. If I did not use the true philosophy, I would also have good reason for trouble because of ambition, the fifth element of our nobility. Not because I would not be among the most advantaged and among the first ones, if I compare the honors, the titles and reputation of mine to the general ones of those of my age. But since so many and extraordinary honors and comforts fall to our house, and since I am not sharing them to such an extent as

[20]*EN*, XII:156–158.

my grandfather, father and all my brothers do, it can seem, and I know that it seems so to many persons, that the reason for this lies in certain deficiencies of mine. But since I know perfectly the reason for this difference, I neither complain nor is my satisfaction decreased by this. It would be foolish ungratefulness if I complained about the fortune of my house and that person who grounds his happiness in the foolish and intemperate opinion of the common people is crazy.

And free from endless responsibilities and disturbing duties related to the honor of our country, I enjoy the freedom and use my time in conformity to my taste and needs. And although I do not share that extraordinary veneration, which does not fit well to my mind, I live however without envy and denial.

I tell you these things about myself so that, since you are wise and very prudent, once you see the source of my pleasures, you will be able from the same source to allow a new brook of happiness to spring forth, which is very abundant, thanks to the machinery of your prudence. And to console me greatly, you will advise me about each new inundation that happens. And about this issue it is enough now.

I have added and changed particulars to my instrument daily to measure the temperature in such a way that I could easily tell you the whole story of my invention or, better said, improvements, should I have the chance to do so orally in your presence. But since, as you write and as I now believe, Your Most Excellent Lordship is the first author and inventor of it, then I believe that the instruments made by you and by your very good craftsman are much better than mine. I would therefore like to ask you to write to me at the first occasion of the kinds of works you have made and I will write to you what has been done here. And should we touch on some aspects of this issue in every letter, I will write to you some imperfect speculations of mine that will be perfected by your mind and intelligence without any effort and with pleasure. The person who claims himself to be the inventor of these instruments is not only inappropriate, but completely unable to explain it to me according to my wishes and needs, in the same way as I vainly tried to have him understand the reasons for the effects one sees in some of my (I will say) compound and multiplied instruments.

It is impossible to find Apelle's book here and the Venetian librarians did not go to the last fair at Frankfurt. If Your Most Excellent Lordship gives me some indications, I will try to serve you.

Certainly only two long lenses of the quality you desire have been made and, as I heard, these are exquisite and in the possession of the Most Illustrious Lord Vincenzo Gussoni, who was ambassador in Savoy. He had them worked from a piece of glass that he took from a broken mirror, from which he also had worked many other smaller and very good lenses. It is impossible to take them from his hands. He declares that they are considerably better than my lenses. However, Master Antonio, who made them, tells me there is no difference.

The Most Illustrious Lord Agostino da Mula came back from Verona and all his glasses [lenses] have been stolen. I believe that he will have some of them made with extraordinary diligence. I will certainly take the occasion to send one of them to you, since the die is mine. And to close, I pray for every prosperity and happiness from the Lord God for you.

Venice, March 15, 1615

Of Your Most Excellent Lordship

Yours

G. F. Sagredo

G. F. Sagredo to Galileo in Florence. Venice, April 11, 1615[21]

Most Illustrious and Excellent Lord,

I received the letters of Your Most Excellent Lordship of the fourth of the current month with the usual pleasure and consolation. They are mostly written in another hand, which leads me to suspect you are not in very good health. However the last six or eight lines have led me to believe that you are at least at a good point of recovery, which I truly desire. I pray God to grant this to you as soon as possible and for many years.

Concerning your business with Lord Cremonino, I will obtain a good result with the help of the Judge of Witchcraft, who is a true confidant of mine. I will write to him this evening.

Concerning the glass instruments to measure temperatures, the first ones I made were done in the same way as Your Most Excellent Lordship has had them made. But then I multiplied the invention in different ways. I cannot write about all of them in the present letter since I do not have as much free time as I had fifteen days ago because I was at the Council of the Pregadi and have been elected among the five Sages of the Merchandise. But if I divide this speech into many letters, without haste, this will give us occasion to visit each other more often. I do not think my duties will interrupt our usual and mutual correspondence, which lifts our souls and does not aggravate, even though, using up our time, may impede other actions.

I understood your opinion about the way those instruments function [...] and I would even dare to say [it is] also true, if it were not for the reason that in itself it is not evident to the senses. Neither do I believe that, on the basis of those things that are evident to the senses, one can demonstrate it. *But it satisfies the mind much more than the arguments of the Peripatetics: If, because of the external heat, the air that is inside the warmed glass bowl evidently dilates so that it pushes out the water, it is easy to believe that the heat penetrates the glass. Once it has penetrated there in greater or smaller quantity, it requires more or less space. Since it [the space] cannot simultaneously contain the air and the soft and igneous spirit, the air is obliged to exit the space. In addition, when the external environment cools down, it is believable that the igneous spirit, which is overabundant in the bowl, exits until it equilibrates with the environment. Thus, since the space that contained it becomes empty, the air is obliged to follow, and water or wine after it.* But it is clear that one must acknowledge the existence of the vacuum, as I have shown with the following experiment.

At the furnaces of Murano, I had a vase made with a neck of one palm. When it was very hot, I had it closed so that all the air that was sealed inside, full of heat, could no longer exit. After that, cooled down and consequently cleared of igneous spirit, the air remained inside at the same temperature as the environment. I persuaded those who were present that inside there was really not much air. It was

---

[21] *EN*, XII:167–170.

evident to the senses as well that there was no igneous spirit. There were two demonstrations. The first: having enclosed a sparrow-hawk-bell within it, when moved, this did not produce any sound unless it collided against the glass and, consequently, produced an external sound. It was very easy to believe that this only happened because of the absence of air in the above-mentioned vase, and more so when, having broken the mentioned vase, we found that the bell was able to work normally. The second: once I put the vase with the neck into a basin of water, using an iron tool I delicately opened its mouth, through which so much water went upwards into it that it seemed to try to fill the mentioned vase completely. Our impatience though, which was the reason why it broke, prevented us from seeing it filled completely.

Concerning the difference or inequality of the rising of the water or of the wine, even though at the beginning I did an experiment completely similar to yours using a larger pipe, but without wine and regulated by another equivalent measure, I then did it, however, in another way: After having attracted a certain quantity of liquor into the pipe, I took away the small vase from the bottom and made that liquor rise and fall. However, I soon abandoned this method as well as another one, which was bending the top of the pipe at a right angle toward the ball and the bottom of the pipe toward the other side so that, with the small vase placed at this end, the pipe remained at the same level (Fig. 5). But since these two precautions of mine are not ordinarily useful in reference to instruments that have a large pipe, which are certainly the most perfect ones, I abandoned them as imperfect subtleties and more so because in fact, for the experiments I did, as I will perhaps write to you in others [letters] of mine, I do not find that the difference is great. Hence, although I had the intention of using the other precaution written to me by Your Most Excellent Lordship, to decrease the differences between the higher degrees, I did not do this because I was not really able to speculate this rule as a theory. Hence, if Your Most Excellent Lordship will shed some light on it, I will receive this with great pleasure.

The best and most perfect instruments I made were with a neck as broad as a finger, referring to the internal part of the neck over which I had blown at the furnace of Murano a vase whose volume corresponds to three or four glasses, using the mentioned instrument in the way Your Lordship writes. In this way I have had three of them made in different sizes, and which have worked now for almost three years in such harmony with each other that it is marvelous. These I have observed for over almost one year, one, two, three, four, five, six, up to eight times a day, with such correspondence that from those observations I have achieved a table of correspondences and equations among them. First I have seen that they work with the absolute same proportion, during both extreme heat and extreme cold, so that each time I see one of them I guess, by using the table, the degree of the other two, sometimes with a variation of give or take two or three degrees.

Fig. 5

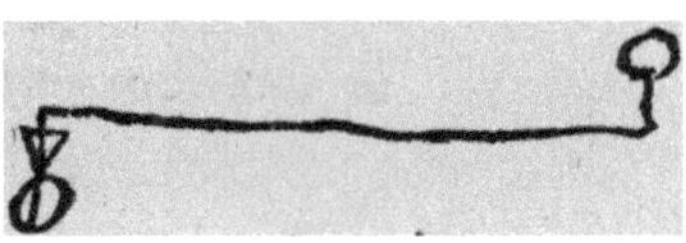

This also happens to those who, leaving Florence, make a pilgrimage to St. Giacomo of Galicia and being on horseback, either on a whim or from necessity, they go for some time at a gallop in front of the companion or they stop behind him two shots of an harquebus away. But each evening they find themselves in the inn at the same table. *So, changing these instruments a little because of the slightest occurrence, they change more or less depending upon whether they are more or less exposed to those occurrences, either because they are close to the apertures of the room, or to the persons, or to the lights, etc. Moreover, since some of them have thicker and some thinner glass, it is conceivable that not all of them change over the same period, but, should some alterations in the temperature of the surroundings occur, the thinner one is the first to sense and show it. Concerning the instruments with a very thin neck, as those of Your Very Excellent Lordship, you should accept that the viscosity of the water and of the wine also causes variation. Therefore I decided to use instruments of such a size, that, when one takes away the lower vase, the neck empties out.* Another time I will write to you of some other particulars and in closing I kiss you hand.

Lord Gagio is here in the room and he disturbs me. I do not want him to see what I am writing. These folios of mine will appear to you a little confused because my mind is busy with other tasks.

In Venice, April 11, 1615

Of Your Most Excellent Lordship

All yours

Giovan Francesco Sagredo

With the next mail I will write to you about that young person whom you propose to me

[In Galileo's hand]

Fig. 6

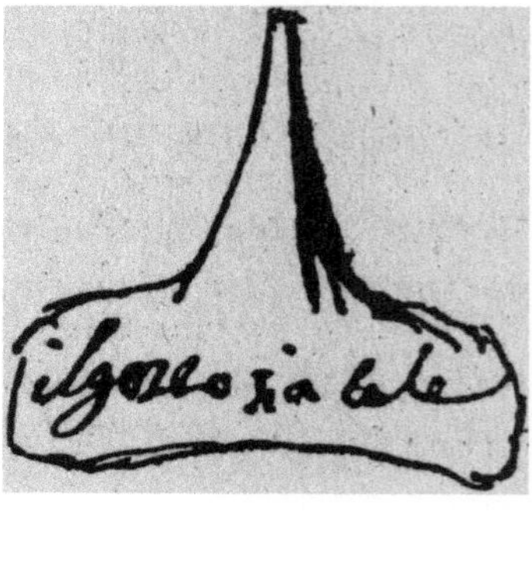

Hold an empty decanter over the fire and from the mouth (which has to be very narrow) observe by means of an air valve whether the igneous spirit exits continuously (Fig. 6).

Fig. 7

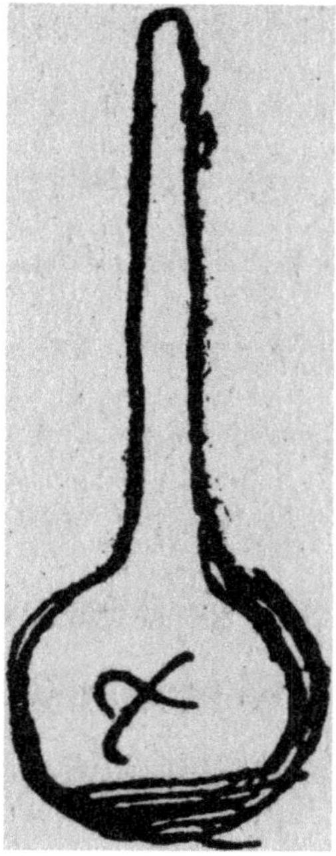

Introduce into decanter X a very small quantity of wine, ink, quicksilver, etc. Then place it over the fire, observe whether the mentioned [things] are consumed, etc., or what it makes (Fig. 7).

B. Castelli to Galileo in Florence. Pisa, May 24, 1617[22]

Most Illustrious and Excellent Lord and Very Cultivated Master,

Angelo di Matteo from Capraia, the carrier, left yesterday morning with the wares of Your Lordship, that is, mattresses and linen. I gave him the order according to what you told me here.

I assemble two telescopes with four rods, as Your Lordship can see in the figure (Fig. 8), so that if telescope A remains immobile, telescope B can move back and forward but always parallel to telescope A. Once this instrument was ready, I targeted the Sun that appeared to me while I was here at the Monastery of St. Girolamo from behind the bell-tower of St. Catharine. I placed the telescope toward the Sun in the same way used to observe the sunspots, and I did it in such a way that one telescope could catch the whole Sun and if I had moved it toward the other telescope, part of the Sun would have been covered by the bell-tower. At the same time, the other telescope could show only a very small portion of the Sun, since it was completely covered by the bell-tower. Once I did this, I concluded that between the place where I was and the bell-tower there was a space-interval of about one hundred and six times the space between the mouths of the telescopes. Since I know that the explanation of this is very well known to Your Lordship, I will not be long

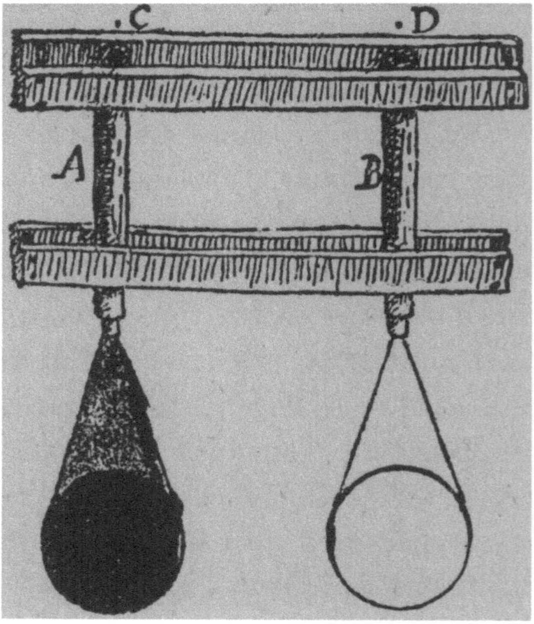

Fig. 8

[22]*EN*, XII:319–320.

in telling it to you. I wish only to tell you that since the mentioned space was three braccia, I found the distance between me and the bell-tower to be three hundred and twenty braccia. This is really a gallant thing and very pleasurable and it will allow very great distances to be measured, such as for example, concerning islands on the sea, etc. But since I am at the end of the folio, I deliver everything to the censorship of Your Lordship and stop here professing myself to be as ever your servant.

Pisa, May 24, 1617

Of Your Most Illustrious and Excellent Lordship

Very Affectionate Servant and Disciple

Don Benedetto Castelli

Brother Bonaventura revers Your Most Excellent Lordship and begs you to keep him in your good favor

Addressed to: The Most Illustrious and Excellent Lord and Cultivated Master

The Lord Galileo Galilei, First Philosopher of His Highness

Florence

Galileo to Leopold of Austria in Innsbruck. Florence, May 23, 1618[23]

Very Serene Lord and Very Cultivated Master,

I am still in the same poor health in which Your Very Serene Highness found me when I was favored and honored by your infinite benignity, so much above my merit. And to these corporal troubles of mine another more unpleasant trouble of mind has been added, which is that I have been unable and am still unable, at least in part, to satisfy the requests of Your Highness to compile some discourses about those problems that I consider pleasurable for you, as I had a mind to do. Therefore it is necessary that I beg you very humbly to allow a deferral, which I need in order to obey your orders more completely. In the meantime I hope you find pleasure with the few things I send to you with the present letter. These are two telescopes, one longer than the other. The larger one is able to serve Your Highness and the members of your family for the observations of celestial things. It is the very same glass with which I have been observing for three years and, if I am not mistaken, it should turn out to be excellent for you. The smaller one is easier to use and is good for discoveries on land, even though in reference to these the longer one will also display the objects larger and more distinct, but you should target them with a little more effort.

I also send you another smaller telescope, shaped from a block of brass. This does not display any finery, because it can serve Your Highness only as a model and example in order to have another one manufactured that fits better to the form and the size of your head or of whoever shall use it. One cannot fix that instrument and device without the actual presence of the head and the eyes of the particular person who shall use it, because of the different positions—higher or lower, more or less inclined to the right or to the left—and these are almost indivisible differences. Your Highness does not lack craftsmen who serve you exquisitely on the basis of this model. I beg you for my sake to keep it as secret as you can.

Enclosed I send you the printed Solar Letters. Moreover, together with the present, you receive a short discourse of mine on the reason for the ebb and flow of the sea. I had to make this discourse in Rome a little over two years ago, ordered by the Most Illustrious and Reverend Lord Cardinal Orsino, while the lords Theologians were thinking of the prohibition of the opinion of Copernicus and of the opinion of the mobility of the Earth. Such an opinion is written in the mentioned book and defended as being true by me at that time, until those Lords decided to prohibit the book and to declare the mentioned opinion as being false and adverse to the Holy Writings. Now, since I know how convenient it is to obey and believe the determinations of the superiors, because they are blessed with a higher knowledge to which the poor quality of my mind does not arrive, I consider this writing I send to you, which is about the mobility of the Earth, that is, one of the physical

[23] *EN*, XII:389–392.

arguments produced to confirm that mobility. I consider it to be a piece of poetry or a dream, and as such Your Highness receives it.

However, as the poets sometimes succumb to some of their whims, in the same way I hold this whim of mine a little in esteem. Because I had written it and showed it to the above-mentioned Lord Cardinal and to someone else, I have then placed some copies into the hands of other great Lords. This is because, if it happens that someone else, distant from our Church, would like to attribute my whim to himself, as has happened to me with a lot of other inventions, the statement of great persons can show that I am the first to dream this fantasy. This fantasy that I am sending you is really only a sketch, because it was written by me in haste and when I hoped that Copernicus would not have been judged to be mistaken, eighty years after the publication of his work. My intention was to greatly extend this beyond this topic with more ease and time, by furnishing it with other proofs and by giving order to it, and by giving it another shape and disposition. But only one celestial voice woke me up and turned all my indistinct and concealed fantasies into fog. Your Highness accepts this with pleasure as it is. If the divine piety allows me to recover so that I am able to work a little, you can expect from me some other and more relevant things. In the meantime be sure that I recognize myself so much highly obligated for your infinite courtesy that I think it is impossible to free myself from such a great obligation. In addition I am always ready to strive for your every request in order to show myself to you as a grateful servant.

And here, very modestly bowing in front of you, with every reverence I kiss your robes and beg you to recommend me to your Very Serene sister and my Lady, when you have the occasion to do so, and with this devotion I revere both Your Highnesses. And may the Lord God grant you the best of happiness.

From Florence, May 23, 1618

Of Your Serene Highness

Very Modest and Obligated Servant

Galileo Galilei

G. F. Sagredo to Galileo in Florence. Morocco, August 4, 1618[24]

Most Illustrious and Excellent Lord,

You surely received my letters of the past week and have learned from them of my good health and of the payment of fifty scudi made by Lord Cremonino. I also wrote something about Germini and something else about that friend of yours. It would be a long task to write all the details about the deficiencies of Germini, as well as of the other friend of yours. So great is the variety of the discourses made and so important the topic, that it is better to keep silent than to say a little about it, without reliable grounds. Each of the persons involved shapes his own particular opinion and the universal one is a mixture of all these. I do not deny having shaped mine, which is very firm and resolute on some points. Speculations are considered by everyone in his own personal way, but I do not consider those. Used to the mathematical discursive form, when somebody tells me that some of his propositions are true, it does not matter whether they are really true or false. I shape the necessary consequences from those, and sure of not being mistaken in the syllogistic form, I do not allow the opponent to deny the conclusions. And if he does, I do not speak to him anymore.

Due to the arrival of Lord Zaccaria, I have been almost completely released from all business. And to restore myself from past efforts, these last days I have sunk into idleness which make me seem negligent in reference to you. The pleasure I receive from your letters and from writing to you is the same as what you recieve from mine, and this makes me sure that we are not unimportant to each other. Your remoteness makes me really unhappy and we could remedy it if you came to these regions to take care of yourself. Do you not remember what Ruzzante told about Padova and its citizens? That the dead persons come to Padova with coffins up their arses and after a few days they are resurrected and become as healthy as a fish. Please make this experience and do not offend the authority of such a great author, who spoke solidly about it on the basis of experience. I promise you that your life will return to you and also to your friends. Take care of your health; do not drink much in order to drink for longer. Remember that you are a gentleman and that gentlemen need to live for at least a hundred years to do long penance and so, enter paradise. Then, since I also declare myself, although unworthy, a gentleman, I beg you not to send me sheep cheese and salty biscuits because they truly are the *petra scandali* and, endangering the health, make me eat additional food after the meal. I am satisfied with abstaining voluntarily from some pleasures in order to enjoy others for longer.

Concerning the telescopes, I get the usual pleasure from them even though my tasks have prevented me from touching them for twenty months. However, I noticed what I wrote in other letters to Your Most Excellent Lordship, and this is that by extending the telescope a little after the last glass with a bit of pipe, in order to cover it from the light, one sees much clearer and more distinctly. Observing the paintings

---

[24] *EN*, XII:403–406.

with the short ones, I discovered an admirable effect in finding out that those imitating the natural environment mislead the eye by representing life marvelously. And if there is some light and shadow made on the painting, and if the painting is nevertheless good, those lights and shadows appear as a beauty spot or something similar accidentally put over it. In conclusion, it seems to me that with this telescope one increases at the same time the imperfections and the best qualities of the painting.

I also observed that the reflections of the concave glass impede the sight sometimes and especially inside the house while looking at some paintings, when the mentioned glass is close to some window or another light and, covered either by hand or by hat or by something else, the sight becomes double. Moreover, since the paintings increase their qualities, if seen through these short telescopes, the same happens to the true bodies: women observed with it from a good site and not too far away appear to be much nicer and more beautiful. I will be glad if you write to me about the experiments you are able to perform from these details. In order to decrease the light reflected inside the telescopes and which cause dimmed sight I found a good remedy with the last telescope. I placed a protection, like a small, perforated circular surface, at a convenient distance and with a convenient size.

On the matter of the glasses, what Your Most Excellent Lordship writes is true because the masters of this city, after having repaired a lot of their forms and having prepared the method of polishing them very well, have only difficulty in finding good glasses. *Experience has shown that the more or less white color does not matter very much. The [glass] vesicles, called «puleghe» by these [workers] of Murano are not very disturbing, only the [glass] spirals, which are spiral [glass] rods that one can often see in the lenses and which originate from the mixture of different glasses. One has therefore to find a way to make homogeneous glass, very similar in all its parts, because it is conceivable that in [reference to] the variety of glasses there is a diversity of hardness that consequently causes the rays, which should run straight through the lens, to refract. Once refracted it follows paths different from the ones intended and [also] different from each other. This is the reason why one sees double and blurry images.* For this reason, about one month ago, I made the proof and, at a furnace in Murano, had a small pan of glass melted, prepared in my presence and in my room. I then had a lot of the best, well-crushed ash brought and, using a very thin leaky sieve, I extracted only 100 pounds from 200 pounds of it, and from this the half again through another sieving. I let this pass for a third time and so reduced it to sixteen pounds of the thin one. I did the same with triturated gravel from Tesino, reducing it to fifteen pounds. I stirred these two very thin and almost impalpable materials and let them pass through the sieve four times so that the mixture corresponded to a good selection. I then sent it to Murano to make the mixture and, once made, it was ground by the millstone for the colors and then sieved twice and finally put into the small pan. But since these actions resemble alchemistic ones, the devil made the small pan overturn and we were not able to see the hoped-for experiment. The worst is that the furnaces will go out today and will stay inactive until October. The manganese goes into the glass and therefore I went again to the glassmaker, who took care of it, instructed by myself in order to obtain the necessary homogeneity. During the next week, I will meet these main persons of

Murano and will write to you about some other details of this topic so that you can do the experiment there. I promise you that if the matter turns out well, I will have two made here in a carefully way by Master Antonio and also by others, and every sort of shape so that very selected ones will not be lacking.

I am in the villa and will be in Venice this evening. I will give all regards on your behalf. Lord Master is doing very well; the same goes for Lord Veniero and Mula and Lord Cavalli. Veniero is always in the Collegio, both as sage of the Council and as sage of the Terraferma, with infinite activity and torment. Lord Mulla, after having been sage of the Terraferma without interruption, was greatly discouraged having now been made sage of the Council. Lord Cavalli married and had a child one month ago, who was baptized at St. Giminiano with many godparents, and in particular four from the Collegio, Lord Count of Levestein with four colonels, his clients, and me, I believe, as the last one. Moreover, as one of the fathers of the ring, I had double the salary in form of sweet things.

I forgot to tell you that the glass, in order to be purified, should stay for at least one month in the furnace with powerful fire, according to the use of the furnaces for glass. But I do not think it is difficult, the most important thing being the homogeneity, not yet achieved until now. But because I see the second folio is full, I will stop and I kiss the hand of Your Most Excellent Lordship.

In the villa of Morocco, August 4, 1618

Of Your Most Excellent Lordship

All Yours

G. F. Sagredo

G. F. Sagredo to Galileo in Florence. Venice, August 18, 1618[25]

Most Illustrious and Excellent Lord,

When I was at the villa during the past week, the small barrel with the sheep cheese and the small sausages reached me in good condition. I thank Your Most Excellent Lordship truly wholeheartedly for this demonstration of love. But because it is superfluous, I beg you it is the last favor of this kind, promising you that if I should need these or other similar favors from those regions, I will ask you for them without any respect.

I spoke with Master Antonio about coming there to serve you for two or three months, but no reason has been able to persuade him. On the other hand, Baci, although he is overloaded with duties because of his considerable family, was not completely contrary to this and asked for the price and he told me that he would be satisfied to serve Your Most Excellent Lordship, with whom he did not want to deal like a merchant.

I am waiting with the greatest desire for Your Most Excellent Lordship's answer to what I wrote to you about the method of purifying glass, to whose compound manganese is added too. One adds it after having made the mixture. And because it is not so simple to stir, it can disturb our plans greatly. However, I hope to be able to find the correct remedy for this too; but because they will not light the fire here until next October, I think I must wait for that time in order to do the experiment.

I send to Your Most Excellent Lordship four lenses for short telescopes; one is less than one fourth and the others are their double. The junctions [focal points?] cannot be more acute than those I sent to you.[26] It is late; I stop and I pray to the Lord God for happiness and gaiety for Your Most Excellent Lordship.

In the villa, August 18, 1618

Of Your Most Excellent Lordship

All yours

Giovan Francesco Sagredo

Addressed to: Most Illustrious and Observed Lord

Most Excellent Lord Galileo Galilei

With a small package

Florence

---

[25] *EN*, XII:407–408.

[26] Probably there was an attached illustration, now lost.

G. F. Sagredo to Galileo in Bellosguardo. Venice, October 27, 1618[27]

Most Illustrious and Excellent Lord,

Today I received the letters of Your Most Excellent Lordship of the 20th of the current month and I immediately summoned Master Alvise dalla Luna, who promised to visit me tomorrow morning. *I will speak to him about what you have written to me and whether he will see fit to do some experiments in order to make clear whether, taking great care, it is possible to perfect the matter of making the desired glasses [lenses]. I placed on my to-do list an item [to make] a pan of crystal cuttings, an experience I did not repeat and which could perhaps turn out well. Concerning the crushed rock crystal instead of the quartz pebbles, this is a thought I had myself. I even have most of the material prepared already, and I wanted to mix it with salt of tartar;* but since I find something to do every day, the time passes; but it too will be done soon. I did the proof of working the glasses with the lathe and of cleaning it during the last week in vain, and I lost the small hope I had about this detail, even though I wish to make another attempt.

Master Ludovico dalla Luna, uncle of Master Alvise, has a great desire to move there next July to do some work. He is an independent person with good fame in Murano, experienced in the arts and very approachable. I keep him in this good disposition because I believe he is a man who will give much pleasure to Your Most Excellent Lordship and to His Highness.

I will send Your Most Excellent Lordship my portrait as soon as Bassano finishes it, together with a copy for you that will be made by the hand of the brother of the Knight and then retouched by himself. And you will do me the pleasure (pardon me if this proposal is as one of an usurer) of sending your portrait to me, done by the hand of one of your most famous painters, so that the pleasure from the beauty of the painting will be added to the one I have seeing your image. And in closing, I kiss the hands of Your Most Excellent Lordship.

In Venice, October 27, 1618

Of Your Most Excellent Lordship

All Yours

Giovan Francesco Sagredo

Addressed to, by other hand: Most Illustrious and Observed Lord

Most Excellent Lord Galileo Galilei

At Bellosguardo. Florence

[27] *EN*, XII:417–418.

G. F. Sagredo to Galileo in Florence. Venice, November 3, 1618[28]

Most Illustrious and Excellent Lord,

I met Master Alvise dalla Luna, from whom I received little knowledge in reference to the very pure glass that you desire. Hence I really do not rely very much on his method.

I have a certain mixture of rock crystal here in the house and I am thinking of collecting a good quantity of it. This will be useful at least for making mirrors, especially circular ones, worked from the die of the lenses of fourteen fourths, with which I derive much pleasure from observing wax figurines inside, which, increased by the mirror, show their naturalness. And to get such an effect it is convenient that the crystal does not have any glass vesicles [*pulega*], otherwise the images are concealed.

Master Alvise greatly praised a workman there, who makes glasses using a flame from a small pipe. It would be very pleasant for me to have some of his smaller works, which Your Most Excellent Lordship considers to be some of the best and most complicated works he can do, because we do not have good men here of this truly beautiful art, and when I need to have something done, it is better for me to supervise them and teach them many things.

Bartolucci did not show up. When you request it, the money from Lord Cremonino is ready, to whom I wrote in order to provide the rest. I am waiting for the answer and will report it to Your Most Excellent Lordship.

I am insisting that Bassano finishes my portrait, but he works so slowly and is so disturbed by everybody that I must have the patience of Job. I do not know whether it will be so difficult for you to get yours from the painters there, among whom, I know there is one, called Bronzino, who is very famous, of whom I have not seen any work. If his quality is diligence, I am not so curious, but if it is naturalness and resemblance, I would like to see some of his works very gladly in order to find out whether they are as good as the other Bassani ones and those of the Knight.

We are experiencing a bad influenza here and therefore there are a lot of persons ill and a number of persons have died which is double that of the last years, which gives me reason to think. But, God be praised, I am doing very well.

To my content, I heard from Master Alvise dalla Luna of the good health of Your Most Excellent Lordship, about whom I was greatly troubled. It seemed to me that you always indicate some health problems in all of your letters. I pray to the Lord God that you thrive for a long time and that He inspires you to come here and stay for a couple of months at Murano, where we could do wonderful experiments. And this would truly be the right way of finding the perfect glass you desire, otherwise I am afraid I am working in vain.

A friend of mine had two hundred pounds of rock crystal fragments brought here to me in my house now and he is giving me hope of letting me have another two

---

[28] *EN*, XII:418–420.

hundred at a good price. And in closure I heartily recommend myself to Your Most Excellent Lordship.

In Venice, November 3, 1618

Of Your Most Excellent Lordship

All Yours

Giovan Francesco Sagredo

Four persons from the house of D. Giovanni were hung yesterday because they attacked some Venetian bullies with their small barrels. Two of them died doing the deed and the other two were taken alive, and all four were hung as an example to the others. D. Giovanni is in Padova; he was not aware of the incident until this morning. The people have a bad opinion of this family because it is said that they always have an harquebus.

Addressed to, by another hand: Most Illustrious and Very Honored Lord

Most Excellent Lord Galileo Galilei

Florence

G. F. Sagredo to Galileo in Florence. Venice, December 22, 1618[29]

Most Illustrious and Excellent Lord,

To what Lord Zaccaria mentions to Your Most Excellent Lordship in the enclosed letters I could add many things because, since Germini's pretence, mixed with a thousand promises and simulations, was aimed at becoming the owner of our businesses. Deceived by the false opinion we had of his qualities, we praised him several times to Your Most Excellent Lordship, without understanding that all the love and humility he showed to us was only the result of his greed. He leaves as a debtor because of his fraudulent actions, although he had decided to leave under the pretext of being requested in his country. But the truth is that he realized he would never have obtained the supreme command of our businesses, since we understood he is unsuitable. He went away without the favor of my brother Lord Zaccaria and went to your city without a single line of my brother. He visited me, pretending to be willing to leave immediately for your city, but then he first tried to find a master. It could be that he leaves today and therefore, for prudence's sake, I wanted to advise Your Most Excellent Lordship, so that you know what the situation is.

In the furnace at Murano there is a new pan; I believe it will be ready in one month. I send to Your Most Excellent Lordship one of my lenses, which did not turn out well. The color is green because of the lack of cobalt oxide [*zafaro*]. It has glass spirals because it has been made with a form similar to the bottom of an artichoke. However it appears to be made of rock crystal, clear and without glass vesicles. And in closing, I kindly kiss the hand of Your Most Excellent Lordship.

Venice, December 22, 1618

Of Your Most Excellent Lordship

Yours

G. F. Sagredo

---

[29] *EN*, XII:429.

G. F. Sagredo to Galileo in Bellosguardo. Venice, March 30, 1619[30]

Most Illustrious and Excellent Lord,

I preferred to wait to write to Your Most Excellent Lordship until this holy week because I could not take the quill in my hand during the last two unhappy and damned weeks, since all that has happened has been bad and unpleasant. First I have been unexpectedly attacked and oppressed by a cold so annoying and disturbing that I could not take pleasure in anything and everything annoyed me, except for the good wine with which I recovered the fever. Then Arno, my very beautiful hound that Your Most Excellent Lordship sent me fifteen months ago, which was my perpetual droll, the most joyful and pleasant among all the dogs I ever saw, was suddenly rabid. First it bit a foreign gentleman and two servants, then the house dog, a marmot, a beech-marten and, finally, a lynx of mine, with which it normally played a perpetual hunt in my rooms, where he could freely move and where he usually kept me company in front of the fire, more domestically than any cat I ever saw. The bitten men are in part recovered and in part still carry the signs. The beech-marten and the house dog are doing well. The marmot died, also because of another ridiculous accident and, for safety, we chained the lynx. Although the dog was only small, once it was chained up, it broke the chains. Finally, we let it run free in a closed room and after ruining the door it died. On top of all my troubles, since I hoped to console myself with the output of the work on crystal which was being prepared at Murano, those wretched Muranesi ruined everything. They broke the pan with the pieces of crystal and worked only twelve sheets so twisted and thin that one cannot polish them. But the worst is that they wasted a great vase of rock crystal before it was fired and, without my permission or awareness, they made only four mirrors out of it and wasted all the rest. It cost me fifty scudi and was enough to make ten mirrors of one braccio each. If they had shown the same characteristics that one could infer from the beginning, I thought I could earn a hundred sequins for each of them, since their color and transparency was comparable to true rock crystal. A small square piece was saved before it was purified and placed in the color mill. I had seven lenses made from it by Bacci, which are all of bad quality. The other seven were made by the spectacle-maker Armano, those of six fourths are mediocre but those of six are better, and the other seven by Master Antonio, among which two of six fourths are good and the rest of three fourths as well. Therefore, one can see that although this matter has not that perfect and extraordinary quality we are searching for, it is at least much better than the ordinary one, of which one wastes two thirds since one third of the lenses is ordinarily less than mediocre. I believe that the reason those of Bacci are bad lies in his working method. The fact that those of six fourths are not always good I am sure is a consequence of the way in which a mirror-maker incorrectly polishes the square piece on one side. This case convinces me that it was

[30] *EN*, XII:446–449.

done on purpose by those rubbers of Murano, perhaps because they are afraid that, had I been able, as one believed, to make mirrors of marvelous beauty, this would compete with their business. This unexpected disgrace has daunted me in such a way that I had definitely decided to stop this activity. But still I want to search for rock crystal and make another great and diligent test, taking on the responsibility for the life of those rascals, who, once they see me very altered because of this first prank, would perhaps not dare to do it a second time.

I received twenty-five silver scudi from Cremonino the day before yesterday and he wrote believing the account to be settled. I will write to him what is needed to get the rest. One cannot find Bortolucci; if I knew where he was, I would send him this money.

Concerning Germini, although I have no doubts about his incompetence and ineptitude, I was not surprised however to read what you wrote to me because in the beginning I was also taken in by his speech and by the extraordinary exhibition of his qualities. But in the end, the experience made it clear that everything was an artifact procured in favor of his particular interest. I am sure he deceived his master telling him a variety of names and many things he saw in our manufacture and his master eventually considered him a volcano with a doctor title. But when he notices he is lost because of just a few duties, unable to distinguish between the good or bad work of the smiths, unaware of what casting is, [*cotizzare*], and how to make a fascine of iron, which represents the advantage of the masters, how the mouth of the bellow should be fixed in order to do a good job, how one masters the fire, how one distinguishes the sorts of coals, which of them are good to cast, which to boil, which ones for weak iron and which for the hard ones and for steel, the qualities of the woods to make coals, which wood is well seasoned and with which moon they must be cut, how one makes a heap of coal, how one sets fire to it, the smog, the cover, the wind, and all the consequences of these particulars, I do not know whether they will continue to praise him. I do not count his intolerable disobedience, his negligence in providing for and foreseeing what is required, and his lack of prudence in handling the workers. I do not change my opinion according to what Your Most Excellent Lordship writes to me, that is, that the craftsmen there are not at the same high level as those here. I only say this to you, to tell the truth, that Germini is as much an expert on the handling of a forge and on a shop of ironware as the German pupils of Pignano are in mathematics. And yesterday one of his pupils from Verona, who visited me, praised him as a great mathematician and as a great mind who knows everything, only because he heard him speaking dangerously about spheres, cylinders, cones, parabolas, parallelepipeds, eccentrics, epicycles, ecliptics and a thousand other very new words, not understood by the audience. They thought he is the best mind in the world, since he could speak so easily about these things. I also had another steward from Pistoia who was more intelligent and acute than our workers, in the same way that gold is better than copper. But this is not relevant. It is enough if that gentleman pays for the fraudulent actions against us and at the right time. I write all these things manifestly, since they are true. I would also like you to relate this to Germini on my behalf adding that, after he left, the new steward had the iron fished out [of the water], scattered everywhere since one year because of Germini's disobedience. He has already recovered eleven jaws and a box of steel and this gives us hope of recovering the rest too, if the great amount of mud that has settled on it over time will allow this.

With many difficulties, the knight Bassano finally finished the head in my portrait. I am very afraid that he will not paint the robes because I am not favored by his lady, who knows that I went behind her back. In order to have it soon, I am thinking of letting his brother Girolamo make a copy of it with the consular robe, similar to one he already made seven years ago and which I appreciate. And then I send it to you.

I pray God that, with the new season, Your Most Excellent Lordship will recover his former health so that I can take pleasure from that healthy status you desire and you can enjoy the portrait in a better mood.

Before the painter leaves, about whom you wrote me, I will try to meet him and to speak with him so that he can communicate to you my wish concerning the paintings, from which since one year I receive great satisfaction. I stop here, wishing you a good Easter, a better Ascension, and the best Christmas and New Year's Day for many years. And I kiss your hand.

Venice, March 30, 1619

Of Your Most Excellent Lordship

Yours

G. F. Sagredo

Addressed to (by another hand): The Most Illustrious and Cultivated Lord

The Most Excellent Lord Galileo Galilei

To Bellosguardo

Florence

G. C. Lagalla to Galileo in Florence. Rome, July 30, 1621[31]

My Most Illustrious and Excellent Lord and Very Respected Master,

There was a long silence between Your Most Excellent Lordship and myself because of the respect that I have, since I do not want to disturb you and to oblige you to answer. Now I am forced to break it and to recur to the favor of Your Lordship because of these works of mine that I am writing. One of them—*De immortalitate animorum ex Aristotelis sententia*—is already finished. In this work I not only keep the soul immortal like Aristotle, but also as an informing and multiplied form so that the eternity of the world and the eternity of the species are saved, without leaving the Aristotelian frame and the natural light. I hope it will be ready at the end of September and I will send one copy to Your Lordship through the Lord Ambassador so that I am favored with your judgment and in this way you do me the favor of presenting it to His Highness the Lord Cardinal de Medici as a sign of my devotion to His Highness and the Serene House.

I am writing some pamphlets on philosophy and among them "De simpatia et antipathia." And I need to think about the remora that hinders the ship in its movements. I try to reduce the cause of this effect, not to an occult cause, but to the obstacle that it could present to the ship, since the ship is in equilibrium within a liquid element where the smallest hindrance can cause the greatest effect, as we can see in the steelyard how each small difference of weight along the line lifts up a great quantity and greatly varies the motion in the center. And this can easily happen with the remora, in part because of the slowness of its fluid, by means of which it adheres so strongly to the keel or to the rudder of the ships, since it is a kind of conch or sea-snail, as Plinius says, half a foot in size, and because it has the fins of the conch projected outside and scattered in such a way that it seems to have feet, as Aristotle says, one can assume that it can cause hindrance to the motion of the ships in the water, the more because Plinius accredits the same effect to every kind of conch. Therefore, before I write this thought of mine, I wanted to kindly ask you for your opinion. So please do me the favor of considering it and see whether one can determine it with mathematical reasons. If Your Lordship approves it, I will write it based on your authority. And pardon me for the annoyance that I am giving you and I beg you not to tire yourself, but to favor me with your ease since your good health is for me more important than any other thing. Kissing your hands, I remain your servant.

From Rome, July 30, 1621

Very Devoted Servant of Your Most Illustrious and Excellent Lordship

Giulio Cesare La Galla.

[31] *EN*, XIII:72–73.

G. B. Guazzaroni to Galileo in Aquasparta. Todi, April 20, 1624[32]

Most Illustrious, Excellent and Very Cultivated Master,

I became aware now thanks to Lord Captain Matalucci that Your Lordship will stay here in Acquasparta and that you will leave for Rome next Monday. Thus, it seemed convenient to me to revere you with these four lines and to remind you of me as your hearty servant now and during the time you spend in Rome and also to learn something, if you are pleased with this.

I remember, when I enjoyed your presence during that short time you had in Todi, that you resolved some doubts of mine concerning your famous telescope, useful and tasteful to your eyes and to the world. Once we also spoke about mirrors and I asked you for the point at which concave mirrors kindle and I told you della Porta's opinion, according to which the point is along the semi-diameter or center of the sphere of which the mirror is a section, and Magini's opinion, according to which the point lies in the fourth part of the diameter. You answered me that the point at which the fire is lit varies according to the variety of the spheres, of which the mirrors are sections. Consequently, the point at which the fire is lit is not at the centre or at the fourth part of the diameter but it changes, and you also told me that it is not a point but a region. It seems to me that Your Lordship told me this, if I am not wrong. And now I have a new doubt after I examined what you said using a concave mirror of steel that I have. And I find that the point at which the fire is lit is most powerful along a segment of the diameter different from those stated in the mentioned opinions; the fire is also set a little bit above and under the mentioned segment, though less powerful but in a greater space and therefore in conformity with Your Lordship's opinion. And the fire is set also at other points outside the diameter, which is the perpendicular line from the mirror, though it is not powerful. But actually, the mentioned point along the perpendicular to the surface, or center of the mirror, is really one single point so that from the mirror a cone is constituted, whose base is the same mirror and the point at which the fire is lit is the top of the cone. And the point where the fire is most powerful is not a region but almost a point. The cone is also around this point and also around the diameter, but the fire is weaker and it never becomes more powerful between that point and the mirror along the perpendicular, but only in the opposite direction farther away from the mirror.

Please tolerate my ignorance but I would like to explain it to you in more detail. In order to decrease the confusion I attach here (Fig. 9) the perfect spherical concavity of my mirror. If we start from the three points A, B, C, on the basis of the ordinary rule I find the two lines DE and FG and, thanks to them, the diameter BH (I do not draw these lines completely because I do not want to use more than half a sheet). Now, to be short I leave aside the numbers and only say that the fourth part of the diameter is the point I, which, according to Magino, is the point of the fire. According to della Porta, the point is much farther away outside of the sheet, since

[32] *EN*, XIII:172–174.

Fig. 9

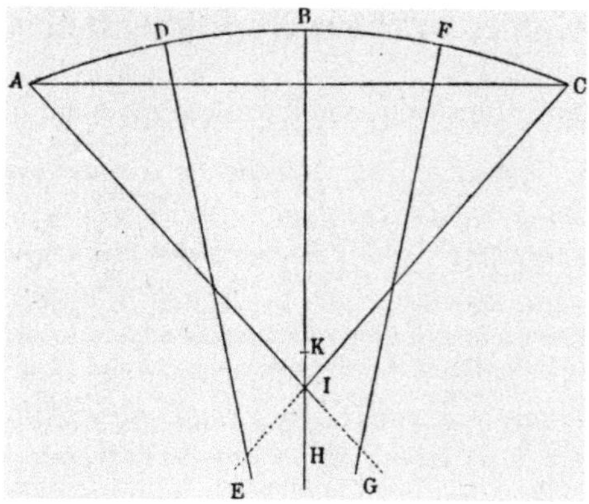

it is on the same semi-diameter behind the point H. But I observe that the real point of the most powerful fire is at K along the diameter, and it is shaped from the base AC and the cone ACK. Around the mentioned diameter, the fire is set also on the right and on the left side and at H, over the mentioned point K and somewhere else but less powerful; and the cone, or conoid, whose base is AC is also visible. And the mentioned mirror of mine is not parabolic or something like it.

I will wait for Your Lordship to tell me something about this issue, and if it is not from Acquasparta, it can be from Rome and at your ease. I am sorry about the Most Illustrious Lord Cesarini's death and I wish Your Lordship a good trip and perfect health and I very affectedly revere Your Lordship and our Most Illustrious and Excellent Lord Prince Cesi, a wonder of our age.

Todi, April 20, 1624

Of Your Most Illustrious and Excellent Lordship

Very Devoted Servant

Gio. Batta Guazzaronio

Galileo to F. Cesi in Rome. Bellosguardo, September 23, 1624[33]

Most Illustrious and Excellent Lord and Very Cultivated Master,

I send to Your Excellence a small «occhialino» [microscope] to observe the smallest things as if they were very close, and I hope you will like it and use it for no short time, for this is what happened to me. I am sending it to you quite late because I was not able to perfect it earlier, having found it difficult to produce perfect crystals [lenses]. One adheres the object to the movable circle which is at the base, and one moves it [the base] in order to watch the entire object, since what you can see at any one time is only a small part [of the object]. And since the distance between the lens and the object must be very precise, if you observe an object with a texture [which has a height, that is, a non-negligible third dimension] one has to be able to bring the lens closer or further away, depending on the part that one is observing. Therefore the foot of the small cannon [the tube], or what we prefer to call its slide, is made movable. Moreover, one has to use it when the air is clear, and in sunlight it is even better because the object should be very much illuminated. I contemplated an infinite number of small animals with enormous admiration, among them the flea is very horrible, the mosquito and the moth are very beautiful. And with great satisfaction I have seen how flies and other small animals adhere as they walk over mirrors and also upwards. But Your Excellence will have the chance to observe thousands and thousands of details, of which I beg you to keep me informed about the most curious ones. Finally one can contemplate the greatness of nature, and how keenly nature works, and with such inexpressible diligence.

I answered Ingoli's writing and will send it to Rome in eight days. Now I resumed the work on the flow and ebb and achieved the following proposition: If the Earth is at rest, it is impossible that flow and ebb take place; and if it moves following the movements assigned to it, it is necessary that flow and ebb take place with all their consequences.

Father Grassi is now a very good friend of Lord Mario Guiducci, who wrote to me that the mentioned Father *non aborret a motu terrae*, since the mentioned Lord Mario resolved his main doubts and now he seems to tend toward my opinions to such an extent that it would not be surprising if he became devoted to me. This is what the same Guiducci wrote to me. I am indebted to the Most Illustrious and Excellent Madame Princess because of the telescope I did not yet send. Your Excellence, please help me in assuring her that I will pay my debts and the delay with great interest. And the reason for this delay is that I did not yet find an object which seems to me worthy of Your Excellence, as I wish, hope and, certainly, as it will happen. I would like to speak with you about many particulars, but the variety of tasks hinders me. For this it would be necessary that I come to stay with you for an entire month with a calm mind and without any other external annoyance.

---

[33] *EN*, XIII:208–209.

Meanwhile favor me by keeping me in your favor and reverently kiss the robe of Your Madame Princess on my behalf, as I do to yourself heartily and devotedly.

From Bellosguardo, September 7, 1624

As usual I am the servant of Lord Stelluti

Two parts constitute the microscope and you can lengthen and shorten it as you prefer

Of Your Most Illustrious and Excellent Lordship

Very Devoted and Obligated Servant

Galileo Galilei Linceo

G. di Guevara to Galileo in Florence. Teano, November 15, 1627[34]

My Most Illustrious and Cultivated Lord,

Although I wrote to you twice, I have not received any letters from Your Lordship for many months. I had asked you to send me some news of your health, which together with every other prosperity, I desire very much, because of both public benefit and for my own consolation as well, since I very much enjoy the light of your doctrine, which you kindly communicated me on more than one occasion. I believe that some of your letters and some of mine were lost because of the place where I live; the only letters to arrive are the ones sent with mail transport to Benevento and not everyone knows this. In fact it is enough to send the letters to Rome, to the address of the Pope, to be sure that I receive them. I would like to tell this to Your Lordship so that you know how to command me should you want to give me reason to serve you.

Now I really beg you to give me some notes on the subject we discussed together last year in Florence, which is the twenty-fourth mechanical problem of Aristotle. Please tell me whether there are authors who tested Aristotle's solution again, and tell me also on which fundament they base their arguments because those few authors whom I read do not criticize him. Please write me also what you told me about your idea to solve that difficulty of the mentioned problem more clearly and in a different way. At that time many things occupied my mind and I was so alienated by my studies that I did not appropriately catch what Your Lordship told me quickly and thus I cannot remember every detail now. Please do me this favor, as you already did many times, and please do it without too many words and ceremonies and therefore with the same confidence I am using with you now and I will remain very much obligated to you. I very humbly revere my Lord the Very Serene Grand Duke and I devotedly kiss the hand of Your Lordship from a place which is similar to your Bello Sguardo, where I often imagine to be and so to enjoy at least what I remember of the speeches held with Your Lordship. And I pray to Our Lord God a perfect happiness for you.

From Teano, November 15, 1627

Of Your Most Illustrious Lordship

Lord Galileo Galilei

Whose opinion in reference to the mentioned issue I will consider as much as I have to do; and please remember that you already partly allowed me to use it as my own.

---

[34]*EN*, XIII:377–378.

Very Devoted Servant

G. di Guevara, Bishop of Teano

Addressed to: My Most Illustrious and Cultivated Lord

The Lord Galileo Galilei

Florence

A. Arrighetti to Galileo in Siena. Florence, September 25, 1633[35]

My Most Respected and Illustrious Lord,

I have been unable to disobey what Lord Mario Guiducci ordered me to do on your behalf about the short study I made of your first proposition of mechanics, sent here by Your Lordship to the same Lord Mario. Together with some other demonstrations that depend on it, this will be attached. I am sure you will see the entire thing as having been made for my entertainment, pardoning any weakness and if, since I did not read them again, I have inadvertently made some mistakes while copying them and, consequently, if they are not well ordered. If I hear that you have no difficulties with them and that they are not wrong, I will work on the other demonstration lately sent, since it is possible to find the thickness of the solid in another way, which is unique among all things similar to it, also if its momentum is greater than the resistance of its base and if the opposite takes place.

I will not try to condole with you about your troubles since you know very well how much I must share them because of the infinite obligations that I profess for you. To conclude I confirm to Your Lordship my obligation, begging you to remember me as a servant with infinite obligations to the Most Illustrious Monsignor and to keep me in your good favor.

Postscript. After having closed the letter, I took the decision to send to Your Lordship the demonstration of your last proposition too, which is added here. Awaiting your opinion I revere you once again.

Given a prism or a cylinder of heavy and fragile matter, homogeneous in all of its parts, that is held either at the middle or at one or each of its ends, I say that due to its own weight lengthening the mentioned solid will cause it to collapse at the point where it is held, or at its middle if it is held at each of its ends. And if the mentioned solid is made thicker, keeping the same length, it will be able all the more to hold other weight besides its own. And among the infinite solids similar to the one given, there is only one solid that is between the fragility and the solidity, so that if the smallest bit is added to it, it will break, and for any bit less it will be able to hold, besides its own, some other quantity of weight.

Let AB be the given solid, held in the middle at point C. I say that by lengthening it, what was said above will happen (Fig. 10). Lengthen [the lines] as far as the points E and F, so that the point C is always in the middle. Hence, since by lengthening the mentioned solid, the base always remains the same, the resistance at point C will remain the same as well. But one supersedes such resistance according to the increase obtained by lengthening DF and DE, since the momentum that results from the gravity of the above-mentioned solids DF and DE also increases according to the increase of the above-mentioned solids. It follows from this that the mentioned solid EF will break due to its own weight.

[35] *EN*, XV:279–281.

Fig. 10

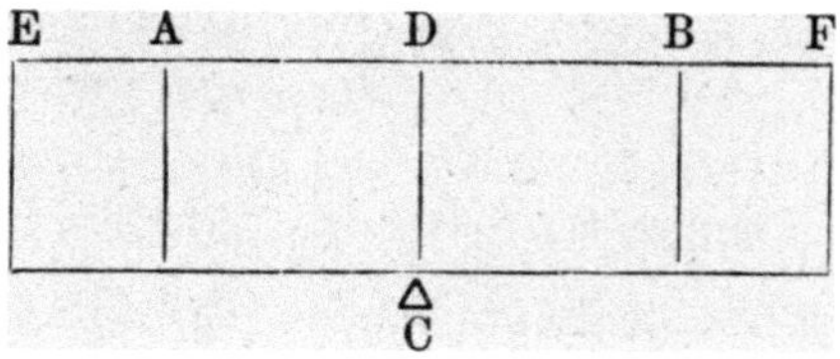

Fig. 11

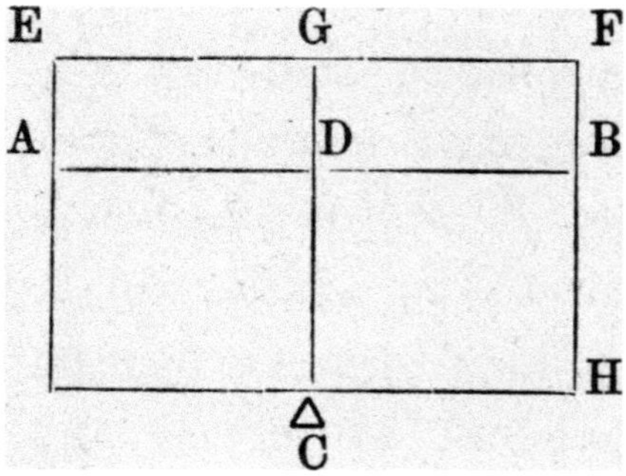

Let the solid AB be increased in reference to its thickness as far as EF, keeping the same length (Fig. 11). I say that the opposite happens, that is, that besides its own weight, the solid will hold some other weight. Because by increasing the mentioned solid, the resistance [of the first solid] to the resistance [of the increased solid] is like the base DC to the base GC, that is like the solid AH to the solid EH, that is, like the momentum of the solid AH to the momentum of the solid EH. But the difference in superseding such resistances increases as much as CG increases, while the length AB remains the same. Hence what one has proposed will happen.

Moreover, I say that using other solids similar to AB, among the infinite ones that one can use there is only one [solid body] which is between fragility and solidity, so that the larger they are, the easier they collapse under their own weight, and the smaller they are, the more they hold some other weight besides their own (Fig. 12). Let the solid AB be in the above-mentioned status and make the solids EF and GH similar to AB, EF larger and GH smaller. Hence, since the resistances in CD, KL and MI have the same proportions to each other as those of the bases CD, KL and MI to each other and, since the momenta of the solids AB, EF and GH have the same proportions to each other as those of the same solids to each other, that is, of the cubes of the same CD, KL and MI, and since the ease in overcoming the resistance remains the same in all the cases, it follows, as one has proposed, that the larger solid always breaks in KL and the smaller one is always able to support some other weight besides its own, and that AB is unique in such a status, as one has proposed. And the same will happen if the mentioned solids are held at one or at each of their ends.

Moreover, if one wants to decrease the thickness of the solid EF so that, keeping it at the same length EF, it has the same status as the solid AB and is also unique in

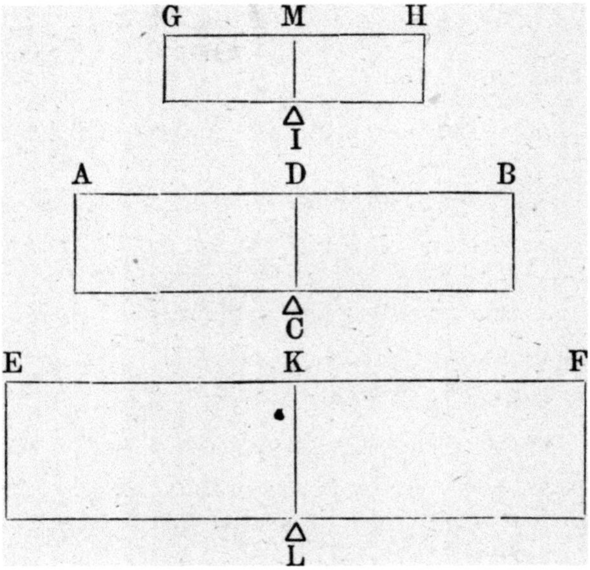

Fig. 12

having such a status among all similar solids, it is sufficient (using the last figure) to find the third proportion of DC and KL, which will be the diameter of the sought cylinder. Therefore, the momentum of the solid AB to the momentum of the solid EF is three times the proportion of CD to KL. And the resistance of CD to the resistance of KL has twice the proportion of the same CD to the same KL, for they are similar solids. And the momentum of the solid EF to the momentum of the found solid (for it has the same height) is twice the proportion of DC to KL, and the resistance of the same EF to the resistance of the found solid is three times the proportion of DC to KL. Hence, the proportion of resistance of the solid AB to the resistance of the solid EF is smaller than the proportion of the momentum of the solid AB to the momentum of the solid EF, in the same way as the proportion of the momentum of the solid EF to the momentum of the found solid is smaller than the proportion of the resistance of KL to the resistance at the base of the found solid. Hence, the resistance of the solid AB to the resistance of the found solid, that is, the resistance at their bases, have the proportion of the momentum of the solid AB to the momentum of the found solid. Hence, the found solid is of the same status as the solid AB. And the same happens if the momentum of the solid AB to the resistance of CD has whatever other given proportion greater or smaller, because the found solid will always be unique in such a status among all solids similar to it.

Galileo to A. Arrighetti in Florence. Siena, September 27, 1633[36]

Most Illustrious and Very Ready Lord Master,

The pleasure with which I read and read again the demonstrations of Your Lordship is greater than the surprise, the pleasure because of the acuteness of the invention and the surprise because it is a work of the mind of Lord Andrea Arrighetti. For a time, the last demonstration had me particularly confused, because of both the unusual text and my worn memory from which fancies escape more easily than they remain impressed. Let this suffice, by the way, to advise Your Lordship to speculate while you are young. The progress of Your Lordship is stately and is above the common geometrical mind, in the way that metaphysics are above pure physics and, keeping Your Lordship among abstract universals, it seems that you despise giving particulars and dealing with persons who are not absorbed in these studies. I repeat to Your Lordship that I had very great pleasure. If you do not disagree, I add this demonstration of yours to the one I made myself in the treatise I am working on. This would be a great pleasure for me. However, in order to make it understandable also to averagely intelligent people, I decrease its stature to my level, but really with some loss of the majesty that Your Lordship lends it, and therefore conclude it in the following way (Fig. 13):

The resistances D, K are to each other as the squares D, K, that is, as the squares K, M, that is, as the prisms E, X, that is as the momenta E, X; the resistances K, M, as the cubes K, M, that is, as the cubes D, K, that is, as the prisms A, E, that is, as the momenta A, E; hence, *ex aequali in proportione perturbata*, the resistances of the prisms D, M are, to each other, as the momenta A, X: but the same prisms are in a similar status.

In reference to myself, I can say that the very pleasant conversation of this very courteous guest of mine really lifts my spirits and the occupation that God gives me about various speculations greatly entertains my mind. And the greatest of all pleasures is to believe that Your Lordship and the other dear friends and masters keep me in their favor and this makes any troubles of mine less disturbing.

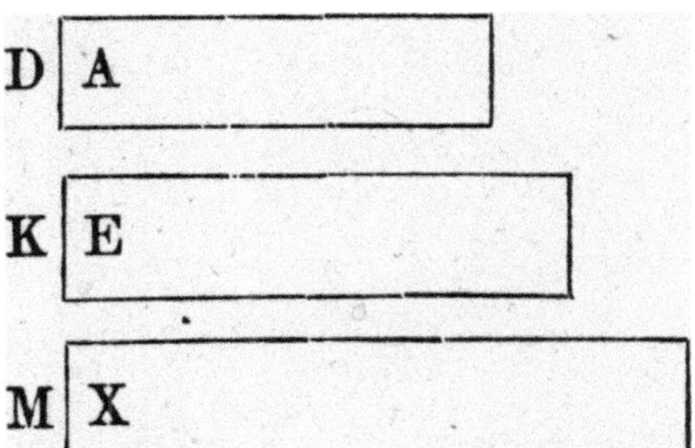

Fig. 13

[36]*EN*, XV:283–284.

N. Aggiunti to Galileo in Florence. Pisa, February 22, 1634[37]

Most Illustrious and Excellent Lord and My Very Cultivated Master,

The truffles Your Lordship gave to me as a gift are so good that I would have received them gladly in whichever place but, especially in this one, they arrived to me as a miraculous pleasure. I thank you endlessly, as Lord Apollonii also does, who glorifies himself because he has been mentioned by you and he obliges me to offer him to Your Most Excellent Lordship, as I do, as the most devoted servant because of your special merit.

We are searching for Lord Rocco's book but we did not yet find it. Once we have it, I am sure that, since the works of Your Most Excellent Lordship have shown us that no one achieved supreme knowledge in the past centuries, the reading of Lord Rocco will show us that the worst knowledge was not achieved then either. The same conviction is confirmed by the reading of Lord Scipione that we are doing now. And if Lord Rocco will turn out to be comparable to Lord Scipione, this will be already relevant.

We read and understood that immense stupid thing about the sunspots that Your Lordship indicated to us. Really, one cannot find anything more stupid and evidently wrong in reference to this issue. When the time arrives, we will use this, like any other thing, as long as you make your notes on that book.

My brother Lodovico proposed to me the following question in the form of a problem (Fig. 14). *How can one make a boat go from one side of a river with a very rapid current to the other without moving anything but the rudder of the said*

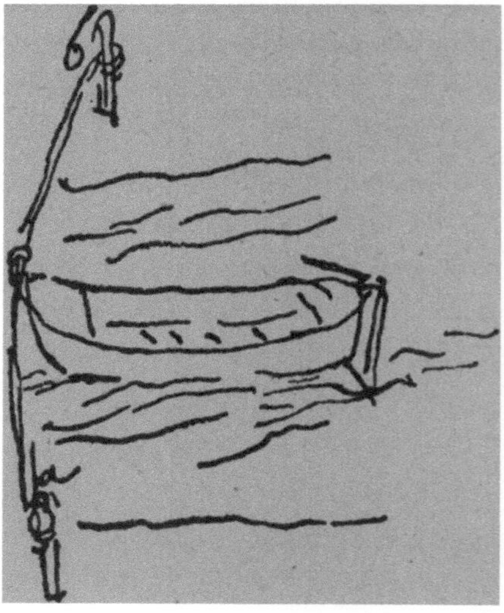

Fig. 14

[37] *EN*, XVI:49–50.

boat? I do not see how the rudder can influence the situation as long as the boat is completely left alone on the current of the river because the rudder and the water are carried with the same velocity and therefore the use of the rudder has no consequence. Therefore, I was also considering if, by somehow blocking the boat, one could solve the problem. And it seems to me that if the boat were fixed at the bow by means of the rope AB, so that it could move along that rope, then it might be that, if one moves the rudder on one side, the bow would move along the rope toward the other side and so the boat would move slowly from one embankment to the other. I do not have time to express my idea more clearly but I think you will understand me also on the basis of these few words. I would like to hear your opinion and since it is very late and G. Battista Pieratti, who will carry the present letter, is about to board the coach and leave with the mail, I close the letter here, since I cannot do otherwise and I pray for you health and prosperity, devotedly kissing your hands.

From Pisa, February 22, 1634

Of Your Most Illustrious and Excellent Lordship

I did not send you the cantucci because I hope to find a better kind. In the end, I was obliged to take what I found. With the first carriage, I will send them to Lord Dino

Very Obligated and Devoted Servant

Niccolò Aggiunti

Addressed to: The Most Illustrious and Excellent Lord and my Very Cultivated Master

The Lord Galileo Galilei, Philosopher and First Mathematician of His Very Serene Highness

Florence

F. Micanzio to Galileo in Florence. Venice, July 8, 1634[38]

Most Illustrious and Excellent Lord, Very Cultivated Lord,

I think that Your Excellent Lordship received the information about the anvil. I ordered my companion to send it to you since I am now hindered.

The Most Illustrious Baitello is here and tells me that Arrisi will send the rest in September; it is 60 scudi in conformity with the agreement and the calculation.

There is a young gentleman who would like to observe the moon but we do not have a telescope because the plague killed the masters and we only have those of bad quality. Thus, I must ask you because you are its inventor and therefore have exquisite things. There is also another person who works as a mechanician following his whims. Oh, if he could stay with Your Lordship for two months, what things he would learn! When he studied your book, he decided to build the Copernican sphere. Yesterday he told me his intention. I do not know whether he will succeed, but he will certainly do something. He also told me that he would like to try to make the dies to work lenses, if he could find out the diameter of the sphere within whose section one makes those lenses. Please give us the necessary hints so that we can make something more than mediocre and give us also those indications which can be helpful for those curious spirits who are better than ordinary.

One cannot leave Roco untouched; the notes should be at the bottom. I am aware of the difficulties you could encounter if you try to publish these and other divine speculations of yours. But if one does not do it, this is like depriving the human intellect of the greatest glory to which it ever arrived in reference to this subject. I would like to have this merit for humanity and so to be the medium of this good thing. If Your Lordship thinks in the same terms I will do it, promising you that the merit will be for that person who really deserves it. Please let me organize this. You can think about it for a little while. It is one of the greatest human pleasures to avenge oneself of the insults, without bearing the fault.

I wait to see if you have recovered the health of your body and the serenity of your mind, for I pray to God for this. And I kiss your hands.

Venice, July 8, 1634

Of Your Most Illustrious and Excellent Lordship

Most Excellent Galileo

Very Devoted Servant

Father Fulgenzio

---

[38] *EN*, XVI:108–109.

A. de Ville to Galileo in Arcetri. Venice, March 3, 1635[39]

Most Illustrious Lord and My Very Cultivated Master,

I already wrote a long time ago to Your Most Illustrious Lordship without receiving an answer. I was afraid to disturb you by writing again, but during these days Lord Father Fulgentio gave some writings of yours to Lord Argoli to read and apologized for his lack of attention, due to other important duties. He gave them to me as well and left them in my hands for only one day. I discussed with him some of my doubts and he begged me with great emphasis to write about these to Your Lordship. To give him this pleasure, I disturb you with my letters for a second time and you will see how I revere your person and admire your rare virtues, and though I am about to write something which could be unpleasant to you, it should not be surprising because your concepts are new, subtle and above the opinion of everyone and therefore doubts and oppositions originate. Moreover, sometimes one proposes and proves what one knows is wrong, as Zeno uses the argument against motion to show the excellence of his mind above those who cannot solve it.

At the beginning of your *Discorso*, it seems to me that you would like to affirm that small-sized machines which work well also work well when they are large, if, when multiplying them, one respects the correct proportion that one must of the instrument and of its parts, and that the deficiency which is in the matter is not a sufficient reason to prove the contrary, since that deficiency is eternal and always the same, of which one can give a reliable rule as in the case of abstract figures. The experience with machines that perform or support force seems to me to show the opposite. For example, one makes a self-moving bridge in order to cross a small canal one foot long with wood one hundredth of a foot thick. But to pass a canal a hundred feet long, one is not able to make a bridge, even though the wood is one foot thick, speaking of the depth of the cantilever, because in this way it is ten thousand times thicker than that which has the side of one hundredth of a foot. The person who would like to make it two hundred feet long does not find any material to support that weight. *One says that, since a timber breaks due to its own weight, hence the matter destroys itself and the machine as well due to its gravity, which does not improve the force. One answers that for this there is a rule. But which one, and with which proportion, and with which matter? Since each material is different—the iron supports very heavy hanging weights, the wood carries them placed upright upon it—then which demonstration is able to show all of the imperfections that can be found in the materials, since there is no science of the singulars? These are all different, and for these differences or affectations, if we do not want to call them imperfections, one cannot give any convenient rule. And not only is there no rule for all of them, there is no single rule for those materials of the same species.* The wood of the white poplar has a different effect than oak and this is also different from the hawthorn. And in the same hawthorn, the wood of the root is different from that of

[39] *EN*, XVI:221–228.

the trunk, and the trunk from the branches, because they are of different rareness or density. The same wood, cut along the grain, does not have the same effect that it has if cut along another one. Wood cut in autumn is different from that cut in spring. In humid weather it stops [the saw] more than in dry weather, then [one has to consider the] knots, grains, the way in which they are joined, if they are old or young, the cleaning, the filing, the right way of working them, and the thousand irregularities that one constantly finds.

Who gave a reliable rule for these? This rule cannot be given. *And while one does not want to call these things defects of the matter, in any case they make the art faulty, as these accidents cannot be recognized in small machines, but become evident in the large ones since they are increased in force and in weight.* Moreover, the forces of the matter do not increase like the quantities and many small sized machines that are very powerful, because the matter is proportioned to such a force, cannot be made larger because of the deficiencies of the matter. *You also observed that a siphon cannot attract for a height of more than 18 feet, no matter how thick and high it is.* Hence, that machine can lift until a certain height but if larger, does not have the same effect. *One must say that this is not a defect of the machine, but of the water.* It does not matter where this defect comes from. It suffices to see that, because of a certain deficiency, the large machine does not perform like the small one, although they are proportioned in reference to all their parts. One says that when one would like to give a rule for the materials, this person does not know of these imperfections or of the deficiencies of the artist. This means that perfect matter, which is not subjected to those irregularities and is an imagined or abstract matter, is then a different issue because one uses abstract figures like cubes, cylinders etc. in the demonstrations. But where is such matter? The same metals of the same kind are different from each other, though they do not have grains, knots, roots and branches, nor do they suffer any humidity or dry context. And these differences make the art imperfect. And for these reasons, one says that a copper sphere does not touch a plane at one point, because, if it is copper, it is completely porous, as you have proved elsewhere, hence, it does not touch at only one point. If the plane is at the bottom, the weight pushes on it; if it comes close, it hits that. Then where does one find those iron tools, lathes and turners, which lathe the sphere and level the plane with that exactness? These accidents and imperfections are inseparable from the art and from the artist and from the matter to such an extent that, when one speaks about a material and artificial thing, this means without that perfect and imagined shape. And what one says about the sphere is valid in reference to every figure or body, regular or irregular, of which one gives the determined shape that is not made randomly. And the art will never come to exactly form what the mind imagines. I do not know whether the reason why nature never makes two similar things is the love of the marvel or the difficulty. It does not make sense to say that it is not a sphere because when one mentions an iron or material sphere, one thinks of a sphere of that material similar, as far as possible, to the ideal or perfect sphere; either one needs to reform the way of speaking and to avoid saying sphere of glass, cylinder of wood, neither any figure, whereas one adds [...] and art. In this discussion, we agree about the fact but we speak in different ways about it, because you do not want one to say what has been commonly said by everyone up to now, for one knows well that a circle of ink is not a perfect circle, nevertheless no one has stopped calling it a circle. The same [is true of] the others things, which, being material, are never perfect.

In your experiment on the fracture of wood or on a line of brass, you see that the same length of the same line, passed through the same draw-plate, does not always break, as one sees in the harpsichords that the same rope at a certain extension does not break while at another one it breaks. I say that at a certain extension it supports a certain weight in a certain moment, but not in another moment. How can one give a rule for such differences?

The trick of descending a rope without injuring the hands I already saw about fifteen years ago; I considered it trivial.

In the instrument to measure the force of the vacuum one has to deduct the force needed because of the friction of the male over the empty cylinder. Then, about this instrument I say that one will never be able to fix it in such a way that the air does not enter as one evidently sees. Moreover, when the air or the water rarefies and when the air cannot rarefy anymore or enter, and when the bucket is heavier so that the instrument cannot hold it, the same instrument breaks completely before the bucket falls, and this experiment will not prove or deny the existence of the vacuum. About this topic I say that one has never proved the impossibility of the vacuum, and that it has not been shown that it exists in nature, and the demonstration that Aristotle does in the fourth [book] of *Physics* in order to prove the impossibility of the vacuum does not conclude anything, whereas one says that the duration of the motion comes not only from the resistance of the body, which does the motion, but also from the nature of the bodies, which can move only for a certain period, even though there is no resistance.

That in the infinite number (if one can say infinite number because it seems a contradiction since every number is finite) there are as many square numbers as there are numbers, this cannot be affirmed in this way, but one can say that there are as many square numbers as there are numbers. This problem is similar to the following: If there were infinite humans, there would be more strands of hair than humans. I would say that both are infinite. But, for me, if one proposes some comparison between infinities, I think one has to say that there is more hair. One says that hence humans are finite. I deny this consequence because infinity is that for which, if you deduct from it, there always remains something to deduct. Deducting humans, some of them will always remain and one never finishes, the same with reference to hair. But in selecting as many humans as one wants to, one always selects more hairs. Since our imagination and reason always lead us to believe that there is more hair than humans, although both are infinite, this happens because the nature of infinity is known by us only through the negation of being finite or of not having an end, since, as long as one keeps that attribute of not having end, it is infinite. Thus, if one says that there are infinite humans and infinite hair, then there is more hair than humans; one does not consider finite one and other and also that one is more infinite than the other. [The operation of] plus, in a finite group [of numbers] denotes the excess of the major to the minor, but in infinity, there being neither major nor minor, [the operation] plus does not denote excess, apart from our way of speaking, that, affirming, can only be within a finite [group]. One thinks that our knowledge, which is finite, imagines that for whichever assumed finite quantity, there is more hair than humans since every assertion involves finite quantities, the assertion being the contrary of the negation. But the negation is the knowledge that we have of infinity and when we say there is as much hair as humans, this affirms an equality and knowledge of the convenience of quantities, and this denotes an end. One says that by affirming that there is more hair than humans, one affirms an inequality and nothing more. It is true. But if we allow that every assertion one makes is in a finite context, it is more convenient to the mind to say that there is more hair than humans, because such is our knowledge.

That the unit is infinite does not seem true to me because the unit, as divisible, means a divisible thing, but not infinite. And this is divisible not into infinite parts equal to a third one or among each other, but into parts proportionally infinitely divisible, as three into thirds and these into others, and into fourths and these into others etc., infinitely, but one does not divide into infinite parts. And this is *continuous* but not infinite because one can be divided into two halves and if one takes a half twice, the whole is taken. [It is] the same with thirds or whatever other part and these divisions are possible. Hence it is not infinite because infinity is not divisible into any part. Then, in infinity, whatever finite quantity one takes as many times as one wants to, one never takes the whole and the infinity always remains. All this is contrary to the unit.

Even though one says that infinity is larger than the finite, one thinks of it without any proportion. And even though the finite increases, it does not get close to infinity, since getting closer or further away has a relation to ends, but infinity does not have any end. In order to say that such an infinity is infinitely larger than the finite, one does not need any proportion or similarity. We can always say one is larger than the other, while the larger contains the smaller and also something more, since from an infinite line one can cut not only one but infinite ones equal to whatever finite line given, and one is able to consider the infinite line longer than the finite one, but without any proportion. That from an infinite line one can cut whichever finite has not only never been denied by anyone, but has been assumed in many geometrical demonstrations.

It seems really incredible to me that Archimedes burnt enemy ships by means of mirrors many miles away, since the joining of rays from the parabola does not happen further away than the fourth of its right side.

It is not possible that, given two concentric circles, of which the larger moves on a plane making a line equal to its circumference, and in the same time the smaller on another plane, the motion of the smaller is made as if in concentric polygons, that is, by jumps and intervals. And it would be difficult to understand if the equinoctial of the first mobile moved on a plane, how one would find those intervals or spaces in a small concentric circle that make a line equal to the circumference of the other. For me, I think the motion of the smaller circle is so: when some parts of the larger circle move on the plane, a proportional part of the smaller one also moves because from any assignable point of the larger one, one can produce semi-diameters to the center, which will determine portions proportional to the concentric circles and the circles are according to the portions. While some part of the larger circle moves on the plane, this part of the smaller one moves. But the smaller one is constantly carried forward with inverse proportion more than the larger one so that the contact that it makes on its plane, and the carriage, balance the portion of the larger one. For example, let the larger circle decouple the smaller one: while the sixth part of the larger moves, the sixth part of the smaller also moves, which will be nothing but the sixtieth part of the larger. But while this sixth part of the smaller moves or rolls, the carriage of this part is made according to the relation of the larger circle to the smaller one, that is, it carries ten times more than the distance along which it moved or rolled, since the carriage, and the rolling, balance the rolling of the larger, which is simple without carriage.

And this does not introduce any contradiction that a body, rolling or turning, is also carried forward, as one allows in epicycles and other motions. Then, in the same demonstration with the polygons, the jumps that you use are carriages that are broken off, as the continuous is broken off along the sides, and the application or contact happens according to the circumference of the polygon. Since the rest is carried more or less, according to its being close to the centre, and what is called trailing is carriage, which, together with the contact of the circumference are equal to the larger circumference [...]. From that it follows that the *continuous* is constituted of divisible parts, infinite and proportional, in such a way that proportional parts of the smaller are smaller than proportional parts of the larger. Thus parts of the smaller circle that walk with the contact, although they are infinitely divisible, have to the parts of the larger that proportion that the smaller circle has to the larger. But parts of the carriage are larger in the smaller circle than in the larger one as the larger circle is to the smaller. In such a way the carriages complement the contacts, as one clearly sees in the example of the polygons and in the centre, which is easily carried and the larger is easily moved. In order to confirm this carriage, in that example where you let the smaller polygon move on the plane, making the line equal to its circumference, one decreases the contact of the larger polygon in such a way that the contacts are equal for both of them. The rest is carriage while they move back, and it is impossible that the contact of one is longer than the contact of the other. Hence the rest is carriage, which is interrupted by jumps in polygons because of the angles between the sides, but in the circle it continues, it being continuous. And those interruptions are impossible in the circle because it would follow that it is a circle and not a circle at the same time, since some parts would be closer to the centre, that is, those that do not touch, and the others more distant, like in the polygons. One cannot say that the sides of the smaller polygon, in the first example, stay at rest according to [...] the length of the side of the polygon [which is inscribed into the larger] circumference, because, though it does not touch the plane, it is carried nevertheless. And for this reason polygons are different from circles, since carriages happen separately from contacts, because the sides are distinct and are not equally distant from the centre in all its parts, but in the circle there is no part that is not equally distant from the centre and no part is interrupted. Hence, they constantly touch the plane and are constantly carried forward and back, according to which the larger or the smaller move with the proportion that circles have, the smaller to the larger for the contact, or the larger to the smaller for the carriage, as one has already said.

When you say that from a straight line one forms a polygon of 100 or 1,000 sides, bending it into a polygon of infinite sides, which is the circle, this cannot happen, because one can well bend a line into a polygon of as many finite sides as one wants, but never infinite. And speaking about bending a straight line into a circle is nothing but making a circumference from a circle equal to a straight line, this has not been demonstrated up to now. And even if a straight line were shaped into a circle, there is no side and one does not demonstrate any division or distinction of parts, because it is certain that no polygon, inscribed into a circle, of as many sides as one wants, is equal to the same circle. Hence, because of the bending one does not obtain a division, either clear or unclear, and going on with the multiplication of the sides one never reaches infinity, the circle being a continuously bent line, where there is no distinction of parts or of sides and the condition that one finds in polygons, of distinct sides, is absent in the circle.

When one says that a sphere touches a plane at one point, it is true. But it is impossible to assign that point, and only with the mind can one understand where the contact is. And one can understand or assign this as separable or indivisible only because it is the negation of further extension, as ends of a line are called points only because, besides those ends, there is nothing more, but, if one chooses another point, we take infinitely divisible parts. So, the contact of the globe is the end of all lines that join in it, but it is not assignable and it does not involve any division or distinction. And in saying one allows the globe to move, then some other point comes; this is not true, because infinite divisible parts come, as there are in the underlying plane. And in this is the difference that one makes the division for assigned parts one after the other, as in the example of the polygons. But in that motion one does not assign any distinct part because one always takes them as being infinitely divisible on the plane as well as on the circle. And saying that the motion takes place along points, first above one, then above another one, and so successively up to the end of the line, it is against the supposition because it is divided into finite parts, trying to divide it into infinite ones. The whole consideration consists of this: that the globe with infinitely divisible parts covers parts of the plane similarly infinitely divisible. The same is done by every plane over a plane, or a side of whichever figure and also of whichever body moving over a plane.

If you want to show that one can make an infinite circle, one well proves that one can make a larger circle and larger than any other given but never infinite; as well as making the angle larger and larger, but when it is made in a straight line, it is no longer an angle. When one cuts a line in the middle, all other lines are equal, since they have to be in the same proportion. Hence it follows that the line, passing through the junction of them, is straight because all are perpendicular and fall at the same point. Hence, they are a straight line and not a circle.

I close here in order not to bore you any longer, begging Your Lordship to pardon me for the boredom that I give you with this badly contextualized discussion. It is permitted to doubt everything in order to better inform oneself and oppositions confirm and make the truth clearer. This is my goal, not to oppose but to make me more competent of your propositions, which, being high and difficult and so sharply proved, can only be understood with difficulty, at least by my weak and inattentive mind because of the constant public businesses for which I am almost constantly traveling. I read books as dogs drink the water of the Nile, since the thirst of knowledge remains always inextinguishable to me like the desire of being able to serve Your Most Illustrious Lordship, whose hands I kiss with extraordinary affection.

From Venice, March 3, 1635

Of Your Most Illustrious and Excellent Lordship

I do not understand the method of deduction of the demonstration that a circle becomes equal to a point

Very Devoted Servant

Knight Antonio Deville

F. Micanzio to Galileo in Florence. Venice, December 1, 1635[40]

Most Illustrious and Excellent Lord, Very Cultivated Lord,

I decided not to answer the letters of Your Most Illustrious and Excellent Lordship until I could collect your pension which matured last September. I had to write, rewrite and quarrel. They finally wrote me that it is in the hands of Lord Baitello, so that Your Lordship can use it as you prefer, that is, the rest of the other installment, which is £ 52, and this one, 140.

That Master Marc'Antonio Mazzoleni died of the plague. There is no one else able to make compasses. It is a strange thing that, although they can be used in such relevant situations, one allows such an invention to perish. And it is also impossible to find the dialogue on its use, for which I am searching ardently.

I do not completely understand what Your Lordship writes in your *Dialoghi* on folio 241. It is not impossible to measure and draw a line greater than any very great circle using the circumference of a small circle and turning it a few times, etc. I asked the experts here but do not understand their answers. I would like to ask you to give me some explanation, if I am able to understand it.

Lord Sigismondo's sphere is observed every day by different people and all of them are very satisfied, especially because of its simplicity and because one sees with the eye all the effects that Your Lordship describes in reference to the sunspots. I could understand them only a little but now I see them very clearly.

I have found a piece of very old mirror, thick and of the highest possible purity, because I would really like to have a telescope and it is the fault of that sent to me by Your Lordship and not of my eyes [...]. Instead of sending it to you, a cat destroyed it. Oh, beast without human reason! I heartily pray for your happiness and I kiss your hands.

Venice, December 1, 1635

Of Your Most Illustrious and Excellent Lordship

Lord Galileo

Very Devoted Servant

Father Fulgenzio

---

[40] *EN*, XVI:355–356.

B. Cavalieri to Galileo in Arcetri. Bologna, March 11, 1636[41]

Most Illustrious and Excellent Lord and Very Cultivated Master,

My long purge, accompanied with constant fluxions, has prevented me up to now from serving Your Most Excellent Lordship in what I have already obliged myself and that I desired so much, remaining for this as mortified as sad because of the gout. I am not yet free, however, but being able to use my hands a little, I immediately wanted to use them to pay this debt. But first I have to say to you that I have retained the letters of the Serene Grand Duke and Very Eminent Lord Cardinal until the current week, having finally sent them only last Wednesday, and this because I wanted them to accompany a letter of mine to the Very Eminent Aldobrandini. I wanted to clarify a little the eventual idea that I would send him letters of recommendation without them being necessary. Since I was not able to write earlier, it was convenient for me to postpone the mail to the mentioned Very Eminent. I know that I am obliged to thank one and the other Serene Master but now I am not fit to write much and therefore, should the occasion arise, I would like to beg you to give my apologies. Hence, I postpone this until I am in a better state of health, hoping that at that time I will also have the answer of the Very Eminent Aldobrandini. The mentioned letters have been very relevant to me, but I believe that I need a very powerful service by the mentioned Very Eminent Aldobrandini and, if possible, it should be made orally (that would be very easy if it is true that the Very Eminent Cardinal of those parts will shortly go to Rome), since that Father Theatine has taken over the protection of this Brother so much, about whom I already wrote and in order to keep me worrying constantly, it is not easy to free myself of him. It is not necessary to do anything else now, but I will inform you of it at the right time.

Finally, in reference to my thoughts about the mirror, etc., I know that if you had given it particular consideration, it would have been easy for you to guess what I think of this, since it just seems to me that you were on track to find it when you indicated to me that you thought it could be a parabolic mirror, if it has a deep base. Hence my thought is this: Let ADG be the parabolic mirror in the above-placed figure (Fig. 15), whose axis XD and focus O are not very distant from the base of the mirror D; and through O one draws BF, perpendicular to XO, that ends at the surface of the mirror in B, F; then as many rays as desired come from the Sun (toward whose centre the axis XD is directed), parallel to the mentioned axis, but, for our example and understanding, we have the two rays HA, LG, which meet the surface of the mirror in its mouth, like in A, G, and the other two MB, NF, which meet the points B, F. Hence, it is evident that these four rays will join each other at the point O, the focus of the mentioned mirror. These, however, will not stop here, but, passing beyond, they will meet the surface of the same mirror again. As HA, LG, that made the first reflections in A, G, will make the second ones in E, C through ER, CS; and the two MB, NF, that made the first reflections in B, F, will

[41] *EN*, XVI:401–404.

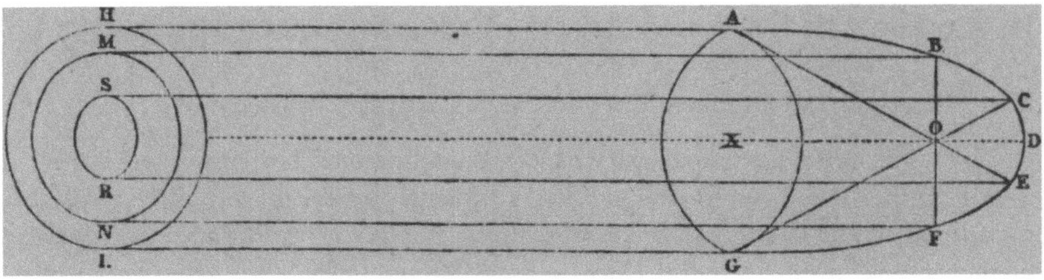

Fig. 15

make the second also in B, F reversely, that is, MB in F through FN, and NF in B through BM.

Thanks to these two reflections of rays one obtains what is our intention, that is, that when the light goes along lines of a certain thickness, parallel to the axis XD, as in the width of the armilla HMNL, the same quantity of light exits through the width of the armilla MSRN. Because the rays, for example, between the two HA, MB, because of their second reflection made after the transit through the focus O, will exit all concentrated between the two ER, FN, reflected by the part EF of the mirror; and the same will happen to the rays between LG, NF, that will exit from BC, concentrated between BM, CS: that is, in conclusion, with this device we will compress the light of the Sun that enters wide and rarefied in the mirror and in the parts AB, GF, reducing it in a smaller space by the second reflection made by the parts BC, EF of that mirror, and also keeping the rays parallel to the axis XD. For this, therefore, it is evident that the closer the focus O is to the base of the mirror (from which it follows that the mirror is more and more concave), the more compressed the light is when it exits through lines parallel to the axis XD: so that we can build such a mirror that reduces it to the narrowness or thinness that we require.

These things are really in conformance with the doctrine of my *Specchio ustorio*, as you will immediately understand, because, though in this operation I use only one mirror, it makes the service of two of them, which are separated by the circle BF; since ABFG is the large mirror and BDF the small one, placed in a way that the focus of the large one, which is O, is joined to the focus of the small one, which is the same O. I consider this junction to conform to the structure taught in my book and therefore very difficult to obtain in practice, although in this way, in my opinion, a great part of the difficulty is removed. But it is also true that in this way I cannot enjoy the comfort of the convertibility of the small mirror BDF to burn everywhere. But as a remedy for this, two things have occurred to me, of which I do not really have a demonstration but only a probable conjecture and one has to wait and see what the experiments will show. The first is that, though it is true that the above-mentioned things happen when the axis of the mirror is directed toward the centre of the Sun, nevertheless, inclining the mirror a little, the rarefaction of the small cannon of light constitutes itself so fast that it also continues to burn (about which I confess to you I have speculated much to find the effect that is made by the rays that enter the mirror obliquely and not parallel to the axis; not being able to understand up to now their effect in the second reflection, as well as in the other conical sections). The other is that, keeping the axis of the mirror toward the centre of the Sun, we could, at the mouth of that mirror, oppose to the exiting small luminous cannon [a bundle of rays] a small flat mirror, movable in every direction and that at any point could just likewise reflect it [the luminous cannon] without changing its thickness. But one has to doubt this because, using three reflections, they weaken the light so much that it is not fit to burn. For this I entrust myself to experience.

This is what I can say to my Lord Galileo so that you remain delighted and can also serve the Serene Grand Duke My Lord. I said, perhaps too daringly, that it seemed to me a beautiful thing, but now I correct myself, entrusting myself to your very fine judgment and selling it to you, or, in order to out it better, offering it to you for what it is worth and nothing more. I will not forget then to make the proof on a small scale. In the meantime, do me the favor of informing me when you receive this because I do not want it to get lost and inform me also about your very valued opinion, sharing it with the Serene Grand Duke, when he returns, and give him my apologies because of the indisposition that I have and my very modest bow. In the meantime, wishing for Your Most Excellent Lordship complete health, I very heartily kiss your hands.

From Bologna, March 11, 1636

Very Obligated Servant of Your Most Illustrious and Excellent Lordship

Francesco Bonaventura Cavalieri

Galileo to L. Reael in Amsterdam. Arcetri, June 5, 1637[42]

From the Villa of Arcetri, June 5, 1637.

Together with the very kind and pleasant letter from Your Most Illustrious Lordship, I received another letter from the Most Illustrious and Literate Lord Martino Ortensio, both sent to me by my very close confidant and diligent friend the Most Illustrious Lord Elia Diodati from Paris. I received these in a period during which I could not read a syllable because of a fluxion of right eye, which decreased my sight as though I were completely blind. Therefore I was obliged to use someone else's eyes. Since this affliction happened because I wrote a lot during the last three months, at this moment I am thus unable to write a single word myself. Thus, according to my sinister accident, in order to give the greatest satisfaction both to Your Most Illustrious Lordship and to Lord Ortensio, I decided to write only to you, but in such a way that my answer is to both of your letters. And this is particularly easy because the questions contained in the two letters are the same.

Your Most Illustrious Lordship communicated to me that you have presented my proposal to the Most Illustrious and Powerful Orders of the United Provinces and that they received it very kindly and decreed it and that I will receive a copy of the report of those Most Illustrious and Powerful Lords by hand of Lord Ortensio. But I still have not received it because the authentication of Lord Cornelio Musch is missing, who is the Gratiario, that is (in our tongue, as I believe), the Chancellor of those Very Powerful Lords. Nevertheless I do not want to miss the chance to give the greatest satisfaction that I can give at the moment to the questions and to the doubts put to me about the normal practice to find the longitude at sea as well as on land in the frame of my invention.

The main doubt, formulated by Your Most Illustrious Lordship, according to what Lord Ortensio tells me, is concerned with the possibility to use the telescope on board the ship because the oscillations caused by the waves would not allow the necessary observations of Jupiter's satellites to be made. The second difficulty, also adduced by Lord Ortensio, is concerned with the lack of telescopes in your country with such perfection that they allow to be distinguished the very small stars orbiting around the planet Jupiter. The same Lord Ortensio asks for tables too and for the method of using them in order to calculate precisely from time to time the movements and therefore the configurations of those small stars. Moreover, he requests the method of building the clock I proposed, which should be so exquisite as to be sufficient for enumerating the parts of time, though very small, without any error in any place and during all seasons of the year.

In reference to the first problem, it is certain that this is the most difficult. But I believe to have found a remedy to this, at least for the ordinary movements of the ship. And this should be enough because, when the ship moves greatly and when there are storms, which normally prevent the sun and the other stars being seen, all

[42] *EN*, XVII:96–105.

observations cease, even the ordinary duties of the sailors. Thus, I think it is possible to transform the status of the person who is supposed to make the observations into a state similar to the tranquility and calm of the sea when the ship moves in the ordinary way. In order to achieve such a goal, I thought to place the observer on a part of the ship so prepared that he does not experience either the oscillations from bow to stern or the lateral ones. And my thought has the following fundament: If the ship were always on calm waters and were not oscillating, it is sure that the use of the telescope would be as easy on board as on land.

Now, I want to place the observer on a small ship placed on the large ship and the small one should contain a certain quantity of water, according to the needs that I will lay down in the following. It is first evident here that the water contained by this small vessel will always remain in equilibrium, without any of its parts rising or falling and so that it will always remain parallel to the horizon, also when the great ship oscillates to the right and the left sides, or to the front and back. In this way, if we build within the small vessel another vessel smaller than the first, floating on the contained water, this last vessel would be on a very calm sea and, consequently, would not oscillate. And this second small ship should be at the place where the observer works. I therefore want the first vessel, which is supposed to contain the water, to be like a great basin in the shape of a semi-sphere and that the smaller vessel is similar to the first one and only smaller to such an extent that the space left between the convex surface of the smaller vessel and the concave one of the other vessel is not larger than one inch. For this reason, a very small quantity of water will be enough to sustain the smaller vessel, as if it were placed on the great ocean, as I demonstrate in my treatise about the things which float on water and, at first sight, this looks truly marvelous and incredible. The size of these vessels should be such that the internal and smaller one is able, without sinking, to sustain the weight of the observer and also the seat and the other devices to which the telescope is fixed. And in order to keep the smaller vessel separate from the containing one in such a way that they never touch each other, so that the smaller one is not influenced by the motion of the ship in the same way that the larger one is, I want some springs to be fixed on the internal and concave surface of the containing vessel, as well on the convex one of the contained vessel, a number of eight or ten, so that they hinder the vessels from touching each other without, however, obliging the smaller vessel to follow upward and downward movements of the sides of the greater vessel. And if we want to use oil instead of water, this would work even better and no great quantity would be needed; two or three barrels of oil would suffice. Your Most Illustrious Lordship and Lord Ortensio could make some tests with two basins of copper, placing a certain quantity of sand into the smaller one so that it can still float in the larger one and, if you stick a small pole into the sand perpendicularly, you move the larger basin inclining it to one side and to the other, you will see that the small pole will always remain perpendicular without inclining, especially if the oscillations of the containing vessel are slow and show a time interval between each oscillation, like that of great ships. And believe, Your Most Illustrious Lordship, that once one begins to speculate on how to practice these operations, men will appear with such a skill, who, with time will get used to practicing these operations without any extra artificial preparation. At the beginning, for use on our galleys, I already made a certain helmet in the shape of a head and when the observer had it on, since a telescope was fixed to this device which was directed in such a way that it always targeted the point targeted by the other eye, without any effort the object that he was looking at was always observed with the telescope as well. One could build a similar machine, which however, is well fixed not only on the head but also on the shoulders and on the bust of the observer, and a telescope of the required size to observe well the small stars of Jupiter could be fixed to this machine so that one eye watches through it the object targeted with the free eye. In this way it would suffice to look at the body of Jupiter to see the stars close to it.

In reference to the second point, which is about the way to find more powerful telescopes than those built in your country, it seems to me that I have already written the last time to you that the power of the one I am using is such that it first shows the disc of Jupiter very sharply (and not blurred) in the same way as the naked eye sees the profile of the moon. And it shows in the same sharp way also its satellites and with the same size the fixed stars of second order when they are observed with the naked eye. Moreover, following the movement of Jupiter with the telescope, one sees the satellites at night before the fixed stars appear and in the morning after they disappear. And following Jupiter with the same telescope, one may see this during the whole day like Venus, the other planets and many of the fixed stars, and here Your Most Illustrious Lordship and Lord Ortensio think of the very great benefit that such a marvelous instrument has for astronomical science. I will certainly send the lenses to Your Most Illustrious Lordship and perhaps already with this present letter if my master, who makes them, has had the time to make a pair of them. And I say this because my Lord the Very Serene Grand Duke, delighted by such instruments, always keeps this man of mine with him and always brings him to all the places and villas to which His Highness moves. Therefore you can be sure about the building method and the power of those devices.

I come now to the second artifact made to increase enormously the precision of the astronomical observations. I am speaking about my time measurer, whose precision is such that it not only gives the exact quantity of hours and prime and second minutes and even third, if we were able to enumerate their frequency. And the precision is such that, building two, four or six such instruments, they would run with such correspondence to each other that one will not differ from the other, even for only one pulsation of the wrist, not only within 1 h, but also 1 day or 1 month. And I take the fundament for such a device from an admirable proposition, which I demonstrate in my book *de motu* which *est sub prelo* of the Lords Elzevirs in Leyden now. And the proposition is the following: Let a circle be erected over the horizon and the perpendicular made from the point of contact, which will therefore be the diameter of the circle, and from the point of contact, which is the lower end of the diameter, one draws as many chords as one wants to, over which one imagines mobiles descending as if they were over inclined planes. The time intervals of their paths over such chords and over the same diameter will all be equal to each other, so that if, for example, one draws from the same point of contact the chords to the circumference with 1, 4, 10, 30, 50, 100, 160 degrees, the mobile descends over these inclinations and lengths with the same times and also along the whole perpendicular diameter. And this happens also in reference to only those sections of the circumference of the two lower quadrants, where, as if they were canals along which a heavy sphere descends, the time taken by the sphere to go along the entire circumference of one section is the same taken by the same sphere when it starts moving 60, 40, 20, 10, 4, 2 or only one degree away from the point of contact. This is a very marvelous accident which one can validate by hanging a small sphere of lead or another heavy matter on a thread tied at the top. Taking it away from the perpendicular line so that it has the inclination of a fourth and letting it free one observes it going back and forth making many oscillations, the first ones are the largest which then continuously decrease until they does not move from the perpendicular line for more than one degree. And since it always moves along the same circumference, one notices that the large, the medium, the small and the smallest oscillations always take the same time.

And if one wants a more convincing test, one hangs two similar spheres from two threads of the same length and taking one of them 80 degrees away from the perpendicular, while the other two or three degrees only, and then letting them free, one person counts the oscillations of one of the two pendulums and another one the oscillations of the other, and you will find that one counts hundreds of the larger oscillations when the other counts hundred of the smallest very precisely.

From this very true and stable principle I deduce the structure of my timekeeper using not a weight hanging from a thread but a pendulum of solid and heavy matter, like for example brass or copper. I give to this pendulum the shape of a section of a circle of twelve or fifteen degrees, whose semi-diameter is two or three spans; the larger it is, the easier it is to use. I make the part of the section along the semi-diameter thick in the middle and thin toward its extreme sides, which end with a very sharp line in order to avoid as much as possible the resistance of the air that retards it. This section is perforated in the center, through which a small pole of iron passes and this has the shape of those around which the *stadera* turns. The small pole ends with an angle and lies over two supports of bronze, so that they do not consume too much because of the frequent moving of the section. Taking the section many degrees away from the perpendicular (if it is well balanced), it will make a very great number of oscillations before it stops. In order to have it working for the necessary time, it is also convenient that the person who works with it gives a powerful impulse to the section so that its oscillations are great. And once the oscillations made during a natural day are counted with patience, measuring the day by means of the revolution of a fixed star, one will have the number of oscillations for 1 h, 1 min and also other minor parts. Once this first experiment with a pendulum of any length is done, one can shorten or lengthen it, so that each oscillation corresponds to a second. Because the lengths of those pendulums have double the proportion to each other as that between their [oscillation] periods, for example: Let a pendulum be four spans long which makes in a given time a thousand oscillations, if we want the length of another pendulum which, in the same time, makes double the number of oscillations, it is required that the length of the pendulum is the fourth part of the length of the first one. Thus, as one can see with the experiments, the number of oscillations of pendulums of different lengths is sub-multiplied from those lengths.

Then, in order to avoid the tedium of the person who has to continuously count the oscillations, there is a very comfortable solution, which is the following: If one makes the section in such a way that from the middle of the circumference of the section a very small and thin stiletto sticks out, and during its movement it strikes a fixed bristle at one if its extremities and the bristle lays over the gears of a wheel, which is as light as paper placed on a horizontal plane close to the pendulum; and if the wheel has a gear made like those of a saw, that is, with one of the two sides perpendicular to the plane of the wheel and the other side obliquely inclined, this will work in such a way that, striking the bristle against the perpendicular side of the gear, it moves it and then the same bristle comes back and shuffles over the oblique side of the gear without moving it and falls at the bottom of the next gear. In this way, at each passage of the pendulum the wheel moves corresponding to the space interval of one of its gears and when the pendulum comes back, the wheel does not move. The motion of the wheel therefore will always be circular in one direction and, if its gears are marked with numbers, one observes the number of gears that have passed, and therefore the number of the oscillations and of the time intervals passed.

Around the center of this first wheel, one can also fix another wheel with a smaller number of gears, which touches another gear wheel greater than it. From the motion of the smallest wheel, we can ascertain the number of the entire revolutions of the first wheel by dividing the number of the gears in such a way that, for example, when the second wheel has completely turned around once, the first one turned around 20, 30, or 40 times or as many times as one wants. However, it is superfluous to tell this to Your Lordships, who have very diligent and ingenious men able to build clocks and other marvelous machines and it is also superfluous because these men will find more acute consequences than I have found from this new fundament that the pendulum, whether it moves along great or small spaces, always makes equal oscillations. Since the problem with clocks is mainly the fact that up to now it has not been possible to make them perform what we call the time of the clock accurately enough to make equal oscillations, with this very simple pendulum of mine, which is not subject to alteration, one has a way of always maintaining the equal measure of time. Now, Your Most Illustrious Lordship together with Lord Ortensio, understand what and how great the advantage is for astronomical observations for which it is not necessary to have the clock run continuously because it is sufficient, in order to measure the hours from noon or from sunset, to ascertain some time data concerned with some eclipses, conjunctions or with some other aspects of the celestial motions.

Concerning the tables of the movements of Jupiter's satellites and the method I used to calculate and produce the ephemerides, at the moment I am unable to satisfy you completely because I am hindered by a fluxion of the right eye which, to my great disappointment, hinders me in writing and reading a single word. In order to establish the squares of such movements, to favor Lord Ortensio, I need to observe again the present constitutions in order to fix their average movements and, besides this, I need to check a great number of observations I made continuously over many years. However, since I cannot make use of my sight, not even a bit of it, it is necessary for me to wait until my fate changes and this could take only a few days.

Concerning that point discussed by the very learned Lord Ortensio, that is, how one can begin to practice my method on land in order to correct the maps and to establish with great precision the longitudes of the isles, harbors and of other fixed places, there is no need for tables or ephemeredes. One needs only two observers, one at rest along the first meridian, which I establish to be the meridian of Amsterdam, and the other who goes from place to place and, in each of them, makes observations for three, four or six nights and watches the conjunctions, the separations and other aspects. They have to note down the times of those aspects and of the noons [of the day when those aspects take place] precisely. These notes should then be sent and compared and they will give the time difference between the two meridians, that is, the sought longitude. Before any other thing, it is therefore convenient that the Most Illustrious and Powerful Lords Orders command that an observatory is assigned and prepared in Amsterdam with all the necessary instruments to make continuous observations and that, for this work, a man is chosen who is a scientist in astronomy, diligent and patient, as I have been for many years in order to discover what I found with truly great effort. I know that for such work there is no lack of appropriate men in your country. However, on the basis of what I could understand of the value of Lord Martino Ortensio, I believe he would not only be very appropriate for this service, but also without equal or, at least, there is no person better than him. If this Lord therefore does not refuse to apply his mind to this enterprise, I will send all that is still missing to Your Most Illustrious and Powerful Lords in order to completely and simply understand each invention of mine.

And since what I am about to add in the following is the most relevant point of this enterprise, I will write further although I already wrote exaggeratedly until here.

Accept therefore, Your Most Illustrious Lordship, that I repeat that not only the beginning of the great arts and enterprises has been weak and requiring industriousness and continuous application of perspicacious minds in overcoming the first apparent difficulties during the time, but that the same happened concerning small and lower arts too. With this I would like to infer that, since I could not appear in front of you with an already established and perfect art because I have never been a sailor and never explored remote places, it is required that Your Most Illustrious and Powerful Lords rely on the opinions of intelligent men and that, if you want to gain the desired goal, you order that such a great enterprise begins without interruption and delay because of those eventual difficulties that could be encountered at the beginning, because they will all be solved since none of the possible difficulties can be as great as the most relevant ones which have been already solved by the human industriousness.

I chose to present my invention to those Most Illustrious and Powerful Lords rather than to whichever absolute prince because, if the latter is not able to understand this whole machinery, it is almost always the case that he relies on the advice of other people who are often not very intelligent. And moreover, because of that tendency, which is seldom separated from the human minds, to refuse to praise someone else more than oneself, it happens that the prince, badly advised, despises the offer and that the donor, instead of praised and thanked, is disturbed and vilified. But a republic where the deliberations depend on the consultation of many people, a small number of persons or even one single person among the powerful ones, and also averagely intelligent of the proposed matters, could be able to convince the other Lords to give their approval and so to support the enterprise. I hope to receive this help from the favor and the authority of Your Most Illustrious Lordship and if it happens that, thanks to your advice, one does start this enterprise, I will feel great happiness, although my very old age does not give me hope of seeing how my studies and my efforts produced and matured the advantage that, thanks to me, the human genus would achieve in reference to these two very great and noble arts, the nautical and the astronomical.

I kept Your Most Illustrious Lordship excessively busy. I ask you to excuse me and to communicate what I wrote to Lord Ortensio and to Lord Blauvio, elected as third among the Lord Commissioners and with reverent devotion I send you my greetings. I humbly bow in front of Your Most Illustrious Lordship and I pray to God for the best of happiness for you.

Galileo to F. Micanzio in Venice. Arcetri, November 20, 1637[43]

Very Reverend Father and My Very Cultivated Lord,

I will shortly answer the very pleasant letter of Your Very Reverend Father, since I am obliged to answer many letters and at the same time I am not able to write one single verse.

Concerning the Elzevirs, I had already written to Your Very Reverend Father that the printed sheets, sent by them to me, are at number 23, although I did not receive the six sheets between the tenth and the seventeenth, but I communicated it to them and they will certainly send them again.

Concerning the parabolic mirror, I always considered it very difficult, if not impossible, to reduce it [a mirror] to such a shape. But, if it is a spherical one and a portion of a very large sphere, the shape around its center differs so little from the parabolic one, that, since the spherical one can be built perfectly, as opposed to the parabolic one, the effect of burning would turn out to be stronger with the spherical one than with the parabolic, even though the latter joins the reflected rays at only one point which the other does not.

I am sorry that I cannot satisfy Your Very Reverend Father and the German gentleman, your friend, in reference to the lenses for a telescope. Since I cannot use them anymore, I gave away two of them of mediocre quality that I had and kept for myself only the ancient ones of mine, the discoverer of the celestial news, and I also destined these to the Grand Duke my Lord. For some months he has been applying his mind obsessively to the manufacture of those lenses and he always takes with him to the villas and everywhere one person who makes them continuously. His Serene Highness always acts as his supervisor and he does not want the master to work for anyone else and His Serene Highness is so jealous of them that it is impossible to have only one of them. Therefore it is better that you excuse me that I do not serve you, although this would be my wish.

When my small pension is in your hands, please do me the favor of keeping it until my nephew Albert, who serves the High Serene of Bavaria and who is here with me now, passes through Venice in order to revere Your Very Reverend Father, during his travels back to Munich. There he would like to purchase one of the violins of Cremona or Brescia because he plays that instrument very gently. And he will use the mentioned small pension to pay for the violin. I believe that it is possible to find those instruments there, although they are built somewhere else. And if there are not and it is required to acquire it outside the city, you may do me the favor of ensuring that a person who has expertise in that art chooses an absolutely perfect one from Brescia.

Apart from the unhappiness because of my sight, the status of the rest of my body is mediocre, although the frequent access of my old troubles because of the shivers with cold still continues. And I very cordially revere you here and I feel

---

[43] *EN*, XVII:220–221.

a very particular pleasure in knowing that the Very Excellent Lord Commissioner Veniero remembers me.

From Arcetri, November 20, 1637

Of Your Very Reverend Father

I commit to you the enclosed one

Very Devoted and Obligated Servant. Galileo Galilei

Credits

Biblioteca Nazionale Centrale of Florence (Concession of the Ministero per i Beni e le Attività Culturali della Repubblica Italiana/Biblioteca Nazionale Centrale di Firenze):
Figure 1 of Chapter 1. Figure 11 of Chapter 3. Figure 1 of Chapter 6.

Biblioteca Riccardiana, Florence:
Figure 10 of Chapter 3. Figure 6 of Chapter 5.

Istituto e Museo di Storia della Scienza, Florence. Fotographs of Franca Principe:
Figures 2, 8, 11 of Chapter 2. Figure 2 of Chapter 5.

Library of the Max Planck Institute for the History of Science, Berlin:
Figures 4, 5, 7, 10, 12, 16, 19 of Chapter 2. Figures 7, 8, 18 of Chapter 3. Figures 1, 2, 3, 4 (right), 5 of Chapter 4. Figures 1, 5 of Chapter 5.

© SUB Carl von Ossietzky Hamburg, Abt. Handschriften:
Figures 12, 13, 14, 15, 16 (left), 17, 19, 20, 21, 22, 23, 24 of Chapter 3.

© Electa: *La ragione e il metodo*, edited by Marco Bona Castellotti et al. 1999/2000: Illustration of the inside cover.

References

Galileo's Works

Galilei, G. 1585–1586. *La bilancetta*. Reprinted in *EN*, I:210–228.

Galilei, G. 1587–1588. *Theoremata circa centrum gravitatis solidorum*. Reprinted in *EN*, I: 179–208.

Galilei, G. 1592–1593a. *Delle macchine*. Hamburg: Staats- und Universitätsbibliothek. Cod. Math. 200b, cc. 15–33.

Galilei, G. 1592–1593b. *Breve istruzione all'architettura militare*. Reprinted in *EN*, II:17–75.

Galilei, G. 1592–1593c. *La Sfera ovvero Cosmografia*. Reprinted in *EN*, II:203–255.

Galilei, G. 1592–1593d. *Trattato di fortificazione*. Reprinted in *EN*, II:77–146.

Galilei, G. 1598–1599. *Le mechaniche*. Reprinted in *EN*, II:145–191.

Galilei, G. 1598–1634. *Ms. Gal. 26*. Florence: Biblioteca Nazionale Centrale di Firenze. Reprinted partially in *EN*, XIX:130–206.

Galilei, G. 1602 ca.–1637 ca. *Notes on Motion. Ms. Gal. 72*. Florence: Biblioteca Nazionale Centrale di Firenze. Electronic edition by the Max Planck Institute for the History of Science: http://www.mpiwg-berlin.mpg.de/Galileo_Prototype/INDEX.HTM/ Accessed July 2009.

Galilei, G. 1606. *Le operazioni del compasso geometrico et militare*. ed. Padova: In casa dell'Autore, Per Pietro Marinelli. Reprinted in *EN*, II:335–424.

Galilei, G. 1609 –. *Ms. Gal. 49*. Florence: Biblioteca Nazionale Centrale di Firenze. Reprinted partially in *EN*, XIX:130–206.

Galilei, G. 1610a. *Le matematiche nell'arte militare*. In Dep.i Canc.a, Ordinario, 19, 1777–1806. Padova: Archivio del Comune di Padova.

Galilei, G. 1610b. *Sidereus nuncius*. Venetiis: Apud Thomam Baglionum. Reprinted in *EN*, III: 54–96.

Galilei, G. 1612. *Discorso intorno alle cose che stanno in su l'acqua*. Firenze: Appresso Cosimo Giunti. Reprinted in *EN*, IV:57–141.

Galilei, G. 1619. *Postille al Dicorso delle Comete di Mario Guiducci*. Reprinted in *EN*, VI: 672–680.

Galilei, G. 1623. *Il Saggiatore*. Roma: Appresso Giacomo Mascardi. Reprinted in *EN*, VI: 197–372.

Galilei, G. 1632. *Dialogo sopra i due massimi sistemi del mondo, tolemaico e copernicano*. In Fiorenza: Per Giovan Batista Landini. Reprinted in *EN*, VII:25–520.

Galilei, G. 1638. *Discorsi e dimostrazioni matematiche intorno à due nuove scienze*. Leida: Appresso gli Elsevirii. Reprinted in *EN*, VIII:39–318.

Galilei, G. 1655. *Discorsi e dimostrationi matematiche, Intorno à due nuove scienze, Attenenti alla Mecanica, & i Movimenti Locali*. Bologna: Per gli HH del Dozza.

Galilei, G. 1657–1667. *MSS Galileani, Vol. IV*. Florence: Biblioteca Nazionale di Firenze. Accademia del Cimento.

Galilei, G. ca. 1590. *De motu antiquiora*. Reprinted in *EN*, I:243–419.

Galilei, G., and L. Danesi. 1649. *Della Scienza Mecanica, e delle Utilità, Che si traggono da gl'Instromenti di quella*. Ravenna: Appresso gli Stamp. Camerali.
Galilei, G., and W. F. Edwards. 1988. *Tractatio de praecognitionibus et praecognitis. Tractatio de demonstratione*. Padova: Antenore.
Galilei, G., and A. Favaro. 1899. Delle meccaniche lette in Padova l'anno 1594 da Galileo Galilei. *Memorie del Regio Istituto veneto di Scienze* XXVI (5).
Galilei, G., and A. Favaro. 1968. *Le opere di Galileo Galilei*. 4th. ed. 20 vols. Firenze: Giunti Barbèra Editore. First edition, 1890–1909.

Primary Literature

Epistolae clarorum virorum selctae de quamplurimis optimae ad indicandam nostrorum temporum eloquentiam. 1586. Coloniae Agrippinae.

Adorno, F., and L. Zangheri. 1998. *Gli statuti dell'Accademia del disegno*. Firenze: L.S. Olschki.

Agricola, G. 1541. *Georgii Agricolae, medici, Bermannus, sive, De re metallica*. Parisiis: apud Hieronymum Gormontium.

Alberti, L. B. 1973. Ludi Matematici. In *Opere Volgari*, edited by C. Grayson. Bari: Laterza, 131–173.

Alexandrinus, P., and F. Commandino. 1588. *Mathematicae Collectiones*. Pesaro, Venetia: Girolamo Concordia.

Anonymous. XVI century. *Fortificationi. De Artiglieria*. Florence: Biblioteca Nazionale Centrale di Firenze. Ms. XIX-62.

Archimedes. 2002a. On the Equilibrium of Planes. In *The Works of Archimedes*, edited by T. L. Heath. New York: Dover Publications, 189–220.

Archimedes. 2002b. On the Floating Bodies. In *The Works of Archimedes*, edited by T. L. Heath. New York: Dover Publications, 253–300.

Archimedes, and G. d. Monte. 1588. *Guidiubaldi è Marchionibus Montis In duos Archimedis aequeponderantium libros paraphrasis scholijs illustrata*. Pisa: Apud H. Concordiam.

Archimedes, and N. Tartaglia. 1543. *Opera Archimedis*. Venice.

Aristotle. 1936. *Aristotle Problems*. In *Aristotle in Twenty-Three Volumes*. Vol. 15 and 16. Cambridge, MA: The Loeb Classical Library.

Aristotle. 1980. *Mechanical Problems*, edited by W. S. Hett. In *Aristotle in Twenty-Three Volumes*. Vol. 14. Cambridge, MA: The Loeb Classical Library, 331–411.

Aristotle. 1987. *Meteorologica*, edited by G. P. Goold. In *Aristotle in Twenty-Three Volumes*. Vol. 7. Cambridge, MA: The Loeb Classical Library.

Aristotle, and B. Baldi. 1621. *Mechanica Aristotelis problemata exercitationes*. Moguntiae: Typis & Sumptibus Viduae Ioannis Albini.

Aristotle, and G. Biancani. 1615. *Aristotelis loca mathematica*. Bononiae: Apud Bartholomaeum Cochium.

Aristotle, and M. E. Bottecchia Dehò. 1982. *MHXANIKA*. Padova: Ed. Antonore.

Aristotle, and M. E. Bottecchia Dehò. 2000. *Problemi meccanici*. Catanzaro: Rubbinetto.

Aristotle, and V. Fausto. 1517. *Aristotelis Mechanica Victoris Fausti industria in pristum habitum-restituita ac latinitate donata*. Paris: Venundatur in aedibus Iodoci Badii.

Aristotle, and A. Guarino. 1573. *Le Mechaniche d'Aristotile trasportate di Greco in Volgare Idioma*. Modena: Andrea Gadaldino.

Aristotle, and G. d. Guevara. 1627. *In Aristotelis mechanica commentarii. Una cum Additionbus Quibusdam Ad eandem materiam pertinentibus*. Roma: Apud Jacobum Mascardum.

Aristotle, and W. K. C. Guthrie. 1939. *On the Heavens*. In *Aristotle in Twenty-Three Volumes*. Vol. 6. Cambridge, MA: Loeb Classical Library.

Aristotle, and W. S. Hett. 1936. *Minor Works*. In *Aristotle in Twenty-Three Volumes*. Vol. 14. Cambridge, MA: The Loeb Classical Library.

Aristotle, and W. S. Hett. 1936. *On the Soul*, edited by W. S. Hett. In *Aristotle in Twenty-Three Volumes*. Vol. 8. Cambridge, MA: Loeb Classical Library, 8–203.

Aristotle, and W. D. Ross. 1936. *Aristotle's Physics. A Revised Text with Introduction and Commentary*. Oxford: Clarendon Press.

Aristotle, and W. D. Ross. 1960. *Aristotle's Physics*. Oxford: Clarendon Press.

Bacon, F. 1620. *Sylva Sylvarum of A Natural History in Ten Centuries*. London: for W. Lee.

Baldinucci, F. 1728. *Notizie de' Professori del Disegno da Cimabue*. Firenze: S. Franchi.

Belluzzi, G. B. ca. 1545. *Il Trattato delle fortificazioni di terra*. Firenze: Biblioteca Riccardiana di Firenze. Ms. cart. Riccardiano n. 2587. Reprinted in Lamberini, D. *Il Sanmarino. Giovan Battista Belluzzi architetto militare e trattatista del Cinquecento*, Firenze: Olschki, Vol. II: 403–449.

Benedetti, G. B. 1585. *Io Baptistae Benedicti diversarum speculationum mathematicarum, et physicarum liber: quarum feriem sequens pagina indicabit.* Taurini: Bevilaqua.

Besson, J. 1578. *Theatrum instrumentorum et machinarum.* Lyon.

Biancani, G. 1620. *Sphaera mundi, seu Cosmographia demonstrativa ac facili methodo tradita.* Bononiae: typis Sebastiani Bonomii.

Biringucci Vannocci, O. After 1562. *Architettura, fortificazioni e macchine.* Siena: Biblioteca Comunale di Siena. Ms. S.IV.1.

Biringuccio, V. 1540. *De la pirotechnica.* Venetia: Rossinello.

Buonarotti, M., and G. Milanesi. 1875. *Le lettere. Pubblicate coi ricordi ed i contratti artistici.* Milano: Le Monnier.

Busca, G. 1601. *Dell'architettura militare di Gabriele Busca.* Milano: presso Bordone e Locarni.

Caprae, B. 1607. *Usus et fabrica circini.* Patavii: Apud Petrum Paulum Tozzium. Reprinted in *EN*, II:425–510.

Cardi da Cigoli, L., and R. Profumo. 1992. *Trattato pratico di prospettiva, Manoscritto Ms 2660A del Gabinetto dei Disegni e delle Stampe degli Uffizi.* Rome: Bonsignori Editore.

Cassini, G. D. 1668. *Ephemerides Bononienses Mediceorum.* Bononiae: Mariae.

Cataneo, G. 1582. *Avertimenti, et essamini intorno a quelle cose, che richiedono a un perfetto bombardiero.* Vinegia: Salicato.

Caus, S. d. 1615. *Les Raisons des forces mouvantes avec diverses machines.* Frankfurt a. M.: Abraham Bacquart.

Cavalieri, F. B. 1632. *Lo specchio ustorio overo trattato delle settioni coniche, et alcuni loro mirabili effetti intorno al lume, caldo, freddo, suono, e moto ancora.* Bologna: Ferroni.

Cennini, C. 1437. *Tratttato della pittura.* Roma: co' torchi di Paolo Salviucci.

Ceredi, G. 1567. *Tre discorsi sopra il modo d'alzar acque da' luoghi bassi.* Parma: Appreso Seth Viotti.

Cesi, F. 1625. *Apiarium.* Romae: Mascardi.

Clavius, C. 1582. *In Sphaeram Ioannis de Sacro Bosco commentarius.* Romae: apud Franciscum Zanettum, ex officina Dominici Basae.

Commandino, F. 1565. *Liber de centro gravitatis solidorum.* Bononiae: ex officina Alexandri Benacii.

Contarini, G. 1592–1593. *Fabrica di galee.* Venezia: Archivio di Stato di Venezia. Archivio Proprio Giacomo Contarini, busta 25. Electronic edition by the Max Planck Institute for the History of Science: http://echo.mpiwg-berlin.mpg.de/content/under_construction/venicearsenal/archive/ Accessed October 2009.

Contarini, G. ca. 1580. *Scritture concernenti a fortificazioni.* Venezia: Archivio di Stato di Venezia. Archivio Proprio Giacomo Contarini, busta 19, registro 1.

Euclid, and T. L. Heath. 1956. *Elements.* 3 vols. New York: Dover Publications.

Euclid, and N. Tartaglia. 1543. *Euclide Megarense philosopho.* Vinegia: Rossinelli.

Fausto, V. 1551. *Victoris Fausti Veneti Orationes Quinque eius amicorum cura quam fieri potuit diligenter impressae.* Venetia: reprinted in *Epistolae.* 1586.

Fibonacci, L. XIII. century. *Lo livero dell'abbecho secondo la oppinione de maiestro Leonardo della chasa degl'figliuogle Bonacie da Pisa.* Florence: Biblioteca Riccardiana. Codex 2186.

Fludd, R. 1617. *Utriusque cosmi maioris scilicet et minoris metaphysica, physica atqve technica historia, in duo volumina secundum cosmi differentiam diuisa.* 2 vols. Oppenheim: De Bry.

Fludd, R. 1626. *Philosophia sacra et vere Christiana Seu Meteorologica Cosmica.* Frankfurt: prostat in officina Bryana.

Fludd, R. 1638. *Philosophia Moysaica: In qua Sapientia & scientia creationis & creaturarum Sacra veréque Christiana ... ad amussim & enucleaté explicatur.* Goudae: Rammazenius.

Galilei, V. 1581. *Dialogo di Vincentio Galilei nobile Fiorentino della musica antica et della moderna.* Fiorenza: Marescotti.

Giorgio Martini, F. d. 1475–1478. *Opusculum de architectura.* London: British Museum. Ms. 197.b.21.

Giorgio Martini, F. d. 1967. *Trattati di archittetura civile e militare.* Milan: Edizione Il Polifilo.

Grassi, O. 1619. *De tribus cometis anni MDCXVIII*. Romae: ex typographia Iacobi Mascardi. Reprinted in *EN*, VI:37–108.

Grayson, C. 1973. *Opere volgari di Leon Battista Alberti*. 3 vols. Bari: Laterza.

Guiducci, M. 1619. *Discorso delle comete*. Firenze: stamperia di Pietro Cecconcelli. Reprinted in *EN*, VI:37–108.

Hero of Alexandria, and G. B. Aleotti. 1589. *Gli artifitiosi et curiosi moti spiritali di Herrone*. Ferrara: per Vittorio Baldini.

Hero of Alexandria, and F. Commandino. 1575. *Heronis Alexandrini Spiritalium liber*. Urbini.

Hero of Alexandria, B. Davanzati, C. Gargiolli, and F. Martini. 1862. *Della natura del voto di Erone Alessandrino*. Firenze: Stamperia del Monitore.

Hero of Alexandria, and A. Giorgi. 1592. *Spiritali di Herone Alessandrino*. Urbino: Appresso Bartolomeo, e Simone Ragusij fratelli.

Hero of Alexandria, and W. Schmidt. 1899. *Druckwerke und Automatentheater*. Leipzig: B. G. Teubner.

Hero of Alexandria, and O. Vannocci Biringucci. 1582. *Libro degli artifizii spiritali over di fiato d'Herone Alessandrino*. Siena: Biblioteca degli Intronati. Ms. L. VI. 44. Electronic edition by the Max Planck Institute for the History of Science. http://echo.mpiwg-berlin.mpg/content/pratolino/sources/ Accessed October 2009.

Horky, M. 1610. *Brevissima Peregrinatio contra Nuncium Sidereum*. Modena: Apud Iulianum Cassianum.

Kepler, J. 1604. *Ad Vitellionem Paralipomena, quibus Astronomiae pars Optica traditur; potissimum de Artificiosa observatione et aestimatione diametrorum deliquiorumque Solis & Lunae*. Frankfurt: Claudius Marnius Johann Aubrius.

Kepler, J. 1610. *Dissertatio cum Nuncio Sidereo*. Pragae: Typis Davidis Franci.

Kepler, J. 1611. *Dioptrice*. Augustae Vindelicorum (Augsburg): Typis Davidis Franci.

Kircher, A. 1646. *Ars magna lucis et umbrae*. Romae: Sumptibus Hermanni Scheus.

Kircher, A. 1673. *Athanasii Kircheri Phonurgia nova, sive, Conjugium mechanico-physicum artis & natvra paranympha phonosophia concinnatum*. Campidona: Per Rudolphum Dreherr.

Lagalla, G. C. 1612. *De Phoenomenis in orbe lunae novi telescopii usu a D. Gallileo Gallileo nunc iterum suscitatis Physica disputatio*. Venetiis: Apud Thomam Baglionum. Reprinted in *EN*, III:311–395.

Leurechon, J. 1624. *Récréation mathématique: composée de plusieurs problémes plaisants et facétieux*. Basel.

Leurechon, J. 1663. *Mathematical Recreations*. London: for William Leake.

Lorini, B. 1609. *Le Fortificazioni*. Venetia: Presso Francesco Rampazetto.

Magalotti, L. 1666. *Saggi di naturali esperienze*. Firenze: Per Giuseppe Cocchini all'Insegna della Stella.

Magini, G. A. 1602. *Theorica speculi concavi sphaerici*. Bononiae: apud Ioannem Baptistam Bellagambam.

Magini, G. A. 1611. *Breve instruttione sopra l'apparenze: et mirabili effetti dello specchio concavo sferico*. Bologna: Presso G. B. Bellagamba.

Manzini, C. A. 1660. *L'occhiale all'occhio. Dioptrica pratica*. Bologna: per l'Herede del Benacci.

Maurolicus, F. 1575. *Opuscola mathematica*. Venetia: apud Franciscum Franciscium Senensem.

Mazzoni, J. 1597. *Jacopi Mazonii in universam Platonis et Aristotelis philosophiam praeludia sive de comparatio*. Venetiis: Guerilius.

Mellini, D. 1588. *Discorso di Domenico Mellini nel quale si prova contra l'oppenione di alcuni non si potere artifizialmente ritrovare, ne dare ad un corpo composto di Materia corrotibile un Movimento, che sia continovo e perpetuo*. Fiorenza: Appresso Bartolomeo Sermatelli (Bernardo Buontalenti's copy is preserved at the Biblioteca Nazionale Centrale of Florence, Fondo Targioni Tozzetti 86).

Mersenne, M. 1636. *Harmonie universelle*. Paris: Chez Sebastien Cramoisy.

Mersenne, M. 1639. *Les nouvelles pensées de Galilei, mathematicien et ingenieur du Duc de Florence*. Paris: Rocolet.

Metio Alcamariano, A. 1623. *Praxis nova geometrica per usum circini et regula proportionalis*. Amstelrodami: Bibliopol.
Moletti, G., and W. R. Laird. 2000. *The Unfinished Mechanics of Giuseppe Moletti*. Toronto: University of Toronto Press.
Montaigne, M. d. 1929. *The Diary of Montaigne's Journey to Italy. 1580 and 1581*. London: Leonard and Virgina Woolf.
Monte, G. d. 1577. *Guidiubaldo e Marchionibus Montis Mechanicorum liber*. Pisauri: Apud Hieronymum Concordiam.
Monte, G. d. 1579. *Planisphaeriorum universalium theorica*. Pisauri: Gerolamo Concordia.
Monte, G. d. 1609. *Problematum astronomicorum*. Venetiae: Bernardo Giunta e Giovanni Battista Ciotti.
Monte, G. d. 1615. *De cochlea libri quatuor*. Venetiae: Deuchinus.
Monte, G. d., and F. Pigafetta. 1581. *Le Mechaniche dell'Illustrissimo Sig. Guido Ubaldo De'Marchesi del Monte*. Venetiae: Appresso Francesco di Franceschi Sanesi.
Neri, A. 1612. *L'arte vetraria*. Reprinted in Barovier Mentasti R. (ed.) Edizioni Polifilo, Verona, 1980.
Neroni, M. 1602. *Atlante*. Firenze: Biblioteca Nazionale Centrale di Firenze. II.I.281.
Pereira, B. 1576. *De Communibus omnium rerum naturalium principiis et affectionibus libri quindecim*. Romae: Venturini Tramezini, Apud Franciscum Zanettum et Bartholomaeum Tosium.
Philo of Byzantium, and F. D. Prager. 1974. *Pneumatica*. Wiesbaden: Dr. Ludwig Reichert Verlag.
Pifferi, F. 1604. *Sfera di Giovanni Sacrobosco tradotta e dichiarata da Don Francesco Pifferi Sansovino*. Siena: appresso Salvestro Marchetti.
Pliny, and H. Rackam. 1952. *Natural History*. Cambridge, MA: Loeb Classical Library, Harvard University Press, 1983.
Porta, G. B. d. 1589. *Magiae naturalis libri XX*. Neapoli: Apud Horatium Salvianum.
Porta, G. B. d. 1593. *De refractione optices parte*. Neapoli: O. C. Salviani, A. Giacomo Giovanni, Pace.
Proclus, and G. Friedlein. 1873. *Im primum Euclidis Elementorum librum commentari*. Leipzig: Teubner.
Puccini, B. 1990. Libri primo della fortificazione di Bernardo Puccini. In *Il principe difeso: vita e opere di Bernardo Puccini*, edited by D. Lamberini. Florence: La Giuntina, 325–350.
Ramelli, A. 1588. *Le diverse et artificiose machine*. Parigi: In casa del Autore.
Redi, F. 1678. *Lettera intorno all'invenzione degli occhiali*. Firenze: per Francesco Onofri Stampatore Granducale.
Ricci, O. ca. 1560–ca. 1590a. *L'uso dell'archimetro*. Firenze: Biblioteca Nazionale Centrale di Firenze. Fondo II, 52 (Nelli, 61 1891), fol. 36–48.
Ricci, O. ca. 1560–ca. 1590b. *Problemi di geometria pratica*. Firenze: Biblioteca Nazionale Centrale di Firenze. Fondo II, 52 (Nelli, 61 1891), fol. 1–35.
Ricci, O. ca. 1560–ca. 1590c. *Problemi geometrici*. Firenze: Biblioteca Nazionale Centrale di Firenze. Fondo II, 52 (Nelli, 61 1891), fol. 48–84.
Ricci, O. ca. 1560–ca. 1590d. *Problemi geometrici*. Firenze: Biblioteca Nazionale Centrale di Firenze. Mss. Gal. 10.
Ricci, O. ca. 1560–ca. 1590e. *Trattato d'architettura*. Modena: Biblioteca Estense. Collezione Campori, CAMP APP 753, Ms. y R. 2.8, cc. 41 ff.
Sanctorius, S. 1612. *Commentaria in artem medicinalem Galeni*. Venetiis: apud Marcum Antonium Brogiollum.
Sanctorius, S. 1630. *Commentaria in artem medicinalem Galeni*. Venetiis: apud Marcum Antonium Brogiollum.
Sanctorius, S. 1634. *De statica medicina*. Venetiis: apud Marcum Antonium Brogiollum.
Sanctorius, S. 1646. *Commentaria in primam fen primi libri Canonis Avicennae*. Venetiis: apud Marcum Antonium Brogiollum.
Sanuto, M., and R. Fulin. 1969–1970. *I diarii. 1496–1533*. 58 vols. Bologna: Forni.

Sarsio, L. 1619. *Libra astronomica ac philosophica, qva Galilaei Galilaei opiniones de cometis a Mario Gvidvcio in Florentina Academia expositæ atque in lucem nuper editae, examinantus.* Perusiae: ex typ. Marci Naccarini. Reprinted in *EN*, VI:111–180.

Sarsio, L. 1626. *Ratio ponderum librae et simbellae: in qua quid e Lotharii Sarsii libra astronomica, quidque è Galilei Galilei Simbellatore, de cometis statuendum sit, collatis vtriusque rationum momentis, philosophorum arbitrio proponitur.* Lutetiae Parisiorum: samptibus Sebastiani Cramoisy, via Iacobaea sub Ciconiis. Reprinted in *EN*, VI:375–500.

Sgrilli, B. S. 1742. *Descizione della regia villa, fontane, e fabbriche di Pratolino.* Florence: Per Tartini e Franchi, stamperia granducale.

Soprani, R. 1674. *Le vite de' pittori, scoltori, et architetti genovesi.* Genoa: per Gioseppe Bottaro e Gio:battista Riboldi Copagni.

Tartaglia, N. 1537. *Nova scientia.* Venetia: Stephano da Sabio.

Tartaglia, N. 1554. *Quesiti et inventioni diverse de Nicolo Tartalea Brisciano.* Venetia.

Tartaglia, N. 1565. *Opuscolum de ponderositate.* Venetia: apud Curtium Troianum.

Telioux, B. 1611. *Mathematica meravigliosa.* Paris: Bibliothèque de l'Arsenal. Ms. 8525.

Theti, C. 1588. *Discorsi sulle fortficationi.* Venetia: Appresso Nicolo Moretti.

Toledo, F. d. 1574. *Commentaria una cum quaestionibus in octo libros Aristotelis De Physica auscultatione.* Köln: Brickmann.

Toledo, F. d. 1575. *Commentaria una cum quaestionibus in II libros De generatione et corruptione.* Venetia.

Valle, P. d. 1622. *Logica. Duobus tomis distincta, quorum primus artem veterem, Secundum novam comprehendit.* 2 vols. Lugduni: Sumtibus Ludovici Prost Haeredi Rouille.

Vasari, G. 1568. *Le vite de' più eccellenti pittori, scultori e architettori.* Firenze: Giunti.

Vitruvius, S. I. Frontinus, and F. G. Giocondo. 1513. *De architectura libri X iterum et Frontinus à Iocundi revisi repurgatique quantum ex collatione licuit.* Firenze: Filippo Giunta.

Vitruvius, M. P., and D. Barbaro. 1567. *I dieci libri dell'architettura di M. Vitruvio.* Venetia: Appresso Francesco de Franceschi Senese. Reprinted by Edizioni Il Polifilo, 1997.

Vitruvius, M. P., I. D. Rowland, T. N. Howe, and M. J. Dewar, eds. 1999. *Ten Books on Architecture.* Cambridge: Cambridge University Press.

Viviani, V. 1654. *Racconto istorico.* Reprinted in *EN*, XIX:597–632.

Zonca, V. 1607. *Novo teatro di machine et edificii.* Padova: Appresso Pietro Bertelli.

Zublero, L. 1607a. *Fabrica et usus instrumenti chorografici.* Basel: Bibliopole Basiliensis.

Zublero, L. 1607b. *Novum instrumentum geometricum.* Basel: Bibliopole Basiliensis.

Secondary Literature

Nel trecentesimo natalizio di Galileo in Pisa, 18 Febbraio MDCCCLXIV 1864. Pisa: Tipografia Nistr.

Actes du XIIe congrés international d'histoire des sciences. 1971. Paris: Librairie Sci. Tech. Blanchard.

La corte il mare i mercanti. La rinascita della scienza. Editoria e società. Astrologia, magia e alchimia. 1980. Florence: Electa Editrice.

Abrahams, P. 2002. The First 300 Years of Binocular Telescopes. http://www.europa.com/~telscope/bintlhst.txt/ Accessed October 2009.

Ackerman, J. S. 1954. Architectural Practice in the Italian Renaissance. *Journal of the Society for Architectural History* 13:3/10.

Andrews, W. J. H. 1996. *The Quest for Longitude.* Cambridge, MA: Harvard University Press.

Arrighi, G. 1965–1967. Un programma di didattica di matematica nella prima metà del Quattrocento (dal codice 2186 della Biblioteca Riccardiana di Firenze). *Atti e Memorie dell'Accademia Petrarca di Lettere, Arti e Scienze di Arezzo* nuova serie (XXXVIII):112–128.

Aymard, M. 1980. L'arsenale e le conoscenze tecnico-marinaresche. In *Storia della cultura veneta*, edited by G. Arnadi and M. Pastore Stocchi. Vicenza: Neri Pozza Editore, 289–315.

Baldini, U. 1992. *Legem impone subactis. Studi su filosofia e scienza dei gesuiti in Italia, 1540–1632.* Rome: Bulzoni

Barclay, R. 1996. *The Art of the Trumpet-Maker. The Materials, Tools and Techniques of the Seventeenth and Eighteenth Centuries in Nuremberg.* Oxford: Oxford University Press.

Becchi, A. 2004. *Q. XVI. Leonardo, Galileo e il caso Baldi. Magonza, 26 marzo 1621.* Venice: Marsillo.

Bedini, S. A. 1986. Galileo and Scientific Instrumentation. In *Reinterpreting Galileo*, edited by W. A. Wallace. Washington D.C.: The Catholic University of America Press, 127–153.

Bennett, J. A. 1986. The Mechanics' Philosophy and the Mechanical Philosophy. *History of Science* 24 (3):1–28.

Bennett, J. A. 1991. The Challenge of Practical Mathematics. In *Science, Culture and Popular Belief in Renaissance Europe*, edited by S. Pumfrey, P. L. Rossi and M. Slawinski. Manchester: Manchester University Press, 176–190.

Bensaude-Vincent, B., and W. A. Newman, eds. 2007. *The Artificial and Natural: An Evolving Polarity.* Cambridge, MA: MIT Press.

Berti, D. 1876. Storia dei manoscritti galileiani della Biblioteca Nazionale di Firenze ed indicazione di parecchi libri e codici postillati da Galileo. In *Atti della R. Accademia dei Lincei, Memorie della classe di scienze morali, storiche e filologiche*, 96–105.

Bertoloni Meli, D. 1992. Guidobaldo dal Monte and the Archimedean Revival. *Nuncius* 7 (1):3–34.

Bertoloni Meli, D. 2006. *Thinking with Objects: The Transformation of Mechanics in the Seventeenth Century.* Baltimore: The Johns Hopkins University Press.

Biagioli, M. 1989. The Social Status of Italian Mathematicians, 1450–1600. *History of Science* 27:41–95.

Biagioli, M. 1993. *Galileo Courtier.* Chicago: The University of Chicago Press.

Biagioli, M. 2006. *Galileo's Instruments of Credit: Telescopes, Images, Secrecy.* Chicago: University of Chicago Press.

Biener, Z. 2004. Galileo's First New Science: The Science of Matter. *Perspectives on Science* 12 (3):262–287.

Blackwell, R. J. 1991. *Galileo, Bellarmine, and the Bible. Including a Translation of Foscarini's Letter on the Motion of the Earth.* Notre Dame, Ind.: University of Notre Dame Press.

Bona Castellotti, M., E. Gamba, and F. Mazzocca, eds. 1999. *La ragione e il metodo. Immagini della scienza nell'arte Italiana dal XVI al XIX secolo.* Milan: Electa.

Boschiero, L. 2007. *Experiment and Natural Philosophy in Seventeenth-Century Tuscany: The History of the Accademia del Cimento.* Dordrecht: Springer.

Bozzi, P., C. Maccagni, L. Olivieri, and T. B. Settle. 1995. *Galileo e la scienza sperimentale*. Padova: Dipartimento di Fisica "Galileo Galilei."

Bredekamp, H. C. 2007a. *Antikensehnsucht und Maschinenglauben. Die Geschichte der Kunstkammer und die Zukunft der Kunstgeschichte*. 3rd ed. Berlin: Wagenbach. First edition: 1993.

Bredekamp, H. C. 2007b. *Galileo der Künstler*. Berlin: Akademie Verlag.

Bruce, M. 2005. Knowing How and Knowing That: Artisans, Bodies, and Natural Knowledge in the Scientic Revolution. *Studies in History and Philosophy of Science Part A* 36 (3):577–585.

Burke, P. 1972. *The Italian Renaissance. Culture and Society in Italy*. New York: Charles Scribner's Sons.

Büttner, J. 2009. *Galileo's Challenges: The Origin and Early Conceptual Development of Galileo's Theory of Naturally Accelerated Motion on Inclined Planes*. PhD (submitted in March 2009), Institut für Geschichtswissenschaften, Humboldt Universität zu Berlin, Berlin.

Büttner, J., P. Damerow, and J. Renn. 2001. Traces of an Invisible Giant: Shared Knowledge in Galileo's Unpublished Treatises. In *Largo campo di filosofare: Eurosymposium Galileo 2001*, edited by J. Montesinos and C. Solís. La Orotava: Fondación Canaria Orotava de Historia de la Ciencia, 183–202.

Büttner, J., P. Damerow, J. Renn, and M. Schemmel. 2003. The Challenging Images of Artillery - Practical Knowledge at the Roots of the Scientific Revolution. In *The Power of Images in Early Modern Science*, edited by W. Lefèvre, J. Renn and U. Schoepflin. Basel: Birkhäuser, 3–38.

Camerota, F. 2000. *Il compasso di Fabrizio Mordente*. Florence: Olschki.

Camerota, M. 1992. *Gli scritti «De motu antiquiora» di Galileo Galilei. il Ms. Gal 71. Un'analisi storico-critica*. Cagliari: CUEC Editrice.

Camerota, M. 2004. *Galileo Galilei e la cultura scientifica nell'età della Controriforma*. Rome: Salerno Editrice.

Caruso, A., and A. C. Crombie. 1983. The Jesuits and Galileo's Ideas of Science and of Nature. *Annali dell'Istituto e Museo di Storia della Scienza di Firenze* VIII (2):3–67.

Casi, F. 1993. *Scienziati aretini dal '400 al '700 - Strumenti*. Edited by C. A. e. Promozioni. Arezzo.

Castiglione, A. 1931. The Life and Work of Santorio Santorio (1561–1636). *Medical Life* 38: 729–785.

Caverni, R. 1972 (Reprint). *Storia del metodo sperimentale in Italia*. Florence: Johnson Reprint Corporation. Original edition: 1891–1901.

Chaldecott, J. A. 1952. Bartolomeo Telioux and the Early History of the Thermometer. *Annals of Science* 8:195–201.

Concina, E. 1988. *L'Arsenale della Repubblica di Venezia. Tecniche e istituzioni dal Medioevo all'età moderna*. Milan: Electa.

Concina, E. 1990. *Navis*. Torino: Einaudi.

Cuomo, S. 1997. Shooting by the Book: Notes on Niccolò Tartaglia's «Nova Scientia.» *History of Science* 35 (2):155–188.

Cuomo, S. 2000. *Pappus of Alexandria and the Mathematics of Late Antiquity, Cambridge Classical Studies*. Cambridge: Cambridge University Press.

Cuomo, S. 2001. *Ancient mathematics, Sciences of Antiquity*. New York: Routledge.

Damerow, P., G. Freudenthal, P. McLaughlin, and J. Renn. 2004. *Exploring the Limits of Preclassical Mechanics. A Study of Conceptual Development in Early Modern Science: Free Fall and Compounded Motion in the Work of Descartes, Galileo and Beeckman*. 2 ed. New York: Springer Verlag.

Damerow, P., J. Renn, S. Rieger, and P. Weinig. 2002. Mechanical Knowledge and Pompeian Balances. In *Homo Faber: Studies on Nature, Technology, and Science at the Time of Pompeii*, edited by J. Renn and G. Castagnetti. Rome: L'Erma di Bretschneider, 93–108.

Daston, L. 2008. Unruly Weather: Natural Law Confronts Natural Variability. In *Natural Law and Laws of Nature in Early Modern Europe: Jurisprudence, Theology, Moral, and Natural Philosophy*, edited by L. Daston and M. Stoilleis. Farnham: Ashgate, 233–248.

Daston, L., and K. Park. 1998. *Wonders and the Order of Nature. 1150–1750*. New York: Zone Books, Distributed by the MIT Press.

De Ceglia, F. P. 1997. *Reazioni Romane. L'idraulica galileiana negli scritti di Giovanni Bardi e Giuseppe Biancani*. Bari: Laterza.

Dempsey, C. 1980. Some Observations on the Education of Artists at Florence and Bologna During the Later Sixteenth Century. *Art Bullettin* 62:552 ff.

Desarnaulds, V., Y. Loerincik, and A. P. O. Carvalho. 2001. *Efficiency of 13th Century Acoustic Ceramic Pots in Two Swiss Churches*. Paper read at Noise-Con 2001, October 29–31 at Portland Maine.

Dezzi Bardeschi, M., ed. 1985. *Il ritorno di Pan. Ricerche e progetti per il futuro di Pratolino*, Pratolino. Laboratorio di meraviglie. Florence: Alinea Editrice.

Dijksterhuis, E. J. 1983. *Die Mechanisierung des Weltbildes*. Berlin: Springer. Original edition: Amsterdam: J. M. Meulenhoff 1950.

Dragoni, G. 1991. *Instrumenta. Il patrimonio storico scientifico italiano: una realtà straordinaria*. Bologna: Grafis Edizioni.

Drake, S. 1958. Galileo Gleanings. V. The Earliest Version of Galileo's Mechanics. *Osiris* XIII:262–290.

Dupré, S. 2000. Mathematical Instruments and the Theory of the Concave Spherical Mirror: Galileo's Optics beyond Art and Science. *Nuncius. Annali di Storia della Scienza* 15 (2): 551–588.

Dupré, S. 2005. Ausonio's Mirrors and Galileo's Lenses: The Telescope of the Sixteenth-Century Practical Optical Knowledge. *Galilaeana: Journal of Galilean Studies* 2:145–180.

Fara, A. 1988. *Bernardo Buontalenti: l'architettura, la guerra e l'elemento geometrico*. Genoa: Sagep.

Fara, A. 1995. *Bernardo Buontalenti*. Milan: Electa.

Fara, A. 1998. *Bernardo Buontalenti e Firenze: archittetura e disegno dal 1576 al 1607*. Florence: Olschki.

Favaro, A. 1886. La libreria di Galileo Galilei. *Bullettino di bibliografia e di storia delle scienze matematiche e fisiche* 19:1–77.

Favaro, A. 1966. *Galileo Galilei e lo studio di Padova*. 2 vols. Padua: Editrice Antenore.

Favaro, A., and P. Galluzzi. 1983. *Amici e corrispondenti di Galileo*. 3 vols. Florence: Libreria Editrice Salimbeni.

Fiocca, A., ed. 1998. *Giambattista Aleotti e gli ingegneri del Rinascimento*. Florence: Leo S. Olschki.

Fredette, R. 1970. *Les De Motu "plus anciens" de Galileo Galilei: prolégomnès*. PhD thesis, Université de Montreal.

Freedberg, D. 2002. *The Eye of the Lynx*. Chicago: The University of Chicago Press.

Gallo, I. 1986. Filosofia e scienza agli albori del seicento: Giulio Cesare Lagalla tra Aristotele e Galilei. *Rassegna Storica Salernitana* 6:55–75.

Gallo, I. 1987. Ancora su Giulio Cesare Lagalla: Medicina e tecnica della dissimulazione. *Rassegna Storica Salerinitana* 8:17–34.

Galluzzi, P. 1979. *Momento. Studi Galileiani*. Rome: Edizioni dell'Ateneo & Bizzarri.

Galluzzi, P. 1991. *Prima di Leonardo. Cultura delle macchine a Siena nel Rinascimento*. Milan: Electa.

Gandt, F. d. 1986. Les Méchaniques attribuées a Aristote et le renouveau de la science des machines au XVI siécle. *Les Etudes philosophiques* 3:391–405.

Gatto, R. 2001. Sull'edizione critica de «Le Mechaniche» di Galileo. In *Largo campo di filosofare: Eurosymposium Galileo 2001*, edited by J. Montesinos and C. Solís. La Orotava: Fondación Canaria Orotava de Historia de la Ciencia, 203–216.

Gatto, R. 2002. *Le mechaniche*. Firenze: Olschki.

Goldthwaite, R. A. 1972. Schools and Teachers of Commercial Arithmetic in Renaissance Florence. *Journal of European Economic History* 1:418–433.

Grayson, C. 1973. *Opere volgari di Leon Battista Alberti*. 3 vols. Bari: Laterza.

Greco, V., G., and G. Molesini. 1997. Galileo e Sagredo: primo enunciato della formula dei fabbricanti di lenti. *Quaderni di storia della fisica* 1:35–38.

Hall, B. S. 1979. Der Meister sol auch kennen schreiben und lesen: Writings about Technology ca. 1400– ca. 1600 AD and their Cultural Implications. In *Early Technologies*, edited by D. Schmandt-Besserat. Malibu: Undena Publications, 47–58.

Helden, A. v. 1996. Longitude and the Satellites of Jupiter. In *The Quest for Longitude* edited by W. J. H. Andrews. Cambridge, MA.: Harvard University Press, 85–100.

Hellmann, G. 1920. Beiträge zur Erfindungsgeschichte meteorologischer Instrumente. In *Abhandlungen der Preussischen Akademie der Wissenschaften*. Berlin: Akademie der Wissenschaften.

Henninger-Voss, M. J. 2002. How the New Science of Cannons Shook Up the Aristotelian Cosmos. *Journal of the History of Ideas* 63 (3):371–397.

Humphrey, J. W., J. P. Oleson, and A. N. Sherwood, eds. 1997. *Greek and Roman Technology. A Sourcebook*. London: Routledge.

Hunt, F. V. 1978. *Origins in Acoustics. The Science of Sound from Antiquity to the Age of Newton*. New Haven and London: Yale University Press.

King, H. C. 1955. *The History of the Telescope*. London: Charles Griffin & Company Limited.

Knowles Middleton, W. E. 1964. *The History of the Barometer*. Baltimore, Maryland: John Hopkins University Press.

Knowles Middleton, W. E. 1966. *A History of the Thermometer and Its Use in Meteorology*. Baltimore: John Hopkins University Press.

Knowles Middleton, W. E. 1971. *The Experimenters*. Baltimore: The John Hopkins University Press.

Kuhn, T. S. 1976. Mathematical versus Experimental Traditions in the Development of Physical Science. In *The Essential Tension. Selected Studies in Scientific Tradition and Change*, edited by T. S. Kuhn. Chicago: The University of Chicago Press. Original edition: *The Journal of Interdisciplinary History*, 7 (1976): 1–31, 31–65.

Kuhn. 1977. *The Essential Tension. Selected Studies in Scientific Tradition and Change*. Chicago: The University of Chicago Press.

Laird, W. R. 1986. The Scope of Renaissance Mechanics. *Osiris* 2:43–68.

Lamberini, D. 1986. The Military Architecture of Giovanni Battista Belluzzi. *Fort. The International Journal for Military Architecture and Fortifications* 14:5–16.

Lamberini, D. 1987. Practice and Theory in XVI Century Fortifications. *Fort. The International Journal for Military Architecture and Fortifications* 15:5–20.

Lamberini, D. 1990. *Il principe difeso. Vita e opere di Bernardo Puccini*. Florence: Editrice La Giuntina.

Lamberini, D. 2007. *Il Sanmarino. Giovan Battista Belluzzi. Architetto militare e trattatista del Cinquecento*. 2 vols. Florence: Olschki.

Lefèvre, W. 2001. Galileo Engineer: Art and Modern Science. In *Galileo in Context*, edited by J. Renn. Cambridge: Cambridge University Press, 11–27.

Lefèvre, W., J. Renn, and U. Schoepflin, eds. 2002. *The Power of Images in Early Modern Science*. Basel: Birkhäuser.

Leitão, H. 2000. *O comentário de Pedro Nuñes à Navegação a Remos*. Lisbon: Edifil/Publicações económicas Lda.

Levere, T. H., and W. R. Shea, eds. 1990. *Nature, Experiments, and the Sciences. Essays on Galileo and the History of Science in Honour of Stilman Drake*. Dordrecht: Kluwer Academic Publishers.

Lindberg, D. C. 1983. Optics in Sixteenth-Century Italy. In *Novità celesti e crisi del sapere*, edited by P. Galluzzi. Florence: Istituto e Museo di Storia della Scienza, 131–148.

Lüthy, C. 2003. The Invention of Atomist Iconography. In *The Power of Images in Early Modern Science*, edited by W. Lefèvre, J. Renn and U. Schoepflin. Basel: Birkhäuser, 117–138.

Maffioli, C. S. 2008. Galileo, Guiducci and the Engineer Bartolotti on the Bisenzio River. *Galilaeana* 5:179–206.

Martellozzo Forin, E. 2005. *La bottega dei fratelli Mazzoleni, orologiai in Padova (1569)*. Saonara (PD): Il Prato.

Meskens, A. 1997. Michiel Coignet's Contribution to the Development of the Sector. *Annals of Science* 54:143–160.

Micheli, G. 1995. *Le origini del concetto di macchina*. Florence: Olschki.

Miniati, M. 1991. L'Istituto e Museo di Storia della Scienza di Firenze. Ricerche e Prospettive. In *Instrumenta. Il patrimonio storico scientifico italiano: una realtà straordinaria*, edited by G. Dragoni. Bologna: Grafis Edizioni, 157–162.

Moll, G. 1831. On the First Invention of Telescopes, Collected from the Notes and Papers of the Late Professor Van Swinden. *The Journal of the Royal Institution of Great Britain* 1:319–332; 483–496.

Montesinos, J., and C. Solís, eds. 2001. *Largo campo di filosofare. Eurosymposium Galileo 2001*. La Orotava: Fondación Canaria Orotava de Historia de la Ciencia.

Napolitani, P. D. 1988. La geometrizzazione della realtà fisica: il peso specifico in Ghetaldi e in Galileo. *Bollettino di storia delle scienze matematiche* VIII (2):139–237.

Newman, W. R. 2001. Corpuscolar Alchemy and the Tradition of Aristotle's «Meteorology,» with Special Reference to Daniel Sennert. *International Studies in the Philosophy of Science* 15 (2):145–153.

Olschki, L. 1919–1927. *Geschichte der neusprachlichen wissenschaftlichen Literatur*. 3 vols. Heidelberg: Max Niemeyer Verlag.

Palmieri, P. 2003. Mental Models in Galileo's Early Mathematization of Nature. *Studies In History and Philosophy of Science Part A* 34 (2):229–264.

Paré, G. M. 1947. *Les idées et les lettres au XIIIe siècle: Le Roman de la Rose*, Université de Montrîal. Bibliothèque de philosophie. Montréal: Centre de psychologie et de pédagogie.

Pedersen, O. 1968. Sagredo's Optical Research. *Centaurus* 13 (2):139–150.

Penzo, G. 1999. *La Gondola. Storia, progettazione e costruzione della più straordinaria imbarcazione tradizionale di Venezia*. Venice: Istituzione per la conservazione della gondola e la tutela del gondoliere.

Petrányi, G. 2000. The Ottoman Coinage of Cyprus (II). *as-Sikka. Online Journal of The Islamic Coins Group* II (3).

Popplow, M. 1998. *Neu, Nützlich und Erfindungsreich. Die Idealisierung von Technik in der frühen Neuzeit*. Münster: Waxmann.

Popplow, M. 2004. Why Draw Pictures of Machines? The Social Contexts of Early Modern Machine Drawings. In *Picturing Machines. 1400–1700*, edited by W. Lefèvre. Cambridge: The MIT Press, 17–48.

Popplow, M., and J. Renn. 2002. Ingegneria e macchine. In *Storia della scienza. Vol. V: La rivoluzione scientifica*, edited by S. Petruccioli. Rome: Istituto della enciclopedia Italiana, 258–274.

Poupard, P., ed. 1984. *Galileo Galilei. 350 anni di storia (1633–1983)*. Rome: Piemme.

Presas i Puig, A. 2002. *Ostilio Ricci, the Practical Education and the Canon of Technical Knowledge at the Beginning of the Italian Renaissance*. Berlin: Max-Planck-Institut für Wissenschaftsgeschichte, Preprint 200.

Promis, C. 1874. *Biografie di ingegneri militari italiani dal secolo XIV alla metà del XVIII*. Vol. 14, *Miscellanea di Storia Italiana*, serie 1 Torino: Fratelli Bocca Librai di S. M.

Rank, L. 1984. *Die Theorie des Segelns in ihrer Entwicklung*. Berlin: Dietrich Reimer Verlag.

Redondi, P. 2004. *Galileo eretico*. 2 ed, Biblioteca Einaudi 175. Turin: G. Einaudi.

Reeves, E. A. 2008. *Galileo's Glassworks: The Telescope and the Mirror*. Cambridge, MA: Harvard University Press.

Renn, J. 2000. Mentale Modelle in der Geschichte des Wissens: Auf dem Wege zu einer Paläontologie des mechanischen Denkens. *Dahlemer Archivgespräche* 6:83–100.

Renn, J, ed. 2001. *Galileo in Context*. Cambridge: Cambridge University Press.

Renn, J., and G. Castagnetti, eds. 2002. *Homo Faber: Studies on Nature, Technology, and Science at the Time of Pompeii*. Rome: L'Erma di Bretschneider.

Renn, J., P. Damerow, and S. Rieger. 2001. Hunting the White Elefant. In *Galileo in Context*, edited by J. Renn. Cambridge, MA: Cambridge University Press, 29–152.

Renn, J., and M. Valleriani. 2001. Galileo and the Challenge of the Arsenal. *Nuncius* 16 (2): 481–503.

Reti, L. 1963. Francesco di Giorgio Martini's Treatise on Engineering and Its Plagiarists. *Technology and Culture* IV:287–298.

Righini, G. 1980. *Le operazioni del compasso militare di Galileo Galilei*. Florence: Istituto e Museo di Storia della Scienza di Firenze.

Rose, P. L. 1971. Materials for a Scientific Biography of Guidobaldo del Monte. *Actes du XIIe congrès internationale des sciences*. Paris: Librairie Sci. Tech. Blanchard, 12 ff.

Rose, P. L. 1975. *The Italian Renaissance of Mathematics*. Geneva: Libraire Droz.

Rose, P. L. 1976. Jacomo Contarini (1536–1595), a Venetian Patron and Collector of Mathematical Instruments and Books. *Physis* XVIII (2):117–130.

Rose, P. L., and S. Drake. 1971. The Pseudo-Aristotelian Questions of Mechanics in Renaissance Culture. *Studies in the Renaissance* 18:65–104.

Russo, L. 2001. *La rivoluzione dimenticata. Il pensiero scientifico greco e la scienza moderna*. Milano: Feltrinelli. English edition: *The Forgotten Revolution: How Science Was Born in 300 BC and Why It Had To Be Reborn*. Berlin: Springer. 2004.

Schemmel, M. 2008. *The English Galileo: Thomas Harriot's Work on Motion as an Example of Preclassical Mechanics*. Dordrecht: Springer.

Schiefsky, M. 2007. Art and Nature in Ancient Mechanics. In *The Artificial and the Natural: An Evolving Polarity*, edited by B. Bensaude-Vincent and W. Newman. Cambridge, MA: MIT Press.

Schmitt, C. B. 1967. Experimental Evidence for and against a Void: The Sixteenth-Century Arguments. *Isis* 58 (3):352–366.

Schneider, I. 1970. *Der Proportionalzirkel. Ein universelles Analogrecheninstrument der Vergangenheit*. Munich: R. Oldenbourg Verlag.

Schürmann, A. 1991. *Griechische Mechanik und antike Gesellschaft: Studien zur staatlichen Förderung einer technischen Wissenschaft*. Stuttgart: Steiner.

Selmi, F. 1864. Un particolare ignoto della vita di Galileo Galilei. In *Nel trecentesimo natalizio di Galileo in Pisa, 18 Febbraio MDCCCLXIV*. Pisa: Tipografia Nisti.

Settle, T. B. 1971. Ostilio Ricci, a Bridge between Alberti and Galileo. *Actes du XIIe congrès internationale des sciences*. Paris: Librairie Sci. Tech. Blanchard, 122–126.

Settle, T. B. 1995. La rete degli esperimenti galileiani. In *Galileo e la scienza sperimentale*, edited by P. Bozzi, C. Maccagni, C. Olivieri and T. B. Settle. Padua: Dipartimento di Fisica "Galileo Galilei," 11–62.

Shapin, S., and S. Schaffer. 1985. *Leviathan and the Air-pump: Hobbes, Boyle, and the Experimental Life: Including a Translation of Thomas Hobbes, Dialogus physicus de natura aeris by Simon Schaffer*. Princeton: Princeton University Press.

Shelby, L. R. 1965. Medieval Masons' Tools. II. Compass and Square. *Technology and Culture* 6 (2):236–248.

Sherwood, A. N. 1997. *Greek and Roman Technology. A Sourcebook*. London: Routledge

Sherwood Taylor, F. 1942. The Origin of the Thermometer. *Annals of Science* 5:129–156.

Smith, P. H. 2004. *The Body of the Artisan: Art and Experience in the Scientific Revolution*. Chicago: University of Chicago Press.

Strano, G. 2009. La lista della spesa di Galileo: un documento poco noto sul telescopio. *Galilaeana* 6:197–211.

Stringa, P. 1982. *Genova e la Liguria nel Mediterraneo*. Genoa: SAGEP Editrice.

Szabó, I. 1987. *Geschichte der mechanischen Prinzipien und ihrer wichtigsten Anwendungen*. Basel: Birkhäuser.

Taddei, G. 1954. *L'arte del vetro in Firenze*. Florence: Felice Le Monnnier.

Tiere, G. 1932. *Cornelius Drebbel*. Amsterdam: H. J. Paris.

Trabucco, O. 2009. Giochi d'acqua e d'aria tra magia e meccanica. Ancora su della Porta e Galileo. *Galilaeana* VI:167–196.

Traina, M. 1975. *Gli assedii e le loro monete*. 3 vols. Bologna: R. Giannantoni.

Turner, G., ed. 1991. *Catalogue of Microscopes*. Florence: Giunti.

Ulivieri, L., and S. Merendoni, eds. 2009. *Pratolino, un mito alle porte di Firenze - Pratolino, a Myth at the Gates of Florence*. Venice: Marsilio.

Valleriani, M. 2001. A View on Galileo's "Ricordi Autografi." Galileo Practitioner in Padua. In *Largo campo di filosofare*, edited by J. Montesinos and C. Solís. La Orotava: Fondaci6n Canaria Orotava de Historia de la Ciencia, 281–292.

Valleriani, M. 2007. From Condensation to Compression: How Renaissance Italian Engineers Approached Hero's «Pneumatics.» In *Übersetzung und Transformation*, edited by H. Böhme, C. Rapp and W. Rösler. Berlin: DeGruyter, 333–354.

Valleriani, M. 2008. Galileo in the Role of the Caster's Assistant: The 1634 Bell of the «Torre del Mangia» in Siena. *Galilaeana* 5:89–112.

Valleriani, M. 2009a. The Transformation of Aristotle's «Mechanical Questions:» A Bridge Between the Italian Renaissance Architects and Galileo's First New Science. *Annals of Science* 66 (2):183–208.

Valleriani, M. 2009b. The Transformation and Reconstruction of Hero of Alexandria's «Pneumatics» in the Garden of Pratolino. In *Pratolino, un mito alle porte di Firenze - Pratolino, a Myth at the Gates of Florence*, edited by L. Ulivieri and S. Merendoni. Venice: Marsilio, 155–181.

Valleriani, M. 2009c. Galileis astronomische Werkstatt. *Sterne und Weltraum* (2):42–51.

Valleriani, M. 2010. Il ruolo della pneumatica antica durante il Rinascimento: l'esempio dell'organo idraulico nel giardino di Pratolino. In *La civiltà delle acque dal Medioevo al Rinascimento*, edited by A. Calzona and D. Lamberini. 2 vols. Florence: Olschki, vol. 2, pp. 613–632.

Valleriani, M. Forthcoming, a. The Garden of Pratolino: Ancient Technology Breaks Through the Barriers of Modern Iconology. In *Ludi Naturae - Spiele der Natur. Transformationen eines antiken Topos in Wissenschaft und Kunst*, edited by N. Adamowsky, R. Felfe and H. Böhme. Munich: Fink Verlag.

Valleriani, M. Forthcoming, b. The War in Ariosto's «Orlando furioso:» A Snapshot of the Passage from Medieval to Early Modern Technology. In *Krieg in Worten / War in Words. Transformationen des Krieges von der Antike bis Clausewitz*, edited by H. Böhme and M. Formisano. Berlin: DeGruyter.

Vallerini, F. 1992. Galileo e il celatone. *Pisa economica* 1:33–38.

Van Dyck, M. 2005. The Paradox of Conceptual Novelty and Galileo's Use of Experiments. *Philosophy of Science* 72 (5):864–875.

Van Helden, A. 1977. *The Invention of the Telescope, Transactions of the American Philosophical Society* v. 67, pt. 4. Philadelphia: The American Philosophical Society. Reprinted in 2008.

Van Helden, A. 1983. Galileo and the Telescope. In *Novità celesti e la crisi del sapere*, edited by P. Galluzzi. Florence: Istituto e Museo di Storia della Scienza, 149–158.

V6rin, H. 2001. Galil6e et Antoine de Ville. Un courrier sur l'id6e de mati6re. In *Largo campo di filosofare*, edited by J. Montesinos and C. Solís. La Orotava: Fondaci6n Canaria Orotava de Historia de la Ciencia, 307–321.

Wallace, W. A. 1984a. Galileo e i professori del Collegio Romano alla fine del secolo XVI. In *Galileo Galilei. 350 anni di storia (1633–1983)*, edited by P. Poupard. Roma: Piemme, 76–97.

Wallace, W. A, 1984b. *Galileo and His Sources. The Heritage of the Collegio Romano in Galileo's Science*. Princeton: Princeton University Press.

Wallace, W. A. ed. 1986a. *Reinterpreting Galileo*. Washington D.C.: The Catholic University of America Press.

Wallace, W. A. 1986b. Reinterpreting Galileo on the Basis of His Latin Manuscripts. In *Reinterpreting Galileo*, edited by W. A. Wallace. Washington D.C.: The Catholic University of America Press, 3–28.

Wallace, W. A. 1990. The Dating and Significance of Galileo's Pisan Manuscripts. In *Nature Experiment and Sciences*, edited by T. H. Levere and W. R. Shea. Dordrecht: Kluwer, 3–50.
Westfall, R. S. 1983. Galileo and the Accademia dei Lincei. In *Novità celesti e crisi del sapere*, edited by P. Galluzzi. Florence: Istituto e Museo di Storia della Scienza, 189–200.
Westfall, R. S. 1985. Science and Patronage. *Isis* 76 (1):11–30.
Willach, R. 2001. The Development of Lens Grinding and Polishing Techniques in the First Half of the 17th Century. *Bullettin of the Scientific Instrument Society* 68:10–15.
Williams, M. R., and E. Tomash. 2003. The Sector: Its History, Scales, and Uses. *Annals of the History of Computing* 25 (1):34–47.
Wilson, C. 1951. *The Invisible World.* Princeton: Princeton University Press.
Zangheri, L. 1987. *Pratolino, il giardino delle meraviglie.* 2 vols, Documenti inediti di cultura toscana. Florence: Gonnelli.
Zilsel, E. 2000. *The Social Origins of Modern Science*. Dordrecht: Kluwer. Original edition: *Die sozialen Ursprünge der neuzeitlichen Wissenschaft*. Frankfurt am Main: Suhrkamp. 1976.

Name Index

Illustration Index

MIX
Papier aus verantwortungsvollen Quellen
Paper from responsible sources
FSC® C105338

If you have any concerns about our products,
you can contact us on
ProductSafety@springernature.com

In case Publisher is established outside the EU,
the EU authorized representative is:
Springer Nature Customer Service Center GmbH
Europaplatz 3, 69115 Heidelberg, Germany

Printed by Libri Plureos GmbH
in Hamburg, Germany